# Exploring
# Anatomy & Physiology
## in the Laboratory
### Second Edition

Erin C. Amerman

# MORTON
PUBLISHING

925 W. Kenyon Avenue, Unit 12
Englewood, CO 80110

www.morton-pub.com

## Book Team

| | |
|---|---|
| Publisher: | Douglas N. Morton |
| President: | David M. Ferguson |
| Acquisitions Editor: | Marta R. Martins |
| Project Manager: | Melanie Stafford |
| Editorial Assistant: | Rayna Bailey |
| Production Manager: | Joanne Saliger |
| Production Assistant: | Will Kelley |
| Illustrations: | Imagineering Media Services, Inc. |
| Cover Design: | Imagineering Media Services, Inc. |

Printed in the United States of America

10   9   8   7   6   5   4   3   2

ISBN-13:  978-161731-056-0

Library of Congress Control Number:  2012951978

# Preface

I first started writing lab procedures for my students 12 years ago in response to frustration my students and I were feeling with the anatomy and physiology lab and the accompanying lab manual. My students wanted—and needed—focused activities, clear objectives and explanations, and exercises that enabled them to appreciate the often-missed "big picture" of A&P. None of the available lab manuals provided this, so I was left with no choice but to write my own lab procedures. The results of my efforts were well worth it, as my students were engaged and active the entire lab period. Their lab and class grades improved.

In 2003, I met with David Ferguson, now the president of Morton Publishing Company, and he offered me a dream opportunity: to share my exercises with students and instructors throughout the country. The result of that meeting was the text *Exercises for the Anatomy and Physiology Laboratory*, a simple black-and-white manual with focused activities. My goal with this book was to solve the teaching problems in the A&P lab and to enhance the experience for my colleagues and our students.

*Exercises* was enthusiastically received, suggesting that we had indeed provided instructors and students something they had been needing. We were so encouraged that we set out to produce an expanded, full-color version of the exercises that included more explanations, new activities, a complete art program, and new pedagogy. This book became *Exploring Anatomy and Physiology in the Laboratory*, or *EAPL*. Like its predecessor, it was warmly received.

Over the life span of the first edition of *EAPL*, we received a great deal of wonderfully helpful feedback from our adopters. When it came time to write the second edition, we assembled that feedback to determine what was working so that we could enhance it, and what was not working as well as intended so that we could improve it. The result of that process is the book you are holding in your hands. As you peruse the book, please notice the following changes, which were a direct result of your feedback:

▌ **Vastly improved art program.** There are more than 300 new and improved figures in *EAPL 2/e*. We replaced most of the histology images with higher-quality micrographs, added new photos and illustrations, and improved the quality of more than 150 of the existing illustrations.

▌ **More critical thinking questions.** The "Check Your Understanding" critical thinking questions have always been a popular feature, so between one and three new questions have been added to nearly every unit in the book.

▌ **New activities.** Several new activities have been added to this edition, including a blood cell differential exercise, a basic ECG interpretation, an ELISA procedure, and a "big picture" tracing exercise.

▌ **Improved Pre-Lab Exercises.** Many instructors told us that the pre-lab exercises, although useful, were too long and difficult to grade. In response, we shortened the longer pre-labs so that they are more manageable. We are also offering the pre-lab exercises to instructors in an electronic format so that they may be assigned online and graded much more efficiently.

- **Improved student assessment.** As with the pre-lab exercises, several instructors told us that the end of chapter quizzes were difficult to grade and sometimes confusing for their students. To help fix this, we added leader lines to the labeling questions, changed the format of the matching questions, and added lines after every short-answer question. We also have a number of additional questions available that instructors may assign online.

- **New paper.** Another frequent comment was that the book was just too big. We didn't want to take away any of the book's content, so instead we changed to a different type of semigloss paper. This paper not only makes the photos and illustrations look better but also it decreases the thickness of the book by about 35%.

- There are a number of other improvements we made as well, including providing larger areas for student drawings, adding lines to procedural questions, adding checkboxes to materials lists and exercises, providing bigger tables, and revising several of the tracing exercises so that they will be easier for your students to complete. We hope that you enjoy the second edition of *EAPL*. Please share your feedback with us about ways we can continue to improve this text so that it provides an optimal learning experience for your students.

- **The following pages offer a walk-through of the key features in *Exploring Anatomy & Physiology in the Laboratory*, 2nd edition.**

—Erin C. Amerman

# Be Prepared

**OBJECTIVES** set learning goals to prepare students for what they are expected to know after completing the lab. They also aid in the review of material.

**PRE-LAB EXERCISES** encourage students to actively prepare for the lab by defining key terms, doing labeling and coloring exercises to learn anatomical structures, and reviewing vital material from previous units, saving instructors from having to spend extra time reviewing material from the lecture. These exercises can be completed using information available in the lab manual and will help students move through the lab more effectively. By asking students to draw their own leader lines and write out definitions, the pre-lab exercises are designed to help students retain information and build a deeper understanding of the content.

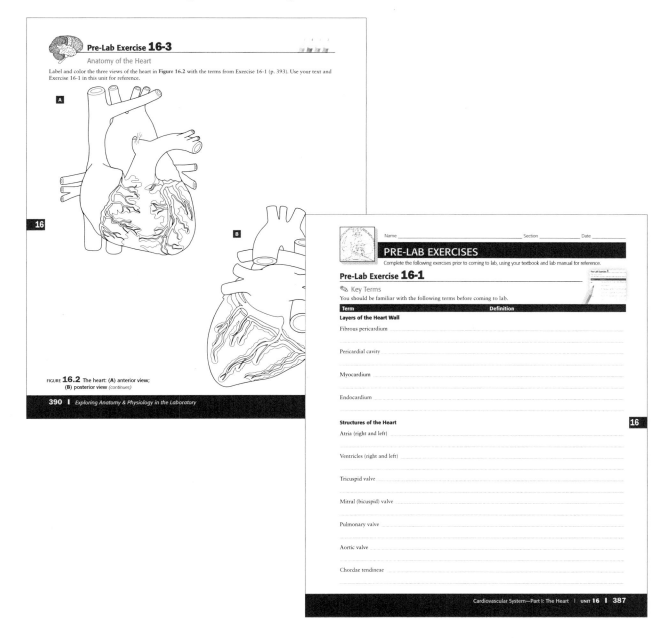

Pre-Lab Exercise **16-3**

Anatomy of the Heart

Label and color the three views of the heart in **Figure 16.2** with the terms from Exercise 16-1 (p. 393). Use your text and Exercise 16-1 in this unit for reference.

**A**

**B**

FIGURE **16.2** The heart: **(A)** anterior view; **(B)** posterior view *(continues)*

**390 |** *Exploring Anatomy & Physiology in the Laboratory*

Name _____ Section _____ Date _____

## PRE-LAB EXERCISES

Complete the following exercises prior to coming to lab, using your textbook and lab manual for reference.

Pre-Lab Exercise **16-1**

✎ Key Terms

You should be familiar with the following terms before coming to lab.

| Term | Definition |
|---|---|
| **Layers of the Heart Wall** | |
| Fibrous pericardium | |
| Pericardial cavity | |
| Myocardium | |
| Endocardium | |
| **Structures of the Heart** | |
| Atria (right and left) | |
| Ventricles (right and left) | |
| Tricuspid valve | |
| Mitral (bicuspid) valve | |
| Pulmonary valve | |
| Aortic valve | |
| Chordae tendineae | |

Cardiovascular System—Part I: The Heart **|** UNIT **16 | 387**

# Be Organized

**MODEL INVENTORIES** provide organized and easily referenced lists of anatomical structures students are responsible for identifying. These lists help students catalog the specimens they see in the lab. The emphasis on examination, description, pronunciation, and writing the names of anatomical structures encourages students to be actively involved in the learning process and allows them to better retain the material.

---

left coronary sulcus to supply the left atrium and posterior left ventricle. When a coronary artery is blocked, the reduced blood flow to the myocardium can result in hypoxic injury and death to the tissue, a condition termed **myocardial infarction** (commonly called a heart attack).

The anatomy of the cardiac veins often varies from person to person, but the following three main veins generally are present:

1. **small cardiac vein,** which drains the inferolateral heart,

2. **middle cardiac vein,** which drains the posterior heart, and

3. **great cardiac vein,** which drains most of the left side of the heart.

All three veins drain into the large **coronary sinus** located on the posterior right atrium. The coronary sinus drains into the right atrium.

## Procedure  Model Inventory for the Heart

Identify the following structures of the heart on models and diagrams, using your textbook and this unit for reference. As you examine the anatomical models and diagrams, record on the model inventory in **Table 16.1** the name of the model and the structures you were able to identify. When you have completed the activity, answer Check Your Understanding questions 1–3 (p. 409).

1. General structures
   a. Mediastinum
      (1) Pericardial cavity
   b. Pericardium
      (1) Fibrous pericardium
      (2) Parietal pericardium
      (3) Visceral pericardium (epicardium)
   c. Myocardium
   d. Endocardium
   e. Apex of the heart
   f. Base of the heart
   g. Anterior interventricular sulcus
   h. Posterior interventricular sulcus
   i. Coronary sulcus
2. Right atrium
   a. Opening of the superior vena cava
   b. Opening of the inferior vena cava
   c. Opening of the coronary sinus
   d. Right auricle
   e. Interatrial septum
   f. Fossa ovalis
   g. Pectinate muscles
3. Left atrium
   a. Opening of the pulmonary veins
   b. Left auricle
   c. Interatrial septum
4. Structures of the ventricles
   a. Right ventricle
   b. Left ventricle
   c. Interventricular septum
   d. Chordae tendineae
   e. Papillary muscles
   f. Trabeculae carneae

5. Atrioventricular valves
   a. Tricuspid valve
   b. Mitral valve
6. Semilunar valves
   a. Pulmonary valve
   b. Aortic valve
7. Great vessels
   a. Superior vena cava
   b. Inferior vena cava
   c. Pulmonary trunk (artery)
   d. Right and left pulmonary arteries
   e. Pulmonary veins
   f. Aorta
   g. Ligamentum arteriosum
8. Coronary arteries
   a. Right coronary artery
   b. Marginal artery
   c. Posterior interventricular artery
   d. Left coronary artery
   e. Anterior interventricular artery
   f. Circumflex artery
9. Cardiac veins
   a. Small cardiac vein
   b. Middle cardiac vein
   c. Great cardiac vein
   d. Coronary sinus

16

TABLE **16.1** Model Inventory for the Heart

| Model | Structures Identified |
|---|---|
| | |
| | |
| | |
| | |
| | |
| | |
| | |

16

# Be Focused

**ILLUSTRATIONS** and **PHOTOGRAPHS** in *Exploring Anatomy & Physiology in the Laboratory, 2e,* were specifically designed to improve student understanding of important concepts and procedural instructions in the laboratory setting. (Many lab manuals simply reproduce artwork and exposition from the related textbook, unnecessarily increasing the redundancy, size, and price of the manual.)

There are 200 new and revised illustrations and 100 new photographs in this edition offering a more detailed and realistic view of human anatomy. Almost all of the histology images are new, taken from the best sources of commercially available slides.

# Be Active

**FOCUSED ACTIVITIES** are the guiding philosophy of this lab manual. Students learn best when they are actively engaged in the laboratory. In this manual, students are asked to be active by describing, labeling, writing, coloring, and drawing. Several new activities have been added to this edition, including a blood cell differential exercise, a basic ECG interpretation, an ELISA procedure, and a "big picture" tracing exercise.

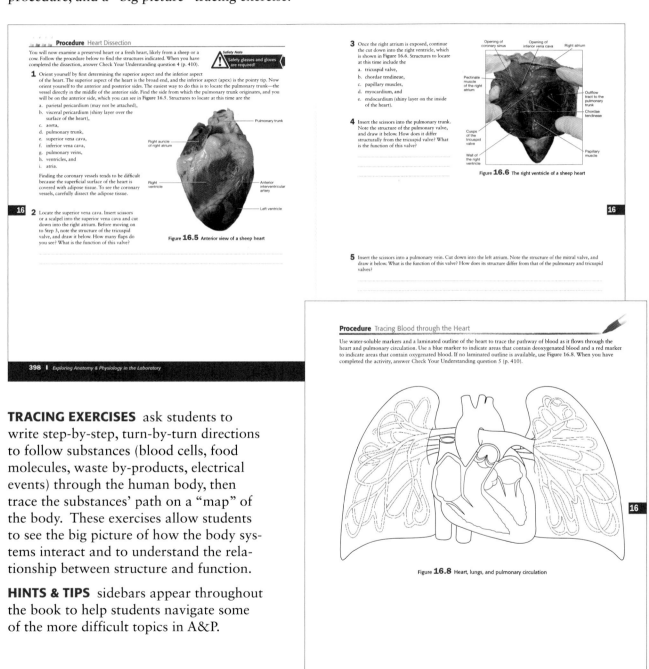

**TRACING EXERCISES** ask students to write step-by-step, turn-by-turn directions to follow substances (blood cells, food molecules, waste by-products, electrical events) through the human body, then trace the substances' path on a "map" of the body. These exercises allow students to see the big picture of how the body systems interact and to understand the relationship between structure and function.

**HINTS & TIPS** sidebars appear throughout the book to help students navigate some of the more difficult topics in A&P.

# Be Sure

**CHECK YOUR PROGRESS** sections completing each unit ensure that students understand key concepts and achieve the learning objectives, signaling that they are ready to move on.

**CHECK YOUR RECALL** "quizzes" consist of labeling, fill-in-the-blank, multiple choice, and sequencing questions that test the students' ability to retain the material they completed in the lab. These sheets can be used as graded lab quizzes and/or to check attendance in the lab.

**CHECK YOUR UNDERSTANDING** "tests" ask students to use the information they have learned to answer critical thinking questions, frequently related to clinical scenarios. The ability to synthesize and apply knowledge is the desired outcome of any A&P course. These questions check students' deeper understanding by challenging them to provide answers they cannot find verbatim in either the lab manual or the textbook. New Check Your Understanding questions have been added to nearly every unit in this edition.

Name _____

Section _____ Date _____

**3** *Matching:* Match the following terms with the correct definition.

_____ Myocardium
_____ Parietal pericardium
_____ Tricuspid valve
_____ Aortic valve
_____ Papillary muscles
_____ Pulmonary veins
_____ Mitral valve
_____ Visceral pericardium
_____ Pulmonary trunk
_____ Coronary sinus
_____ Chordae tendineae
_____ Aorta

A. Located between the left ventricle and the aorta
B. Located between the left atrium and left ventricle
C. Largest artery of the systemic circuit
D. Bring(s) oxygenated blood to the left atrium from the lungs
E. Layer of the heart composed of cardiac muscle tissue
F. Bring(s) deoxygenated blood to the lungs
G. Located between the right atrium and the right ventricle
H. Largest vein of the coronary circulation
I. Outer layer of the serous pericardium
J. Fibrous cord(s) that attach(es) to valves
K. Inner layer of the serous pericardium
L. Fingerlike muscular projections from the ventricles

**4** The veins of the systemic circuit carry _____ blood, and the veins of the pulmonary circuit carry _____ blood.
a. oxygenated; deoxygenated
b. oxygenated; oxygenated
c. deoxygenated; deoxygenated
d. deoxygenated; oxygenated

**5** The pulmonary and aortic valves are known as the
a. atrioventricular (AV) valves.
b. semilunar valves.
c. coronary valves.
d. chordae tendineae.

**6** The right and left coronary arteries are the first branches off the
a. aorta.
b. superior vena cava.
c. pulmonary trunk.
d. pulmonary veins.

**7** *Fill in the blanks:* Cardiac muscle cells are also known as _____. Adjacent cells are joined together by _____, which allow the heart to _____

Cardiovascular System—Part I: The Heart | UNIT **16** | **407**

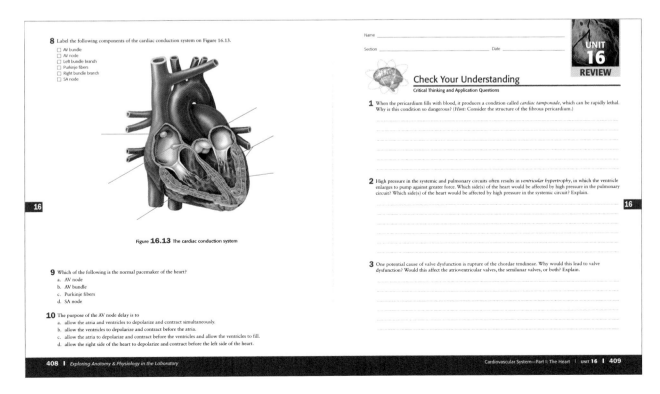

**8** Label the following components of the cardiac conduction system on Figure 16.13.
☐ AV bundle
☐ AV node
☐ Left bundle branch
☐ Purkinje fibers
☐ Right bundle branch
☐ SA node

Figure **16.13** The cardiac conduction system

**9** Which of the following is the normal pacemaker of the heart?
a. AV node
b. AV bundle
c. Purkinje fibers
d. SA node

**10** The purpose of the AV node delay is to
a. allow the atria and ventricles to depolarize and contract simultaneously.
b. allow the ventricles to depolarize and contract before the atria.
c. allow the atria to depolarize and contract before the ventricles and allow the ventricles to fill.
d. allow the right side of the heart to depolarize and contract before the left side of the heart.

Name _____

Section _____ Date _____

UNIT
**16**
REVIEW

## Check Your Understanding

**Critical Thinking and Application Questions**

**1** When the pericardium fills with blood, it produces a condition called *cardiac tamponade,* which can be rapidly lethal. Why is this condition so dangerous? (*Hint:* Consider the structure of the fibrous pericardium.)

**2** High pressure in the systemic and pulmonary circuits often results in *ventricular hypertrophy,* in which the ventricle enlarges to pump against greater force. Which side(s) of the heart would be affected by high pressure in the pulmonary circuit? Which side(s) of the heart would be affected by high pressure in the systemic circuit? Explain.

**3** One potential cause of valve dysfunction is rupture of the chordae tendineae. Why would this lead to valve dysfunction? Would this affect the atrioventricular valves, the semilunar valves, or both? Explain.

## Be Aware

Textbooks are expensive, and the last thing students need is to spend too much for a lab manual. Morton Publishing is committed to providing high-quality products at reasonable prices.

It is our sincere hope that *Exploring Anatomy & Physiology in the Laboratory, 2e*, will provide you the tools necessary for a productive and interesting laboratory experience. We welcome all comments and suggestions for future editions of this book. Please feel free to contact us at eapl@morton-pub.com or visit us at www.morton-pub.com.

## Be Choosy

### MortyPak Options

**Bundle** *Exploring Anatomy & Physiology in the Laboratory, 2e*, with one or more of the following supplemental titles:

- *A Visual Analogy Guide to Human Anatomy*
- *A Visual Analogy Guide to Human Physiology*
- *A Visual Analogy Guide to Human Anatomy and Physiology*
- *A Visual Analogy Guide to Chemistry*
- *A Photographic Atlas of Histology*
- *Photographic Atlas for the A&P Laboratory*
- *A Dissection Guide and Atlas to the Rat*
- *A Dissection Guide and Atlas to the Fetal Pig*
- *A Dissection Guide and Atlas to the Mink*
- *Mammalian Anatomy: The Cat*
- *An Illustrated Atlas of the Skeletal Muscles, 3e*

### Morton CustomLab

In an effort to lower the prices of books, and to provide instructors and students books tailored to their needs, we offer the enhanced Morton CustomLab program. With CustomLab, instructors can remove or combine material from our existing lab manuals or photographic atlases, or may use photographs and illustrations from our extensive online library, to create their own, personalized lab manual. Enrollment minimums apply.

### E-books

As higher education continues to evolve and incorporate technology, we are pleased to offer our titles as e-books for students who prefer an e-format.

# Acknowledgments

Although it is my name on the cover of this text, textbooks are never a solo effort. Many people were integral to the production and development of this book, and I would like to take this brief opportunity to express my gratitude.

First and foremost I would like to thank my family, particularly Elise, my mother Cathy, and Chris. Without your unwavering support, this text would not have been possible. And to Elise, thank you especially for being patient with me being behind my computer screen so often. Also, I can't forget my animals, particularly my cats, who unfailingly managed to be completely in the way of whatever I was doing.

Next I would like to extend my gratitude to the talented book team with whom I was fortunate enough to work: Joanne Saliger, who expertly designed and produced the book; Carolyn Acheson, who skillfully compiled the index; Melanie Stafford and Rayna Bailey, who carefully edited the manuscript; the team at Imagineering Media Services, Inc., who provided the beautiful illustrations; and Photo Researchers, Justin Moore, Mike Leboffe, and John Crawley, who allowed me to use their excellent photos and photomicrographs. I truly appreciate all of your hard work and generosity.

I would also like to thank the following reviewers for their valuable suggestions that helped to shape the contents of this book:

Diana M. Coffman, Lincoln Land Community College

Angela Corbin, Nicholls State University

Dr. Cassy Cozine, University of Saint Mary

Molli Crenshaw, Texas Christian University

Jill E. Feinstein, Richland Community College

Elizabeth Hodgson, York College of Pennsylvania

Steven Leadon, Durham Technical Community College

Eddie Lunsford, Southwestern Community College

Dr. Shawn Macauley, Muskegon Community College

Darren Mattone, Muskegon Community College

Justin Moore, American River College

Michele Robichaux, Nicholls State University

Deanne Roopnarine, Nova Southeastern University

Amy Fenech Sandy, Columbus Technical College, Columbus, GA

Lori Smith, American River College

Valory Thatcher, Mt. Hood Community College

Cathy Whiting, Gainesville State College

The acknowledgments would be incomplete without thanking Doug Morton, who has been kind enough to provide me with another opportunity to publish with his company. And finally, I extend a special thank you to President David Ferguson for his support, patience, friendship, and willingness to cheer for the Florida Gators.

# About the Author

Erin C. Amerman has been involved in anatomy and physiology education for more than 12 years as an author and professor, most recently at Santa Fe College in Gainesville, Florida. She received a B.S. in Cellular and Molecular Biology from the University of West Florida and a doctorate in Podiatric Medicine from Des Moines University. *Exploring Anatomy & Physiology in the Laboratory, 2e,* is her third book with Morton Publishing.

# Contents

# Introduction to Anatomical Terms

**1**

## OBJECTIVES

*Once you have completed this unit, you should be able to:*

- Demonstrate and describe anatomical position.

- Apply directional terms to descriptions of anatomical parts.

- Use regional terms to describe anatomical locations.

- Locate and describe the divisions of the major body cavities and the membranes lining each cavity.

- Demonstrate and describe anatomical planes of section.

- Identify the organ systems, their functions, and the major organs in each system.

Name _____ Section _____ Date _____

# PRE-LAB EXERCISES

Complete the following exercises prior to coming to lab, using your textbook and lab manual for reference.

## Pre-Lab Exercise 1-1

✎ Key Terms

You should be familiar with the following terms before coming to lab.

| Term | Definition |
|------|------------|

**Directional Terms**

Anterior (ventral) _____

Posterior (dorsal) _____

Superior (cranial) _____

Inferior (caudal) _____

Proximal _____

Distal _____

Superficial _____

Deep _____

**Body Cavities and Membranes**

Dorsal body cavity _____

_____

Ventral body cavity _____

_____

Serous membrane _____

_____

**Planes of Section**

Sagittal plane _____

_____

Frontal (coronal) plane _____

_____

Transverse plane _____

_____

# Pre-Lab Exercise 1-2

## Organic Systems

The body has 11 organ systems, each of which contains certain organs, and each with a specific subset of functions. Note that some organs are part of more than one system.

In this exercise you will identify the 11 organ systems, their major organ(s), and the basic function(s) of each system (Table 1.1). Use your textbook and Exercise 1-6 from this unit for reference.

TABLE **1.1** Organ Systems

| Organ System | Major Organs | Organ System Functions |
|---|---|---|
| Integumentary system | | |
| Skeletal system | | |
| Muscular system | | |
| Nervous system | | |
| Cardiovascular system | | |

*(continues)*

TABLE **1.1** Organ Systems *(cont.)*

| Organ System | Major Organs | Organ System Functions |
|---|---|---|
| Respiratory system | | |
| Lymphatic system | | |
| Urinary system | | |
| Digestive system | | |
| Endocrine system | | |
| Reproductive system | | |

# EXERCISES

*The bullet entered the right posterior scapular region, 3 centimeters lateral to the vertebral region, 4 centimeters inferior to the cervical region, and penetrated deep to the muscle and bone, but superficial to the parietal pleura . . .*

Would you believe that by the end of this unit, you will be able to translate the above sentence and also locate the hypothetical wound? Unit 1 will introduce you to the world of anatomy and physiology. We will begin with an introduction to the unique language of anatomy and physiology. Like learning any new language, this may seem overwhelming at first. The key to success is repetition and application: The more you use the terms, the easier it will be for them to become part of your normal vocabulary.

From the terminology we will move on to the organization of the body into body cavities and organ systems. Once you have completed this unit, return to the above sentence, and challenge yourself to locate the precise position of the bullet wound on an anatomical model.

## Exercise **1-1**

### Anatomical Position

In the study of anatomy and physiology, most anatomical specimens are presented in a standard position termed anatomical position. In anatomical position the specimen is presented facing forward, with the toes pointing forward, the feet shoulder-width apart, and the palms facing forward, as shown in **Figure 1.1**. This presentation of anatomical specimens creates a standard point of reference that facilitates communication among scientists and healthcare professionals.

 **Procedure** Demonstrating
Anatomical Position

Have your lab partner stand in a normal, relaxed way, then adjust his or her position so it matches anatomical position. When you have completed this exercise, answer Check Your Understanding question 1 (p. 27).

FIGURE **1.1** Anatomical position

# Exercise **1-2**

## Directional Terms

Another method that makes communication easier and less prone to errors is to use certain terms to define the location of body parts and body markings. For example, when describing a wound on the chest, we could say:

▌ The wound is near the middle and top of the chest; or

▌ The wound is on the right *anterior* thoracic region, 4 centimeters *lateral* to the sternum, and 3 centimeters *inferior* to the acromial region.

The second option is precise and allows the reader to locate the wound exactly. Note that these descriptions are referring to a figure in anatomical position.

Review the definitions of the directional terms that you completed in Pre-Lab Exercise 1-1 and in **Figure 1.2**. Use these terms to fill in the correct directional term in the following practice procedure.

## **Procedure** Directional Terms

Fill in the correct directional term for each of the following items. Note that in some cases, more than one directional term may apply.

The elbow is _____ to the wrist.

The chin is _____ to the nose.

The shoulder is _____ to the sternum (breastbone).

The forehead is _____ to the mouth.

The skin is _____ to the muscle.

The spine is _____ to the esophagus.

The nose is _____ to the cheek.

The spine is on the _____ side of the body.

The arm is _____ to the torso.

The knee is _____ to the hip.

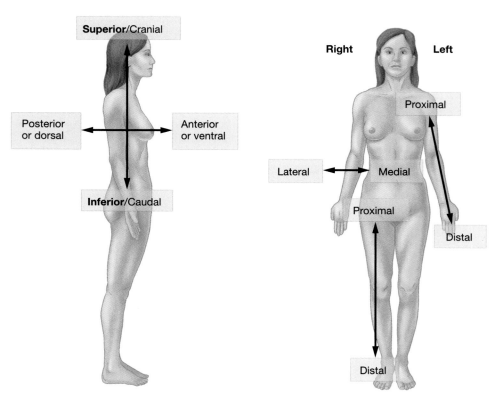

FIGURE **1.2** Directional terms

# Exercise 1-3

## Regional Terms

**MATERIALS**

❏ Laminated outline of the human body
❏ Water-soluble marking pens

You may have noticed in the previous exercise that we used the anatomical terms *thoracic region* and *acromial region* instead of using generic terms such as "chest" and "shoulder." This is a standard practice, again intended to make descriptions as specific as possible and to reduce the potential for errors in communication. For example, "shoulder" could consist of quite a large anatomical area, whereas the "acromial region" refers to one specific location on the shoulder.

The following regional terms, illustrated in **Figure 1.3**, are among the more common terms you will encounter in your study of anatomy and physiology. Note that most of these terms are adjectives rather than nouns. This means that the term is not complete unless it is paired with the term "region." For example, we cannot say, "The wound is in the antebrachial"; we instead must say, "The wound is in the antebrachial *region*."

The following list may look daunting, but you are probably familiar with several of the terms already. For example, you likely know the locations of the "oral," "nasal," and "abdominal" regions. Watch for other terms that you may know.

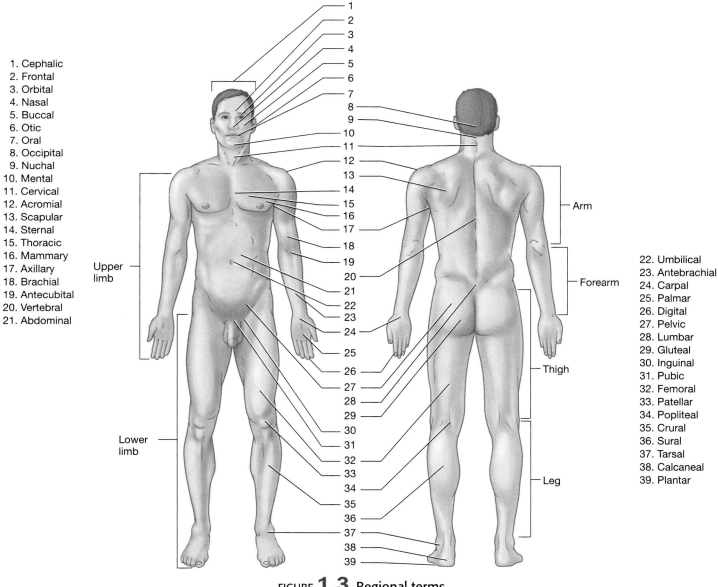

1. Cephalic
2. Frontal
3. Orbital
4. Nasal
5. Buccal
6. Otic
7. Oral
8. Occipital
9. Nuchal
10. Mental
11. Cervical
12. Acromial
13. Scapular
14. Sternal
15. Thoracic
16. Mammary
17. Axillary
18. Brachial
19. Antecubital
20. Vertebral
21. Abdominal

22. Umbilical
23. Antebrachial
24. Carpal
25. Palmar
26. Digital
27. Pelvic
28. Lumbar
29. Gluteal
30. Inguinal
31. Pubic
32. Femoral
33. Patellar
34. Popliteal
35. Crural
36. Sural
37. Tarsal
38. Calcaneal
39. Plantar

FIGURE **1.3** Regional terms

## **Procedure** Labeling Body Regions

Use water-soluble markers to locate and label each of the following regions on laminated outlines of the human body. If outlines are unavailable, label the regions on **Figure 1.4**. When you have completed this exercise, answer Check Your Understanding questions 2–4 (pp. 27–28).

### Adjectives

| | | | | |
|---|---|---|---|---|
| Abdominal | Cervical | Mental | Plantar | |
| Acromial | Cranial | Nasal | Popliteal | |
| Antebrachial | Crural | Nuchal | Pubic | |
| Antecubital | Digital | Occipital | Scapular | |
| Axillary | Femoral | Oral | Sternal | |
| Brachial | Frontal | Orbital | Sural | |
| Buccal | Gluteal | Otic | Tarsal | |
| Calcaneal | Inguinal | Palmar | Thoracic | |
| Carpal | Lumbar | Patellar | Umbilical | |
| Cephalic | Mammary | Pelvic | Vertebral | |

### Nouns

Arm
Forearm
Leg
Lower limb
Thigh
Upper limb

FIGURE **1.4** Anterior and posterior views of the human body in anatomical position

# Exercise 1-4

## Body Cavities and Membranes

### MATERIALS

❑ Human torso models

❑ Fetal pigs (or other preserved small mammals)

❑ Dissection kits/dissection trays

The body is divided into several fluid-filled cavities, each of which contains specific organs. In this exercise you will identify the body cavities and the organs contained within each cavity.

As you can see in **Figure 1.5**, there are two major body cavities, each of which is subdivided into smaller cavities, as follows:

1. **Dorsal (posterior) cavity.** As implied by its name, the dorsal body cavity is largely on the posterior, or dorsal, side of the body. It contains two smaller cavities:

   a. **Cranial cavity.** The cranial cavity is the area encased by the skull. It contains the brain and the special sense organs such as the eyes and the organs for hearing.

   b. **Vertebral** (or **spinal**) **cavity.** The vertebral cavity is the area encased by the vertebrae. It contains the spinal cord.

2. **Ventral (anterior) cavity.** The ventral body cavity is largely on the anterior, or ventral, side of the body. It has two main divisions: the **thoracic cavity**, which is superior to the diaphragm, and the **abdominopelvic cavity**, which is inferior to the diaphragm. Within the thoracic and abdominopelvic cavities are smaller subcavities formed by thin sheets of tissue called **serous membranes**. Cells of these membranes produce a thin, watery fluid called **serous fluid** that lubricates organs so they move with a minimum of friction. Serous membranes are composed of two layers: an outer **parietal layer**, which is attached to the surrounding structures, and an inner **visceral layer**, which is attached to the organ or organs of the ventral cavity. Between the parietal and visceral layers is a thin potential space that contains serous fluid. This potential space is also referred to as a cavity.

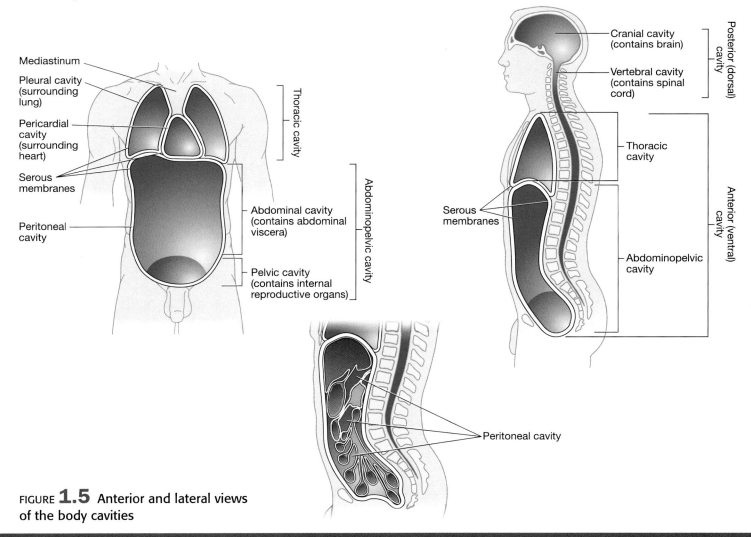

FIGURE **1.5** Anterior and lateral views of the body cavities

The following are the divisions of the ventral cavity:

a. **Thoracic cavity.** The thoracic cavity encompasses the area encased by the ribs. We find the following smaller cavities within the thoracic cavity:

   (1) **Pleural cavities.** Each pleural cavity surrounds one of the lungs. The pleural cavities are located between two serous membranes called the **pleural membranes.** The **parietal pleura** is attached to the body wall, and the **visceral pleura** is attached to the surface of the lung.

   (2) **Mediastinum.** The area between the pleural cavities, called the mediastinum, contains the great vessels, the esophagus, the trachea and bronchi, and other structures. It houses another set of serous membranes that form the **pericardial cavity,** which surrounds the heart. The pericardial cavity is between the **pericardial membranes;** the **parietal pericardium** is attached to surrounding structures, and the inner **visceral pericardium** is attached to the heart muscle.

b. **Abdominopelvic cavity.** The abdominopelvic cavity encompasses the area inferior to the diaphragm and extends into the bony pelvis. It contains the **peritoneal membranes;** the **parietal peritoneum** is attached to the body wall and surrounding structures, and the inner **visceral peritoneum** is attached to the surface of many of the organs in the cavity. Between these two layers of peritoneal membranes we find the peritoneal cavity. Note that some organs are posterior to the **peritoneal cavity;** such organs are said to be **retroperitoneal.** The abdominopelvic cavity is divided into two smaller cavities:

   (1) **Abdominal cavity.** The area superior to the bony pelvis, called the abdominal cavity, houses many of the organs of the digestive, lymphatic, and urinary systems.

   (2) **Pelvic cavity.** The cavity housed within the bony pelvis, the pelvic cavity, contains certain organs of the reproductive system as well as certain organs of the digestive and urinary systems.

We often divide the abdominopelvic cavity into four quadrants, the right upper, right lower, left upper, and left lower quadrants. We can also divide the abdominopelvic cavity into nine regions based on a series of lines drawn over the surface of the abdomen. The regions are listed and labeled in **Figure 1.6.**

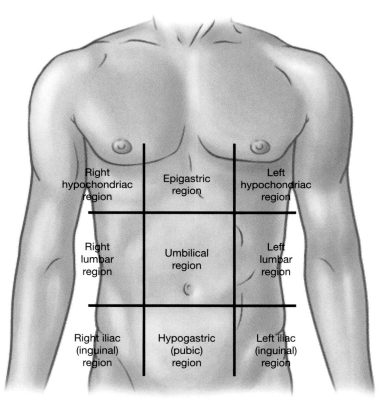

Right hypochondriac region

Epigastric region

Left hypochondriac region

Right lumbar region

Umbilical region

Left lumbar region

Right iliac (inguinal) region

Hypogastric (pubic) region

Left iliac (inguinal) region

FIGURE **1.6** Regions of the abdominopelvic cavity

## **Procedure** Body Cavities

In this procedure you will use a human torso model or a preserved small mammal such as a cat, a fetal pig, or a rat to examine the body cavities. If you are using a preserved small mammal, you may use the following procedure to open the body cavities. Note that you will not open the animal fully in this procedure to preserve structures for future dissections.

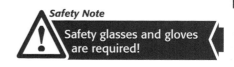

**Safety Note**

Safety glasses and gloves are required!

1. Place the animal on a dissecting tray with its dorsal (back) side facing you. Use your scalpel to make a shallow, +-shaped incision. Make the first incision through the skin along the nape of the neck. Make the second incision from the anterior part of the skull down along the animal's midline to about 4 inches inferior to the first incision. Ensure that your incisions are shallow so you do not damage structures deep to the skin. List the organs you are able to see in Table 1.2.

2. Gently peel back the skin using a blunt dissection probe, and note the appearance of the skin and the exposed muscles, bones, and joints. If your instructor wants you to expose the brain and spinal cord, use either a scalpel or a saw to carefully cut through the skull and a section of the vertebral column.

3. When you have finished your examination of the animal's dorsal side, close the skin, and wrap the area with wet towels soaked with a preservative solution.

4. Flip the animal over, and place it in the dissecting tray with its ventral side facing you. Use a scalpel to make a shallow incision along the animal's midline from the superior neck down to its groin. Make a second incision across the animal's chest and a third incision across the animal's abdomen.

5. Using a blunt dissection probe, peel back the skin of the abdomen carefully to expose the abdominopelvic cavity, and examine its contents.

6. Gently peel back the skin of the chest and neck incisions. Use scissors or a scalpel to carefully cut through the sternum, expose the thoracic cavity, and examine its contents.

When you have opened your preserved small mammal or human torso model, locate and identify each cavity, and do the following:

1. List the organs you are able to see in Table 1.2. See Figures 1.7 and 1.8 for reference.

2. Mark each region of the abdominopelvic cavity with a pin or marking tape (if working with a torso). Note which organs are visible in each region.

TABLE **1.2** Body Cavities and Regions of the Abdominopelvic Cavity

| Cavity | Organ(s) |
|---|---|
| **Dorsal Cavity** | |
| **1.** Cranial cavity | |
| **2.** Vertebral cavity | |
| **Ventral Cavity** | |
| **1.** Thoracic cavity | |
|    a. Pleural cavities | |
|    b. Mediastinum | |
|       (1) Pericardial cavity | |
| **2.** Abdominopelvic cavity | |
|    a. Subdivisions | |
|       (1) Abdominal cavity | |
|       (2) Pelvic cavity | |
|    b. Regions | |
|       (1) Right hypochondriac region | |
|       (2) Epigastric region | |
|       (3) Left hypochondriac region | |
|       (4) Right lumbar region | |
|       (5) Umbilical region | |
|       (6) Left lumbar region | |
|       (7) Right iliac region | |
|       (8) Hypogastric region | |
|       (9) Left iliac region | |

Larynx

Trachea

Blood vessels

Heart

Lungs

Diaphragm

Liver

Small intestine

Large intestine

FIGURE **1.7** Ventral view of the fetal pig

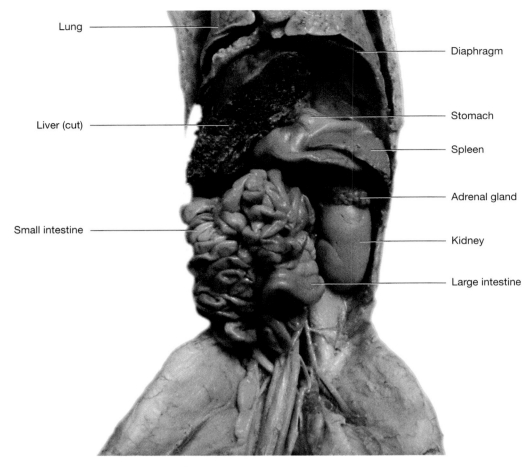

Lung

Diaphragm

Liver (cut)

Stomach

Spleen

Adrenal gland

Small intestine

Kidney

Large intestine

FIGURE **1.8** Abdominopelvic cavity of the fetal pig

## **Procedure** Serous Membranes

**Part 1:** Serous membranes are best examined on a preserved specimen such as a fetal pig or a cat, because their structure is difficult to appreciate on a model. As you dissect the fetal pig or cat, look for the serous membranes listed in **Table 1.3**. Take care not to tear the fragile membranes, which consist of just a few layers of cells. As you identify each membrane, record in the table where you found the membrane and the structure to which the membrane is attached (the lungs, heart, abdominal wall, etc.).

TABLE **1.3** Serous Membranes

| Membrane | Cavity | Structure |
|---|---|---|
| Parietal pleura | | |
| Visceral pleura | | |
| Parietal pericardium | | |
| Visceral pericardium | | |
| Parietal peritoneum | | |
| Visceral peritoneum | | |

**Part 2:**

Draw in the body cavities on your laminated outlines of the human body or **Figure 1.9**. Label the serous membranes surrounding the cavity where applicable.

FIGURE **1.9** Anterior view of the human torso

## **Procedure** Applications of Terms, Cavities, and Membranes

Assume that you are acting as coroner, and you have a victim with three gunshot wounds. In this scenario your "victim" will be your fetal pig or a human torso model that your instructor has "shot." For each "bullet wound," state the anatomical region and/or body cavity in which the "bullet" was found and any serous membranes involved, and describe the location of the wound using at least three directional terms. As coroner, you have to be as specific as possible, and keep your patient in anatomical position!

| | |
|---|---|
| Shot **1** | |
| Shot **2** | |
| Shot **3** | |

---

# Exercise **1-5**

## Planes of Section

### MATERIALS
- ❏ Various anatomical models demonstrating planes of section
- ❏ Modeling clay
- ❏ Knife or scalpel

Often in science and in the medical field, it is necessary to obtain different views of the internal anatomy of an organ or a body cavity. These views are obtained by making an anatomical section along a specific plane. The commonly used planes of section, shown in **Figures 1.10** and **1.11**, are as follows:

1. **Sagittal plane.** A section along the sagittal plane is parallel to the body's longitudinal axis and divides the body part into right and left parts. The sagittal section has two variations:

   a. **Midsagittal sections** divide the body part into equal right and left halves.

   b. **Parasagittal sections** divide the body part into unequal right and left parts.

2. **Frontal plane.** The frontal plane, also known as the **coronal plane,** is also parallel to the body's longitudinal axis. It divides the body part into an anterior (front) part and a posterior (back) part.

3. **Transverse plane.** The transverse plane, also known as a **cross section** or the **horizontal plane,** is perpendicular to the body's longitudinal axis. It divides the body part into a superior (or proximal) part and an inferior (or distal) part.

4. **Oblique section.** An oblique section cuts at an angle and is intended to allow examination of structures difficult to see with standard angles. Note that an oblique section is not illustrated in Figure 1.10.

Note that although there is only a single midsagittal plane, there are a near infinite number of possible parasagittal, frontal, transverse, and oblique planes. In your study of anatomy and physiology, you will see many different examples of these planes of section.

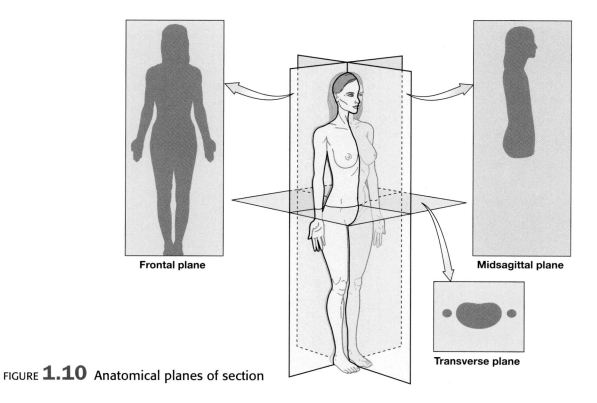

**Frontal plane**

**Midsagittal plane**

**Transverse plane**

FIGURE **1.10** Anatomical planes of section

FIGURE **1.11** Planes of anatomical section visible on computed tomography scans of the brain: (**A**) frontal section; (**B**) midsagittal section; (**C**) transverse section

## **Procedure** Sectioning Along Anatomical Planes

Use a scalpel to cut a ball of modeling clay in each of following anatomical planes. Before you make your cuts, you may wish to mold your clay into the shape of a head and to draw eyes on the head to denote anterior and posterior sides. Use **Figures 1.10** and **1.11** for reference. When you have completed this exercise, answer Check Your Understanding questions 5 and 6 (p. 29).

1. Sagittal
   a. Midsagittal
   b. Parasagittal
2. Frontal
3. Transverse
4. Oblique

## Procedure Identifying Examples of Anatomical Planes of Section

Identify at least two examples of each plane of section listed in **Table 1.4** on anatomical models that your instructor provides. In addition, state which organs are visible in the section.

TABLE **1.4** Anatomical Models and Planes of Section

| Examples of Midsagittal and/or Parasagittal Sections | |
|---|---|
| **Model Name or Description** | **Organ(s) Visible** |
| 1. | |
| 2. | |
| **Examples of Frontal Sections** | |
| **Model Name or Description** | **Organ(s) Visible** |
| 1. | |
| 2. | |
| **Examples of Transverse Sections** | |
| **Model Name or Description** | **Organ(s) Visible** |
| 1. | |
| 2. | |

# Exercise 1-6

## Organs and Organ Systems

### MATERIALS
- ❏ Human torso models
- ❏ Fetal pigs (or other preserved small mammals)
- ❏ Dissection kits/dissection trays

The human body has 11 organ systems, each with specific organs and functions (Figure 1.12). In this exercise you will examine the organ systems and identify their major organs.

## Procedure Organs

Identify the following organs on your preserved mammal specimen or human torso models. Check off each organ as you identify it, and record the organ system to which it belongs in Table 1.5. Remember that some organs may function in more than one system.

- ☐ Adrenal glands
- ☐ Blood vessels
- ☐ Bones
- ☐ Brain
- ☐ Esophagus
- ☐ Gallbladder
- ☐ Heart
- ☐ Intestines
- ☐ Joints
- ☐ Kidneys
- ☐ Larynx
- ☐ Liver
- ☐ Lungs
- ☐ Muscles
- ☐ Pancreas
- ☐ Skin
- ☐ Spinal cord
- ☐ Spleen
- ☐ Stomach
- ☐ Testes (male) or ovaries (female)
- ☐ Thymus
- ☐ Thyroid gland
- ☐ Trachea
- ☐ Urinary bladder

TABLE **1.5** Organs and Organ Systems

| Organ System | Major Organ(s) |
|---|---|
| Integumentary system | |
| Skeletal system | |
| Muscular system | |
| Nervous system | |
| Cardiovascular system | |
| Lymphatic system | |
| Respiratory system | |
| Digestive system | |
| Urinary system | |
| Endocrine system | |
| Reproductive system | |

# **Procedure** Organ Systems

The body's organ systems are illustrated in **Figure 1.12**. Fill in the blanks next to each organ system to identify the major organs shown and principal functions of each system. When you have completed this exercise, answer Check Your Understanding question 7 (p. 29).

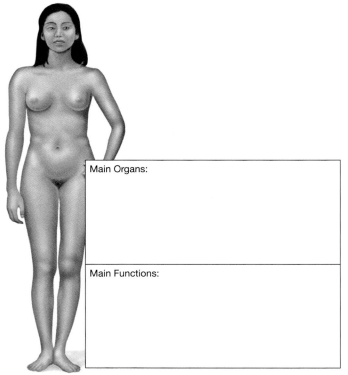

Main Organs:

Main Functions:

**Integumentary System**

Main Organs:

Main Functions:

**Skeletal System**

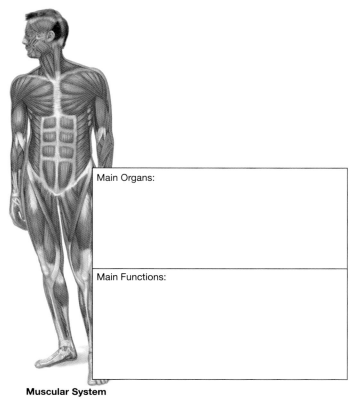

Main Organs:

Main Functions:

**Muscular System**

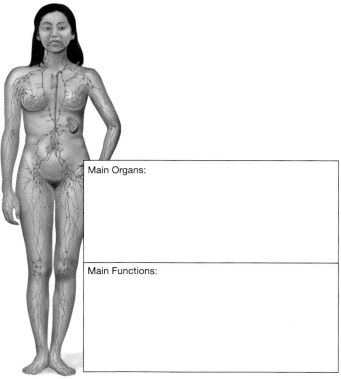

Main Organs:

Main Functions:

**Lymphatic System**

FIGURE **1.12** Organ systems of the body *(continues)*

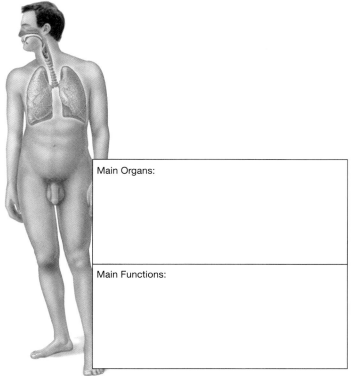

**Respiratory System**

Main Organs:

Main Functions:

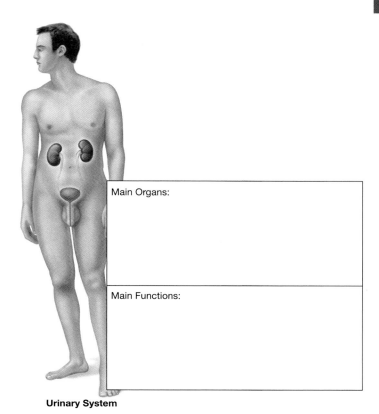

**Urinary System**

Main Organs:

Main Functions:

**Nervous System**

Main Organs:

Main Functions:

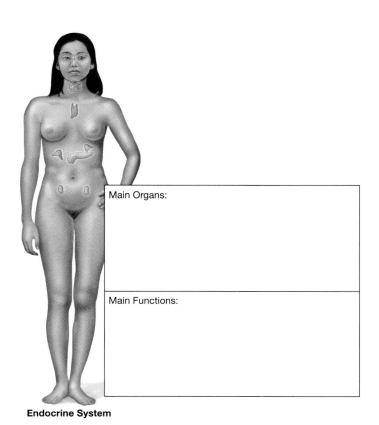

**Endocrine System**

Main Organs:

Main Functions:

FIGURE **1.12** *(cont.)* **Organ systems of the body** *(continues)*

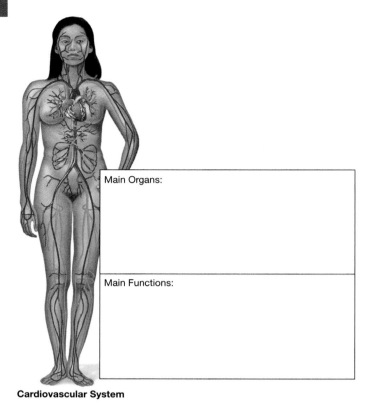

Main Organs:

Main Functions:

**Cardiovascular System**

Main Organs:

Main Functions:

**Digestive System**

Main Organs:

Main Functions:

**Male Reproductive System**

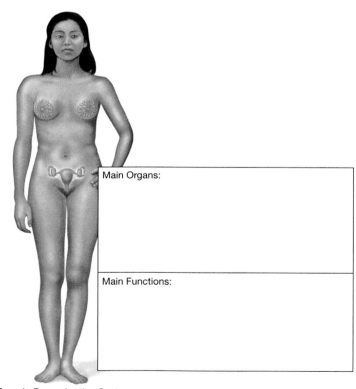

Main Organs:

Main Functions:

**Female Reproductive System**

FIGURE **1.12** *(cont.)* Organ systems of the body

## Check Your Recall

**1** Which of the following best describes anatomical position?

a. Body facing forward, toes pointing forward, palms facing backward

b. Body, toes, and palms facing backward

c. Body facing forward, arms at the sides, palms facing forward

d. Body facing backward and palms facing outward

**2** Match the directional term with its correct definition.

_____ Distal

_____ Lateral

_____ Anterior

_____ Proximal

_____ Inferior

_____ Deep

_____ Superficial

_____ Posterior

_____ Medial

_____ Superior

A. Away from the surface/toward the body's interior

B. Toward the back of the body

C. Closer to the point of origin (e.g., of a limb)

D. Away from the body's midline

E. Toward the head

F. Farther from the point of origin (e.g., of a limb)

G. Toward the body's midline

H. Away from the head/toward the tail

I. Toward the front of the body

J. Toward the surface/skin

**3** Which of the following is an incorrect use of a directional term?

a. The ankle is inferior to the knee.

b. The sternum is superior to the abdomen.

c. The bone is deep to the muscle.

d. The mouth is medial to the ears.

**4** Label the following anatomical regions on **Figure 1.13**:

- ☐ Brachial region
- ☐ Carpal region
- ☐ Cervical region
- ☐ Digital region
- ☐ Forearm
- ☐ Inguinal region
- ☐ Leg
- ☐ Lumbar region
- ☐ Orbital region
- ☐ Otic region
- ☐ Sternal region
- ☐ Upper limb

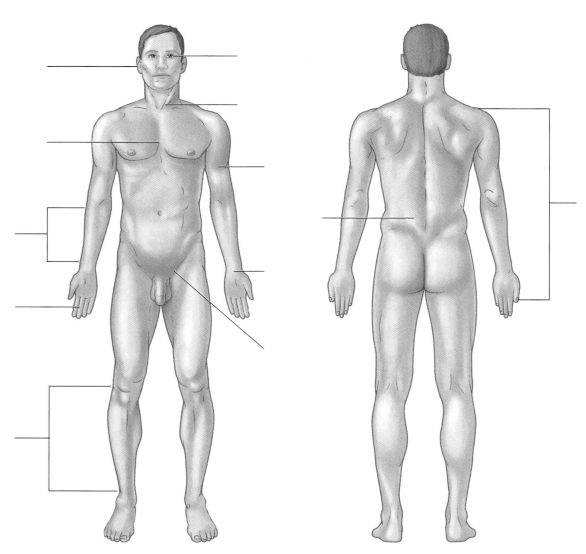

FIGURE **1.13** Anterior and posterior views of the body

**5** Anatomical position and specific directional and regional terms are used in anatomy and physiology to:
   a. standardize units of measure.
   b. provide a standard that facilitates communication and decreases the chances for errors.
   c. provide a standard that is used to develop drug delivery systems.
   d. make students' lives difficult.

**6** Label the following body cavities on **Figure 1.14**, and indicate with an asterisk (*) which cavities are surrounded by serous membranes.

☐ Abdominal cavity          ☐ Dorsal cavity          ☐ Pericardial cavity          ☐ Thoracic cavity
☐ Abdominopelvic cavity     ☐ Mediastinum            ☐ Peritoneal cavity          ☐ Ventral cavity
☐ Cranial cavity            ☐ Pelvic cavity          ☐ Pleural cavity             ☐ Vertebral cavity

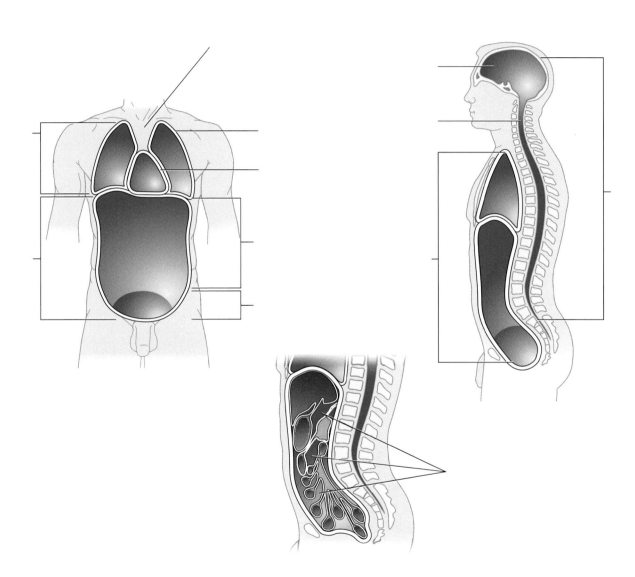

FIGURE **1.14** Anterior and lateral views of the body cavities

**1**

**7** *Fill in the blanks:* A serous membrane secretes _____, which _____ the

organs in certain _____ body cavities.

**8** Define the following planes of section:

**a** Midsagittal plane _____

**b** Parasagittal plane _____

**c** Frontal plane _____

**d** Transverse plane _____

**e** Oblique plane _____

**9** The following organs belong to the _____ system: esophagus, gallbladder, liver.
   a. integumentary
   b. reproductive
   c. lymphatic
   d. digestive

**10** The following organs belong to the _____ system: pancreas, thyroid gland, adrenal glands.
   a. endocrine
   b. urinary
   c. lymphatic
   d. cardiovascular

# Check Your Understanding

### Critical Thinking and Application Questions

FIGURE **1.15** Figure not in anatomical position

**1** Figure 1.15 is not in anatomical position. List all of the deviations from anatomical position.

_____

_____

_____

_____

_____

_____

**2** Locate the following wounds, and mark them on the body in **Figure 1.16**.

**a** The wound is located in the left inferior, posterior lumbar region. It is 5 centimeters superior to the gluteal region and 3 centimeters lateral to the vertebral region.

**b** The wound is located on the right anterior, medial crural region, 10 centimeters proximal to the tarsal region.

FIGURE **1.16** Anterior and posterior views of the body

**1** **3** You are reading a surgeon's operative report. During the course of the surgery, she made several incisions. Your job is to read her operative report and determine where the incisions were made. Draw and label the incisions on **Figure 1.17**.

**a** The first incision was made in the right anterior cervical region, 3 centimeters lateral to the trachea. The cut extended vertically, 2 centimeters inferior to the mental region to 3 centimeters superior to the thoracic region.

**b** The second incision was made in the left anterior axillary region and extended medially to the sternal region. At the sternal region the cut turned inferiorly to 4 centimeters superior to the umbilical region.

**c** The third incision was made in the left posterior scapular region. The cut was extended medially to 2 centimeters lateral to the vertebral region, where it turned superiorly and progressed to one centimeter inferior to the cervical region.

FIGURE **1.17** Anterior and posterior views of the body

**4** Describe the location of the wounds on **Figure 1.18** using correct regional and directional terms.

**a** _____
_____
_____
_____
_____
_____

**b** _____
_____
_____
_____
_____
_____

A

B

FIGURE **1.18** Anterior and posterior views of the body

**5** What may be some real-world applications of anatomical sections? (*Hint:* Think of the medical field.)

_____

_____

_____

_____

_____

**6** Which anatomical section(s) would provide a view of the internal anatomy of *both* kidneys?

_____

**7** The type of anatomy we are studying in this lab manual is called systemic anatomy, which means that we cover the organs related to a specific organ system. Some, however, choose to study anatomy from a regional point of view (e.g., the abdominal region or the thoracic region). Find at least two organ systems that contain organs in different regions of the body, and state where the organs are located in the body.

> *Example:* The nervous system has organs in the cranial cavity (the brain), the spinal cavity (the spinal cord), the thoracic and abdominopelvic cavities (spinal and cranial nerves), and the upper and lower limbs (spinal nerves).

_____

_____

_____

_____

_____

_____

# Chemistry

**2**

## OBJECTIVES

*Once you have completed this unit, you should be able to:*

- Demonstrate the proper interpretation of pH paper.

- Describe and apply the pH scale.

- Describe the purpose and effects of a buffer.

- Explain and demonstrate the purpose of an enzyme.

- Describe the differences in solubility of ionic, polar, and nonpolar covalent molecules in different solvents.

- Apply laboratory techniques to perform aspirin synthesis and DNA extraction.

**2**

# PRE-LAB EXERCISES

Complete the following exercises prior to coming to lab, using your textbook and lab manual for reference.

## Pre-Lab Exercise **2-1**

✎ Key Terms

You should be familiar with the following terms before coming to lab.

| Term | Definition |
|------|------------|
| pH | |
| Acid | |
| Base | |
| Buffer | |
| Chemical reaction | |
| Enzyme | |
| Polar covalent bond | |
| Nonpolar covalent bond | |
| Ionic bond | |
| Cation | |
| Anion | |

Valence electron _____

Solute _____

Solvent _____

Solubility _____

Deoxyribonucleic acid (DNA) _____

# Pre-Lab Exercise 2-2

## The pH Scale and –logarithms

The **pH scale** is a numerical scale from 0–14 that represents the hydrogen ion concentration of a solution. Solutions with a pH of 0 have the highest hydrogen ion concentration, and solutions with a pH of 14 have the lowest hydrogen ion concentration. This is because the pH isn't the actual hydrogen ion concentration; instead, it is a negative logarithm of a solution's hydrogen ion concentration. This actually isn't as difficult as it sounds. Grab your calculator and enter the following:

–log(0.01)      =      _____

–log(0.001)     =      _____

–log(0.0001)    =      _____

–log(0.00001)   =      _____

Notice the general trend here: *As the number gets smaller, its negative log gets bigger.*

   Here's a quick example using hydrogen ions: Solution X has a hydrogen ion concentration of 0.05 M. Solution Y has a hydrogen ion concentration of 0.0002 M.

▮ Which solution has more hydrogen ions? (*Hint*: It's just the solution with the bigger number.) _____

▮ What is the –log of solution X? _____

▮ What is the –log of solution Y? _____

As you can see, the solution with the *higher* hydrogen ion concentration has the *lower* –log. This means that:

▮ Solution X is the more _____ (acidic/basic) solution and has a _____ (lower/higher) pH.

▮ Solution Y is the more _____ (acidic/basic) solution and has a _____ (lower/higher) pH.

One of the more challenging concepts to grasp in chemistry is **chemical bonding**, a topic that pops up again and again in A&P. We revisit it in cytology, the nervous system, blood, respiration, and digestion—to name just a few.

First let's do some basics. Use your textbook and this unit to answer the following questions.

1. Do metals donate or accept electrons? What do they become after they donate/accept electrons?

2. Do nonmetals donate or accept electrons? What do they become after they donate/accept electrons?

3. Electron shells:

   How many electrons can go in the first shell? _____

   How many electrons can go in the second shell? _____

   How many electrons can go in the third shell? _____

4. How many electrons does each element require in its valence shell to become stable? _____

Okay, now let's apply all of this to an example.

## Example 1: Ionic Bond

Let's take a look at sodium (Na) and chlorine (Cl). Draw what I am describing, and you will see it better.

A sodium atom has how many protons? _____

A sodium atom has how many electrons? _____

How many electrons will go in the first shell? _____

How many in the second shell? _____

How many in the third? _____

Now draw this out on the diagram in **Figure 2.1**, and take a look at it, in particular the third (valence) shell.

We know that Na requires eight electrons in its valence shell to become stable. But how many does it have?

So, to fill this shell, will it be easier for sodium to steal seven more electrons from another atom, or will it be easier for sodium to give up that one electron and get rid of that third shell? Sodium is simply going to *give away* that last electron. This means that it will lose an electron (negative charge) but will keep the same number of protons (positive charges). What will the sodium ion's overall charge be now?

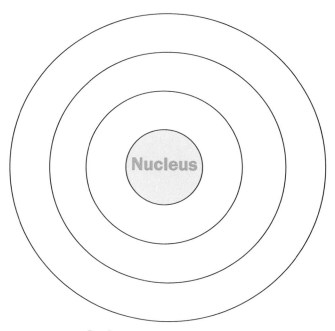

FIGURE **2.1** An orbital model of an atom

Now let's look at chlorine:

A chlorine atom has how many protons? _____

Therefore, we know that a chlorine atom has how

    many electrons? _____

How many electrons will go in the first shell? _____

How many in the second shell? _____

How many in the third? _____

Draw this in **Figure 2.2.**

We know that a chlorine atom requires eight electrons in its valence shell to become stable. Will it be easier for chlorine to steal one electron from another atom, or will it be easier for chlorine to give away seven electrons? It will be *much* easier for chlorine to take an electron from another atom than to try to pawn off seven electrons. This means that chlorine will gain an electron (negative charge) but will keep the same number of protons (positive charges). What charge will chlorine (now called a *chloride ion*) have now?

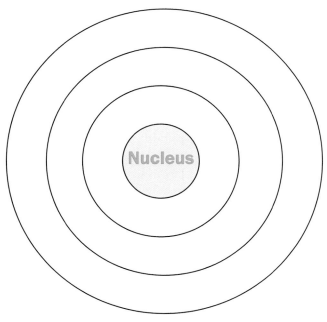

FIGURE 2.2 **An orbital model of an atom**

Let's tie this in with what we learned about metals and nonmetals earlier.

We said that metals _____ electrons to become _____ .

Is sodium a metal or a nonmetal? _____

Looking at our example above, did sodium donate or accept an electron? _____

We also said that nonmetals _____ electrons to become _____ .

Is chlorine a metal or a nonmetal? _____

Looking at our example above, did chlorine donate or accept an electron? _____

Why do we care? Well, when sodium donates an electron to become a positively charged **cation**, and chlorine accepts an electron to become a negatively charged **anion**, this creates an attraction. (Think of it like "opposites attract.") This attraction leads the two ions to *bond* to one another, forming an **ionic bond**. See how simple that is?

## Example 2: Polar Covalent Bond

Now let's tackle another difficult bonding concept, the **polar covalent bond**. The classic example of a polar covalent bond is represented by the two bonds in a water molecule (**Figure 2.3**):

A hydrogen (H) atom has how many protons? _____

A hydrogen atom has how many electrons? _____

How many electrons will go in the first shell? _____

How many electrons would hydrogen require to fill up that first shell? _____

An oxygen (O) atom has how many protons? _____

Therefore, we know that oxygen has how many electrons? _____

How many electrons will go in the first shell? _____

How many electrons will go in the second shell? _____

How many electrons would oxygen require to fill up that second shell? _____

In this case, both hydrogen and oxygen are nonmetals, and neither element will give up electrons in this situation. So, instead of one element taking the electrons, they share them. Because oxygen requires two electrons to fill that valence shell, and hydrogen has only one electron to share, oxygen has to form a partnership with two hydrogen atoms, forming $H_2O$. But oxygen has a higher **electronegativity**, which means that it attracts electrons more strongly than hydrogen does. This means that the electrons will spend more time around the oxygen and less around the hydrogen. Thus, a polar covalent bond is born. This means that:

Oxygen will have a partial _____ charge, and

the two hydrogens will have a partial _____

charge, which you can see in **Figure 2.3**.

A **nonpolar covalent bond** involves the same principles, except that the atoms share the electrons equally. As a result, there are no partial positive or negative charges because the electrons don't spend any more time around one or more of the atoms.

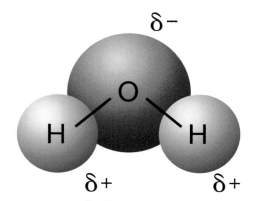

FIGURE **2.3** A water molecule

# EXERCISES

Including a chemistry unit with an Anatomy and Physiology course may seem odd, but consider for a moment the simplest level of organization—the chemical level. Our cells, tissues, and organs, as well as our extracellular environments, are all composed of chemicals that undergo countless chemical reactions every second. So, to be able to understand our anatomy and physiology, we first must understand the most basic structures in our bodies—chemicals.

In the following exercises you will learn chemistry principles as you measure the pH of various solutions, demonstrate the effects of buffers on the pH of solutions, examine the effects of enzymes on chemical reactions, and determine different chemicals' solubilities. Then you will witness chemistry in action by synthesizing aspirin and extracting DNA. It's not every day that you get to say that you made aspirin and extracted DNA!

**Safety Note**

Many of the chemicals you will use in this lab are poisons and can burn your skin, eyes, and mucous membranes. Take care in handling all chemicals. Do not handle any chemicals without safety glasses, gloves, and a lab coat.

## Exercise **2-1**

### pH, Acids, and Bases

#### MATERIALS

- ❏ 3 glass test tubes
- ❏ Test tube rack
- ❏ Graduated cylinder
- ❏ Dropping pipette
- ❏ pH paper
- ❏ Samples of various acids and bases
- ❏ 0.1M hydrochloric acid
- ❏ Samples of various antacid tablets

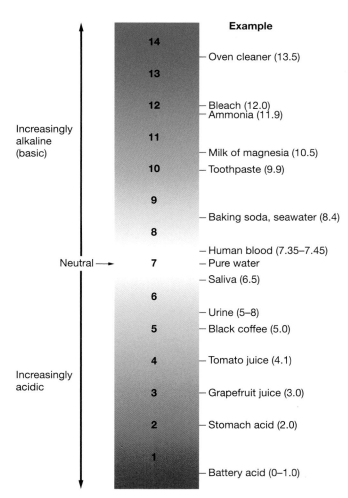

| | Example |
|---|---|
| 14 | Oven cleaner (13.5) |
| 13 | |
| 12 | Bleach (12.0) / Ammonia (11.9) |
| 11 | Milk of magnesia (10.5) |
| 10 | Toothpaste (9.9) |
| 9 | |
| 8 | Baking soda, seawater (8.4) |
| 7 | Human blood (7.35–7.45) / Pure water |
| 6 | Saliva (6.5) |
| 5 | Urine (5–8) / Black coffee (5.0) |
| 4 | Tomato juice (4.1) |
| 3 | Grapefruit juice (3.0) |
| 2 | Stomach acid (2.0) |
| 1 | Battery acid (0–1.0) |

Increasingly alkaline (basic)

Neutral → 7

Increasingly acidic

FIGURE **2.4** The pH scale and examples of solutions with different pH values

The **pH** is a measure of the concentration of hydrogen ions present in a solution. As the hydrogen ion concentration increases, the solution becomes more **acidic**. As the hydrogen ion concentration decreases, the solution becomes more **alkaline**, or **basic**. Notice that the **pH scale**, shown in **Figure 2.4**, ranges from 0 (the most acidic) to 14 (the most basic). As you learned in the Pre-Lab Exercises, the reason for this is that the pH is actually a negative logarithm, so the lower the pH, the higher the hydrogen ion concentration. A pH of 7 is considered **neutral**, which is neither acidic nor basic, because the number of hydrogen ions equals the number of base ions in the solution.

Acids and bases are some of the most important chemicals in the human body: They are found in the stomach, in the blood and other extracellular body fluids, in the cytosol, and in the urine, and are released by cells of the immune system. The normal pH for the blood is about 7.35–7.45. The body regulates this pH tightly, as swings of even 0.2 points in either direction can cause serious disruptions in many of our physiological processes.

## Procedure Reading the pH

A simple way to measure pH is to use pH paper (**Figure 2.5**). To test the pH with pH paper, drop one or two drops of the sample solution on the paper, and compare the color change with the colors on the side of the pH paper container. The pH is read as the number that corresponds to the color the paper turned. When you have completed the activity, answer Check Your Understanding question 1 (p. 49).

**Safety Note**

⚠ Safety glasses and gloves are required!

1 Obtain two samples each of known acids and bases, and record their molecular formulae in **Table 2.1** (the molecular formula should be on the side of each bottle).

2 Measure the pH of each acid and base using pH paper, and record their pH values in **Table 2.1**.

3 Obtain two randomly selected unknown samples, and measure their pH using pH paper.

4 Record the pH values in **Table 2.1**, and determine if this substance is an acid, a base, or is neutral.

FIGURE **2.5** pH paper and indicator strip

TABLE **2.1** Samples of Acids and Bases

| Samples | Molecular Formula | pH |
|---|---|---|
| Acid #1 | | |
| Acid #2 | | |
| Base #1 | | |
| Base #2 | | |
| **Unknown Samples** | **pH** | **Acid, Base, or Neutral?** |
| Unknown #1 | | |
| Unknown #2 | | |

## Procedure pH Applications

Now let's apply the pH scale to physiological systems. The stomach contains concentrated hydrochloric acid, and the pH of the stomach contents ranges between 1 and 3. **Antacids** are a group of medications that neutralize stomach acid to treat a variety of conditions, including *gastroesophageal reflux*, commonly known as *heartburn*. Antacids generally consist of a metal bound ionically to a base. In this activity you will compare the effectiveness of three widely available antacids in neutralizing concentrated hydrochloric acid.

**Safety Note**

Safety glasses and gloves are required!

1 Obtain three glass test tubes, and label them 1, 2, and 3.

2 Obtain a bottle of 0.1M hydrochloric acid (HCl), and record its pH in the blank below.

pH of 0.1M HCl: _____

3 In each tube, place 2 ml of the 0.1M HCl.

4 Add about one-fourth of one crushed Tums® tablet to tube 1.

5 Add about one-fourth of one crushed Rolaids® tablet to tube 2.

6 Add about one-fourth of one crushed Alka-Seltzer® tablet to tube 3.

7 Allow the tubes to sit undisturbed for 3 minutes.

8 Measure the pH of the contents of each tube, and record the values in **Table 2.2**.

9 What effect did the antacids have on the acid?

TABLE **2.2** **Effectiveness of Three Antacids**

| Antacid | Active Ingredient | pH |
|---|---|---|
| Tums® | | |
| Rolaids® | | |
| Alka-Seltzer® | | |

10 Based upon your observations, which antacid is the most effective at neutralizing the acid?

# Exercise 2-2

## 2

## Buffers

### MATERIALS

- ❑ Well plates
- ❑ Stirring rods
- ❑ Dropping pipettes
- ❑ 0.1M hydrochloric acid
- ❑ Buffered solution
- ❑ Distilled water
- ❑ 0.1M NaOH
- ❑ pH paper

A **buffer** is a chemical that resists large or dramatic changes in pH; a solution to which a buffer has been added is called a **buffered solution**. When acid is added to a buffered solution, the buffer binds the added hydrogen ions and removes them from the solution. Similarly, when a base is added to a buffered solution, the buffer releases hydrogen ions into the solution. Both effects minimize the pH changes that otherwise would occur in the solution if the buffer were not present.

## Procedure  Testing Buffered Solutions

In this experiment you will examine the effects of adding an acid or a base to buffered solutions and nonbuffered solutions. You will use distilled water as your nonbuffered solution, and your instructor will choose an appropriate buffered solution for you to use. When you have completed the activity, answer Check Your Understanding question 2 (p. 49).

**1** Obtain a well plate, and number four wells 1, 2, 3, and 4.

**2** Fill wells 1 and 2 about half-full of distilled water. Measure the pH of the distilled water, and record that value in **Table 2.3** ("pH of Initial Solution").

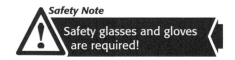

*Safety Note*

**⚠ Safety glasses and gloves are required!**

**3** Fill wells 3 and 4 about half-full of buffered solution. Measure the pH of the buffered solution and record that value in **Table 2.3** ("pH of Initial Solution").

**4** Add two drops of 0.1M HCl to wells 1 and 3. Stir the solutions with a stirring rod (or a toothpick), and measure the pH of each well. Record the pH in **Table 2.3** ("pH after").

**5** Add two drops of 0.1M NaOH (a base) to wells 2 and 4. Stir the solutions, and measure the pH of each well. Record the pH in **Table 2.3** ("pH after").

TABLE **2.3** Buffered and Nonbuffered Solutions

| Well | Contents | pH of Initial Solution | pH After Adding Acid or Base |
|------|----------|------------------------|------------------------------|
| 1 | Water and HCl | | |
| 2 | Water and NaOH | | |
| 3 | Buffer and HCl | | |
| 4 | Buffer and NaOH | | |

**6** Interpret your results. What effect did the buffer have on the pH changes you saw?

# Exercise 2-3

## Enzymes and Chemical Reactions

### MATERIALS

- ❑ 9 glass test tubes
- ❑ Test tube rack
- ❑ Lipase
- ❑ Boiled lipase
- ❑ Bile salts
- ❑ Vegetable oil
- ❑ 0.1M NaOH
- ❑ Ice water bath
- ❑ Warm-water bath (set to 37°C)
- ❑ Distilled water
- ❑ Phenol red

Often when atoms or molecules interact, chemical bonds are formed, broken, or rearranged, or electrons are transferred between molecules. These interactions are called **chemical reactions**. Most chemical reactions can proceed spontaneously, but they often take an extremely long time.

One factor that can alter the rate at which a reaction takes place is *temperature*. Generally, when temperatures increase (up to a point), molecules move faster, and they collide and react at a faster rate. The opposite happens when temperatures are decreased.

Another factor that can affect reaction rate is a substance called a **catalyst** that is added to a reaction to increase its rate. Catalysts are not consumed in the reaction and may be reused after the reaction has completed.

In the body, biological catalysts called **enzymes** speed up essentially all of our chemical reactions. Nearly all enzymes in the body are proteins that work by binding specifically to the reacting components and by reducing the amount of energy required for a reaction to proceed (called the *activation energy*). Note here that an enzyme must bind to the reactants for it to work. This means that if the enzyme is damaged and loses its shape, it will not be able to function. Enzymes may be damaged by the same processes that damage all proteins, including extreme heat and extreme pH swings. An enzyme that has lost its shape as a result of such damage is said to be **denatured**.

## Procedure  Test Enzymatic Activity

In the following procedure, you will be comparing the ability of three solutions—lipase, boiled lipase, and water—to digest vegetable oil at three different temperatures. **Lipase** is an enzyme found in the human digestive tract that speeds up the digestion of dietary fats. Note that you also must add another component, called **bile**, to your mixture. Bile is not an enzyme itself, but it does increase the ability of lipase to facilitate fat digestion.

You will check for the presence of digestion using an indicator called **phenol red**. Phenol red appears pink at an alkaline (basic) pH, changes to an orange-red color at a neutral pH, and changes to a yellow color when the pH becomes acidic.

If fat has been digested, fatty acids will be released that will decrease the pH of the contents of your tube and turn them orange-red or yellow. You may interpret your results in the following way (**Figure 2.6**):

- ❙ Yellow color = the pH is acidic, and significant fat digestion occurred

- ❙ Red-orange color = the pH is neutral, and some fat digestion occurred

- ❙ Pink color = the pH is basic, and no (or limited) fat digestion occurred

When you have completed this experiment, answer Check your Understanding question 3 (p. 49).

1  Obtain nine glass test tubes and label them 1 through 9 with a marker.

**Safety Note**

⚠ Safety glasses and gloves are required!

2  Add 3 ml of vegetable oil to each test tube.

3  Add 8–10 drops of the pH indicator phenol red to each test tube. The oil now should appear pink. If it does not, add drops of 0.1M NaOH (a base) to each tube until the indicator turns pink.

FIGURE 2.6 **Possible results from the lipid digestion experiment**

4  Add the following ingredients to each test tube:

   ▌ Tubes 1, 4, and 7: 1 ml lipase, 1 ml bile

   ▌ Tubes 2, 5, and 8: 1 ml boiled lipase, 1 ml bile

   ▌ Tubes 3, 6, and 9: 2 ml water

5  Place tubes 1, 2, and 3 in an ice bath, and incubate them for 30 minutes.

6  Leave tubes 4, 5, and 6 in your test tube rack to incubate at room temperature for 30 minutes.

7  Place tubes 7, 8, and 9 in a warm-water bath set to 37°C, and incubate them for 30 minutes.

8  After 30 minutes have passed, remove the tubes from the ice bath and warm-water bath, and place them in your test tube rack.

9  Record the color of each tube in Table 2.4, and interpret your results.

TABLE 2.4 Enzymatic Activity of Lipase

| Tube Number | Color | pH (Acidic, Neutral, or Alkaline) | Amount of Digestion that Occurred |
|---|---|---|---|
| 1 | | | |
| 2 | | | |
| 3 | | | |
| 4 | | | |
| 5 | | | |
| 6 | | | |
| 7 | | | |
| 8 | | | |
| 9 | | | |

10 Answer the following questions about your results:

  a  What effect does temperature have on enzyme activity?

  b  Were your results with the boiled lipase different from the results with nonboiled lipase? Why?

  c  Predict what color change you would see in tubes 3, 6, and 9 if you were to leave them in a warm-water bath for several days.

# Exercise 2-4

## Chemical Solubilities

### MATERIALS

- ☐ 10 glass test tubes
- ☐ 10 rubber stoppers
- ☐ Graduated cylinders
- ☐ Test tube rack
- ☐ Distilled water
- ☐ Paint thinner (such as acetone)
- ☐ Table salt
- ☐ Sucrose
- ☐ Iodine crystals
- ☐ Unknown solids

In the Pre-Lab Exercises you learned that three main types of bonds are present between atoms:

1. **Ionic bonds**. An ionic bond forms when one or more electrons is transferred from a metal to a nonmetal. In this transfer, the metal becomes a positive cation, and the nonmetal becomes a negative anion. These positive and negative charges attract and result in the ionic bond.

2. **Covalent bonds**. Covalent bonds form when two or more atoms share electrons. This results in more stable atoms with eight electrons in their valence shells (or two electrons for atoms with five or fewer protons). The two types of covalent bonds are:

   a. **Polar covalent bonds**. A polar covalent bond results from the unequal sharing of electrons between atoms with unequal electronegativities. Essentially, one atom is "greedier" (more electronegative) than the other, and the electrons spend more time around the "greedier" atom. This creates what is known as a **dipole**—a molecule with a partially positive pole and a partially negative pole ("dipole" literally means "two poles").

   b. **Nonpolar covalent bonds**. The electrons are shared equally between the atoms involved in a nonpolar covalent bond because the two atoms have nearly the same electronegativities. As a result, they have no partial positive or negative charges.

Chemicals are not limited to having only ionic, polar covalent, or nonpolar covalent bonds. Certain chemicals are **amphipathic,** meaning that they have both polar and nonpolar parts. Many important biological molecules, including phospholipids and bile salts, are amphipathic.

Determining the types of bonds present in a molecule is important, because molecules will generally only mix well with other molecules that have similar bonds. In general, the following rule can be used to determine whether one molecule will mix with another molecule:

*Like interacts with like.*

This means that chemicals with charged ends (polar covalent and ionic molecules) prefer to interact with chemicals that also have charged ends (other polar covalent and ionic molecules). Chemicals without charged ends (nonpolar covalent molecules) prefer to interact with chemicals that also have no charged ends (other nonpolar covalent molecules). This is something you have seen in action if you have ever tried to combine oil (nonpolar covalent compound) and water (polar covalent compound). In their natural conditions, oil and water don't mix because of their types of chemical bonds.

The degree to which a solid substance, called a **solute,** will mix with another substance, called a **solvent,** is known as its **solubility.** A solute that dissolves completely in a solvent is said to be highly *soluble* in that solvent. Conversely, a solute that does not dissolve in a solvent is said to be *insoluble* in that solvent. The rules for which solutes are soluble in which solvents are simple:

- ▌ Polar covalent and ionic solutes are soluble in most polar covalent solvents.

- ▌ Nonpolar covalent solutes are soluble in most nonpolar covalent solvents.

---

## HINTS & TIPS

### How to Determine the Types of Bonds in a Molecule

Following are some simple rules for determining whether a compound is a polar covalent, nonpolar covalent, or ionic compound.

- ℹ Any metal bound to a nonmetal is an ionic compound.

- ℹ All compounds that contain only C and H are nonpolar covalent.

- ℹ Any molecule containing only identical atoms (e.g., $H_2$) is a nonpolar covalent molecule.

- ℹ A compound containing a significant amount (about one-fourth or more of the total atoms) of O, N, or P is a polar covalent compound.

- ℹ Any compound containing mostly C and H is overall a nonpolar covalent compound. Note that many nonpolar covalent compounds do contain elements such as O or N, but these atoms are in the minority.

## Procedure Test Chemical Solubilities

You now will determine whether solutes are polar covalent, nonpolar covalent, or ionic based upon the type of solvent in which they dissolve. In this exercise, you will have two solvents—one that is polar covalent and one that is nonpolar covalent. You will attempt to dissolve three known samples of solute and two unknown samples of solute in these solvents, after which you will use these solubilities to determine the types of bonds that may be present in your unknown samples.

After you test them, your instructor may wish to give you the molecular formulae of your unknown samples so you may determine the types of molecules more precisely. When you have completed the activity, answer Check Your Understanding question 4 (p. 49).

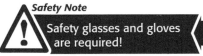

**Safety Note**
Safety glasses and gloves are required!

1  Obtain 10 glass test tubes, a test tube rack, and 10 rubber stoppers. Number the test tubes 1 through 10.

2  Add 2 ml of distilled water to tubes 1–5. Add 2 ml of paint thinner to tubes 6–10.

   a  Which solvent is the polar solvent? _____

   b  Which solvent is the nonpolar solvent? _____

3  Add a pinch of table salt to tubes 1 and 6. What is the molecular formula for salt? _____

4  Add a pinch of sucrose to tubes 2 and 7. What is the molecular formula for sucrose? _____

5  Add a pinch of iodine crystals to tubes 3 and 8. What is the molecular formula for iodine?

6  Add a pinch of unknown substance A to tubes 4 and 9.

7  Add a pinch of unknown substance B to tubes 5 and 10.

8  Place a stopper firmly in each of the tubes. Shake each tube for several seconds until you are certain that no further change will take place in the solute.

9  Examine each tube, and determine if the solute dissolved or remained separate. Record your results in Table 2.5.

TABLE **2.5** Solubilities

| Tube | Dissolved in Polar Solvent? |
|------|------------------------------|
| 1 (salt + water) | |
| 2 (sucrose + water) | |
| 3 (iodine + water) | |
| 4 (unknown A + water) | |
| 5 (unknown B + water) | |
| **Tube** | **Dissolved in Nonpolar Solvent?** |
| 6 (salt + paint thinner) | |
| 7 (sucrose + paint thinner) | |
| 8 (iodine + paint thinner) | |
| 9 (unknown A + paint thinner) | |
| 10 (unknown B + paint thinner) | |

10  Based on your results, answer the following questions:

   a  What type of chemical bonds are present in salt? _____

   b  What type of chemical bonds are present in sucrose? _____

   c  What type of chemical bonds are present in iodine? _____

   d  What type of chemical bonds are present in unknown A? _____

   e  What type of chemical bonds are present in unknown B? _____

# Exercise 2-5

## Aspirin Synthesis

### MATERIALS

- ☐ Well plates
- ☐ Willow bark (generally *Salix alba*)
- ☐ Ethanol
- ☐ Glacial acetic acid
- ☐ Ferric chloride
- ☐ Dropping pipettes

In this exercise you will use the principles you learned in the previous exercises to make one of the most commonly used over-the-counter anti-inflammatory drugs—**aspirin**. Aspirin is derived from the chemical **salicylic acid**, found naturally in willow bark. The "all natural" salicylic acid, however, contains a chemical group called a *phenol*, which makes it highly irritating to the stomach.

Fortunately, we can remove the phenol group and replace it with another chemical group called an *acetyl* group, which is much less irritating to the stomach. This simple substitution changes the drug from salicylic acid to **acetylsalicylic acid**, the chemical name for aspirin.

## Procedure  Synthesize Aspirin

The first step in aspirin synthesis is to isolate salicylic acid from its source, willow bark. This is done by soaking the willow bark in ethanol. Salicylic acid is soluble in the slightly polar ethanol, which pulls it out of the willow bark and into solution. In the second step, the phenol group is removed and replaced with an acetyl group. This is accomplished by adding strong acetic acid to the ethanol-salicylic acid solution. When you have completed the activity, answer Check Your Understanding question 5 (p. 50).

1   Obtain a well plate, some willow bark, and ethanol.

2   Fill the large well with willow bark, and cover the bark with ethanol. Label two smaller wells as 1 and 2.

> **Safety Note**
>
> ⚠ Safety glasses and gloves are required!

3   Soak the bark in ethanol for 15 minutes.

4   After 15 minutes, use a pipette to remove several drops of the ethanol-salicylic acid solution, and place equal amounts into well 1 and well 2.

5   Add two drops of iron chloride (also called ferric chloride) into well 1. Iron chloride reacts with phenol groups to turn a dark purple-brown color. If no phenol groups are present, the solution remains orange.

Color of solution in well 1: _____

Does the solution in well 1 contain salicylic acid? How do you know?

_____

6   Add 15 drops of glacial acetic acid to the solution in well 2, and stir for 15 seconds. You have now synthesized aspirin!

7   Into well 2, add two drops of iron chloride.

Color of solution in well 3: _____

Was your synthesis of aspirin successful? How can you tell?

# Exercise 2-6

## DNA Extraction

### MATERIALS

- ☐ 50 ml beaker
- ☐ Test tube
- ☐ Glass stirring rod
- ☐ Graduated cylinder
- ☐ Pea mixture
- ☐ Isopropyl alcohol
- ☐ Meat tenderizer
- ☐ Liquid detergent
- ☐ Wooden applicator stick

The chemical **deoxyribonucleic acid**, or **DNA**, is located in the nucleus of nearly all cells (mature red blood cells are an exception). DNA contains the instructions for building every protein in an organism. It is the small differences in DNA from one organism to another that account for the vast diversity we see on this planet. DNA extraction and analysis have become commonplace in the search for diagnoses of genetic diseases, our evolutionary history, and many other topics in science and medicine. DNA analysis requires specialized equipment, but DNA extraction is surprisingly simple and can be performed using things you have around the house.

## Procedure  DNA Extraction

In this exercise, we will be extracting DNA from dried green peas. Although this obviously is not animal DNA, keep in mind that it is composed of precisely the same nucleic acid bases as animal DNA, and that the plant cell membrane structure is identical to that of animal cells. The first part of the extraction involves removing the DNA from the cell and the nucleus using a mild detergent. Then you will use enzymes to degrade the proteins around the DNA and free up the nucleic acids. Finally, you will take advantage of DNA's weak polarity to extract it from strongly polar water into weakly polar isopropyl alcohol. When you have completed the activity, answer Check Your Understanding questions 6 and 7 (p. 50).

1. Obtain about 10 ml of the blended pea mixture (which contains only dried peas and water), and pour it into a 50 ml beaker.

2. Add 1.6 ml of liquid detergent to the pea soup. Stir gently with a glass stirring rod, and let the mixture sit undisturbed for 5–10 minutes.

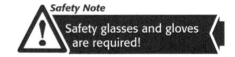
**Safety Note**
Safety glasses and gloves are required!

3. After 5–10 minutes, fill about one-third of a test tube with the pea-soup mixture.

4. Add a pinch of meat tenderizer (which contains enzymes) to the test tube, and stir with a glass stirring rod *very gently* (rough stirring will break up the DNA, making it harder to see).

5. Tilt the test tube, and slowly add isopropyl alcohol with a dropping pipette (or directly from a plastic squeeze bottle) so the alcohol forms a layer on the top of the pea mixture. The amount of alcohol you add should be approximately equal to the volume of the pea mixture.

6. Watch as the DNA (it looks like long, stringy stuff) rises to the alcohol layer (**Figure 2.7**).

7. Use an applicator stick to fish out the DNA, and set it on a paper towel. Voila! You have just extracted DNA!

FIGURE **2.7** Extracted DNA in alcohol

## Check Your Recall

**1** *Fill in the blanks:* A solution with a pH less than 7 is considered _____ , a solution with a pH of 7

is considered _____ , and a solution with a pH greater than 7 is considered _____ .

**2** The pH is a measure of a solution's _____ .
   a. hydrogen concentration
   b. oxygen concentration
   c. cation concentration
   d. hydrogen ion concentration

**3** A buffer is a chemical that
   a. keeps the pH of a solution neutral.
   b. resists a large or dramatic change in pH.
   c. adds hydrogen ions to a solution when the pH decreases.
   d. removes hydrogen ions from a solution when the pH increases.

**4** A chemical reaction takes place when:
   a. chemical bonds are formed.
   b. chemical bonds are broken.
   c. chemical bonds are rearranged.
   d. electrons are transferred between molecules.
   e. All of the above.

**5** *Fill in the blanks:* Increasing the temperature generally _____ the rate of reactions, and decreasing

the temperature generally _____ the rate of reactions.

**6** Which of the following is *not* a property of enzymes?
   a. They increase the rate of a reaction.
   b. They are not consumed in the reaction.
   c. They change the products of a reaction.
   d. They are biological catalysts.

**7** An enzyme that has lost its shape

    a. is denatured and no longer functional.

    b. is denatured but still functional.

    c. is not different from a regular enzyme as shape is not important.

    d. can be restored by extreme temperatures or pH swings.

**8** Predict whether each of the following has ionic, polar covalent, or nonpolar covalent bonds.

$O_2$ _____

$MgBr_2$ _____

$C_6H_{12}O_6$ _____

$CH_4$ _____

$I_2$ _____

$NH_3$ _____

**9** *Fill in the blanks:* The degree to which a substance called a _____ will mix with another

substance called a _____ is known as its _____ .

**10** Which of the following molecules would you expect to dissolve in water?

    a. Polar covalent molecules

    b. Ionic molecules

    c. Nonpolar covalent molecules

    d. Both a and b are correct

    e. Both a and c are correct

# Check Your Understanding

### Critical Thinking and Application Questions

**1** Your lab partner argues with you that if you add acid to a solution, the hydrogen ion concentration should increase and, therefore, the pH should increase. Is he correct? What do you tell him?

**2** A hypothetical drug binds all of the buffer molecules in the blood, effectively removing them from circulation. Predict what will happen to the pH of the blood if hydrogen ions are added to it.

**3** You take the boiled lipase from Exercise 2-3 and compare its structure to the normal, unboiled lipase. You note that the two structures differ. Why are they different? How does this help to explain your results?

**4** The blood is largely composed of water. Which types of biological molecules can the blood transport safely throughout the body? Which types of biological molecules can the blood not transport safely throughout the body without modification? Why?

**5** The popularity of herbal medical remedies has surged recently. Most of these remedies claim that, because they are derived from plants, they are "all natural" and "drug-free." Given the aspirin synthesis you just performed, what do you think of such claims? Furthermore, which preparation is less irritating to the stomach: the "all natural" willow bark or the synthetic aspirin?

**6** Detergents have a structure very similar to the phospholipids in the plasma membrane. Explain why you used detergents in the first step of your DNA extraction. (*Hint:* Think about the rule that "like dissolves like.")

**7** The enzymes in meat tenderizer are called *proteases*, which degrade proteins. Why must we have proteases in this extraction? (*Hint:* Think about the structure of DNA. What is in it besides nucleic acids?)

# Introduction to the Microscope

**3**

## OBJECTIVES

*Once you have completed this unit, you should be able to:*

- Identify the major parts of the microscope.
- Define the magnification of high, medium, and low power.
- Demonstrate proper use of the microscope.
- Practice focusing on low, medium, and high power.
- Define depth of focus.

# EXERCISES

**3**

Working with the microscope and slides seems to be one of the least favorite tasks of anatomy and physiology students. But with a bit of help, a fair amount of patience, and a lot of practice, the use of microscopes becomes progressively easier with each unit (and, yes, we do look at slides in almost every unit!).

## Exercise **3-1**

### Introduction to the Microscope

**MATERIALS**
❑ Light microscopes with three objective lenses

The microscopes that you will use in this lab are called **light microscopes** (**Figure 3.1**). This type of microscope shines light through the specimens to illuminate them, and the light is refracted through objective lenses to magnify the image. Light microscopes have the following components:

▮ **Ocular lens.** The ocular lens is the lens through which you look to examine the slide. The microscope may have one ocular lens (a **monocular** microscope) or two ocular lenses (a **binocular** microscope). Many ocular lenses have pointers that can be moved by rotating the black **eyepiece**. The area of the slide visible when you look into the ocular is known as the **field of view**.

▮ **Objective lenses.** These are lenses with various powers of magnification. Most microscopes have low (4×), medium (10×), and high (40×) power objective lenses. Note that sometimes the 4× objective is referred to as the *scan objective*. On such microscopes, the 10× objective is actually called the *low*-power objective. The objective lenses are attached to the **nosepiece**, which allows the operator to switch between objectives. Certain microscopes have a higher power objective (100×), called the **oil immersion lens**, which requires that a drop of oil be placed between the slide and objective lens. Your instructor may have an oil immersion microscope set up for demonstration purposes.

▮ **Stage.** The stage is the surface on which the slide sits. It typically has stage clips to hold the slide in place. The stage on many microscopes is moveable using the mechanical stage adjustment knob. Others require you to move the slide manually.

▮ **Arm.** The arm supports the body of the microscope and typically houses the adjustment knobs.

▮ **Coarse adjustment knob.** This large knob on the side of the arm moves the stage up and down to change the distance of the stage from the objective lenses. It allows gross focusing of the image.

▮ **Fine adjustment knob.** This smaller knob allows fine-tuning of the image's focus.

▮ **Lamp.** The lamp, also called the *illuminator*, provides the light source. It rests on the *base* of the microscope.

▮ **Iris diaphragm.** This adjustable wheel on the underside of the stage controls the amount of light allowed to pass through the slide.

## Procedure Magnification

Light is refracted through two lenses to obtain magnification—the ocular lens and the objective lens.

Magnification of the ocular lens is usually 10× (magnified 10 times). Magnification of the objectives lens varies, but typically is 4× for low power, 10× for medium power, and 40× for high power. (To verify that this is the case for your microscope, look at the side of the objective lens, which usually is labeled with its magnification.) Remember that oil immersion provides even greater magnification at 100×. The total magnification is obtained by multiplying the power of the ocular lens by the power of the objective lens.

Fill in **Table 3.1** to determine total magnification at each power.

FIGURE **3.1** A compound light microscope

TABLE **3.1**

| Magnification of Ocular Lens | Power | Magnification of Objective Lens | Total Magnification |
|---|---|---|---|
| | Low | | |
| | Medium | | |
| | High | | |
| | Oil Immersion | | |

# Exercise 3-2

## Focusing and Using the Microscope

### MATERIALS

- ❑ Light microscope with three objective lenses
- ❑ Lens paper
- ❑ Introductory slides (3 colored threads and the letter "e")

As I'm certain your instructor will point out, microscopes are expensive! Care must be taken to ensure that the microscopes stay in good working condition. Taking proper care of microscopes makes the histology sections of your labs run more smoothly, and also ensures that you stay on your lab instructor's good side! Bearing that in mind, following are some general guidelines for handling the microscopes:

▌ Use two hands to support the microscope when carrying it—one hand to hold the arm and the other hand to support the base.

▌ Gather up the cord so it does not dangle off the lab table after you have plugged in the microscope. This will help to prevent people from tripping over loose cords.

▌ Clean the lenses with lens paper only. Do not use paper towels or cloth to clean a lens, because this will scratch its surface.

▌ Make sure the lowest power objective is in place before you begin.

▌ Get the image in focus on low power, then switch to higher power and adjust with the fine adjustment knob. Be careful not to use the coarse adjustment knob with the high power objective in place, because you could break the slide and damage the lens.

▌ When you are finished with the microscope, turn the nosepiece to the lowest power objective, and remove the slide. Be sure to clean off the objective if you used oil, because oil left on the objective tends to harden. Turn off the power to the microscope, and unplug it. This will decrease the chances of a blown bulb or fuse next time the microscope is used. Before putting the microscope away, wrap the electrical cord around the base, and cover it with a dust cover.

If you follow these general guidelines, you can rest assured that the microscopes (and your grade) will not suffer any harm.

## Procedure  Focusing the Microscope

Now that we know how to handle the microscope properly, let's practice using it.

**1** Obtain a slide of the letter "e."

**2** Examine the letter "e" slide macroscopically before placing it on the stage. How is the "e" oriented on the slide? Is it right side up, upside down, backward, etc.?

**3** Ensure that the nosepiece is switched to low power, place the slide on the stage, and secure it with the stage clips. Move the slide using the stage adjustment knob until the "e" is in your field of view.

**4** Use the coarse adjustment knob to bring the slide into focus slowly. Once it is grossly in focus, use the fine adjustment knob to sharpen the focus. How is the "e" oriented in the field of view? Is it different from the way it was when you examined it in item 2?

**5** Move the nosepiece to medium power. You should only have to adjust the focus with the fine adjustment knob; no adjustment of the coarse focus should be necessary.

**6** Once you have examined the slide on medium power, move the nosepiece to high power. Again, focus only with the fine adjustment knob.

Wasn't that easy?

## **Procedure** Depth of Focus

At times you will look at a slide and see something brown or black and kind of neat-looking with interesting swirls and specks. What is this fascinating discovery you've made? It's dirt on top of the slide. This happens because people tend to focus the objective on the first thing they can make out, which usually is the top of the coverslip on the slide, which has a tendency to be dirty.

These "dirt discoveries" can be avoided by appreciating what is known as **depth of focus**. Also called the *depth of field*, the depth of focus is the thickness of a specimen that is in sharp focus. Thicker specimens will require you to focus up and down to look at all levels of the specimen. This takes practice and skill. Let's get some practice doing this.

**1** Obtain a slide with three colored threads. The threads are located on the slide at varying depths, and you will have to focus on each thread individually.

**2** Examine the slide macroscopically prior to putting it on the stage.

**3** Ensure that the nosepiece is switched to low power, place the slide on the stage, and secure it with the stage clips. Move the slide using the stage adjustment knob until the threads are in your field of view.

**4** Use the coarse adjustment knob to get the slide into focus on low power.

**5** Switch to medium power, and use the fine adjustment knob to sharpen the focus. Which thread(s) is(are) in focus?

**6** Move the objective up and down slowly with the coarse adjustment knob, focusing on each individual thread. Figure out which color thread is on the bottom, in the middle, and on the top, and write the color below:

Bottom    _____

Middle    _____

Top    _____

## HINTS & TIPS

### How to Approach Microscopy

Okay, so you can focus on newsprint and threads, but what about cellular structures and tissue sections? Well, those certainly are more difficult, but if you keep the following hints in mind, the task becomes much simpler:

**ⓘ Always start on low power.** You are *supposed* to start on low power anyway, to avoid damaging the objective lenses. Sometimes, though, students forget, and jump straight to medium or high power. This risks damaging the lenses and also makes it harder on you. Bear in mind that most slides will have more than one histological or cellular structure on each slide. Starting on low power allows you to scroll through a large area of the slide, and then focus in on the desired part of the section.

**ⓘ Beware of too much light.** It is easy to wash out the specimen with too much light. If you are having difficulty making out details, first adjust the focus with the fine adjustment knob. If this doesn't help, use the iris diaphragm to reduce the amount of light illuminating the specimen. This will increase the contrast and allow you to observe more details. It also helps to reduce headaches and eyestrain.

**ⓘ Keep both eyes open.** It is tempting to close one eye when looking through a monocular microscope. Admittedly, keeping both eyes open isn't easy at first, but it helps to reduce eyestrain and headaches.

**ⓘ Compare your specimen to the photos in your lab manual.** Although the slides you are examining will not necessarily be identical to the photos in this book, they should be similar in appearance. Generally speaking, if you are looking at something that is vastly different from what is in this book, you probably should move the slide around a bit or change to a different power objective to find the correct tissue or cell type on the slide. Other good sources for micrographs include atlases, your textbook, and images on the Internet.

**ⓘ Remember that the slides aren't perfect.** Not all slides will clearly demonstrate what you need to see. Some aren't stained adequately or properly. Some are sectioned at a funny angle. Some don't contain all of the tissue or cell types you need to see. What should you do about this? See the next hint for the answer.

**ⓘ Look at more than one slide of each specimen.** This will help you in the face of subpar slides and also will assist you overall in gaining a better understanding of the specimens you are examining.

**ⓘ Draw what you see.** Students tend to resist drawing, often claiming artistic incompetence as an excuse. But even the most basic picture is helpful for two reasons. First, it allows you to engage more parts of your brain in the learning process. The more areas of your brain engaged, the better are the chances you will retain the information. Also, drawing is helpful in that you actually have to look at the specimen long enough to draw it!

**ⓘ Have patience!** It really does get easier. Don't get frustrated, and don't give up. By the end of the semester, you may come to appreciate the microscope and the fascinating world it reveals!

## Check Your Recall

**1** Match the following terms with the correct definition from the column.

_____ Ocular lens      A. The surface on which the slide sits

_____ Iris diaphragm      B. The microscope's light source

_____ Arm      C. Allows fine tuning of the focus

_____ Objective lens      D. The lens through which you look to view the slide

_____ Coarse adjustment knob      E. Supports the body of the microscope

     F. Controls the amount of light that passes through the slide

_____ Stage      G. Moves the stage up and down; provides gross focusing

_____ Fine adjustment knob      H. Lenses of various powers of magnification

_____ Lamp

**2** Magnification of the ocular lens is usually _____.

**3** The lens that provides 100× magnification is the
  a. low power lens.
  b. medium power lens.
  c. high power lens.
  d. oil immersion lens.

**4** Before you begin, you should make sure that
  a. the objective lens is switched to the highest-power magnification.
  b. the lamp is turned to its full brightness.
  c. the ocular lens is switched to the highest-power magnification.
  d. the objective lens is switched to the lowest-power magnification.

**5** How should you carry the microscope?

_____

_____

**6** What is the depth of focus?

_____

_____

**7** What should you do if your slide looks dramatically different from the slide in your lab manual?

3

**8** If you are having trouble making out details on your slide, what can you do?
   a. Adjust the focus with the fine adjustment knob.
   b. Reduce the amount of light with the iris diaphragm.
   c. Both a and b.
   d. Neither a nor b.

**9** What should you use to clean the lenses?

**10** What are the steps you should use to focus the image on the slide?

# Cytology

**4**

## PRE-LAB EXERCISES

**4**

## Pre-Lab Exercise 4-1

✎ Key Terms

You should be familiar with the following terms before coming to lab.

| Term | Definition |
|------|-----------|

**Cell Structures and Organelles**

Plasma membrane _____

_____

Cytoplasm _____

_____

Nucleus _____

_____

Mitochondrion _____

_____

Ribosome _____

_____

Peroxisome _____

_____

Smooth endoplasmic reticulum _____

_____

Rough endoplasmic reticulum _____

_____

Golgi apparatus (or complex) _____

_____

Lysosome _____

_____

Centrosome _____

_____

Cilia _____

_____

Flagella _____

_____

Vesicle _____

_____

**Membrane Transport**

Diffusion_____

_____

Osmosis _____

_____

Tonicity _____

_____

**Cell Cycle and Mitosis**

Cell cycle _____

_____

Interphase _____

_____

Mitosis _____

_____

Label and color the parts of the plasma membrane depicted in **Figure 4.1** with the terms from Exercise 4-1 (p. 65). Use your text and Exercise 4-1 in this unit for reference.

FIGURE **4.1** The plasma membrane

## Pre-Lab Exercise **4-3**

The Parts of the Cell

Label and color the parts of the cell depicted in **Figure 4.2** with the terms from Exercise 4-1 (p. 65). Note that you may not see all structures in this diagram. Use your text and Exercise 4-1 in this unit for reference.

FIGURE **4.2** A generalized cell

# Pre-Lab Exercise 4-4

## The Cell Cycle

Use your text and lab manual to answer the following questions pertaining to the cell cycle and mitosis.

1. Describe the following stages of the cell cycle:

   a. G1: _____

   _____

   _____

   b. S: _____

   _____

   _____

   c. G2: _____

   _____

   _____

   d. M: _____

   _____

   _____

2. Describe the events occurring in the cell in each of the phases of mitosis in **Table 4.1**.

TABLE **4.1** Stages of Mitosis

| Stage of Mitosis | Events Taking Place in the Cell | Cell Appearance |
|---|---|---|
| Prophase | | |
| Metaphase | | |
| Anaphase | | |
| Telophase | | |

# EXERCISES

A basic principle that we will revisit repeatedly in our study of anatomy and physiology is:

*Form follows function.*

This principle may be stated in a variety of ways and is alternatively called the "principle of complementarity of structure and function." Essentially, what this means is that the anatomy (the *form*) is always suited for the structure's physiology (the *function*). This is obvious at the organ level: Imagine if the heart were solid rather than composed of hollow chambers, or if the femur were pencil-thin rather than the thickest bone in the body. These organs wouldn't be able to carry out their functions of pumping blood and supporting the weight of the body very well, would they? But this principle is applicable even at the cellular level, which we will see in this unit.

We begin with an examination of the cell, its structures, and the relationship of structure and function. Then we turn our attention to two critical processes at the cellular level: *diffusion* and *osmosis*. In the final exercise we explore the *cell cycle* and the process of cell division.

Note that this is the first lab in which you will use a *Model Inventory*—something you will use throughout this lab manual. In this inventory you will list the anatomical models or diagrams that you use in lab (if the model is not named, make up a descriptive name for it), and record in a table the structures you are able to locate on each model. This is particularly helpful for study purposes, as it allows you to return to the proper models to locate specific structures.

## Exercise **4-1**

### Organelles and Cell Structures

#### MATERIALS

- ❏ Cell models and diagrams
- ❏ 5 colors of modeling clay
- ❏ Fetal pigs or other preserved small mammals
- ❏ Wooden applicator sticks
- ❏ Methylene blue dye
- ❏ Glass slide (blank)
- ❏ Distilled water
- ❏ Coverslip
- ❏ Light microscope
- ❏ Cell slides (with red blood cells, skeletal muscle, and sperm cells)
- ❏ Colored pencils

Most cells in the body are composed of three basic parts—the plasma membrane, the cytoplasm, and the nucleus.

1. **Plasma membrane.** The plasma membrane is the outer boundary of the cell (**Figure 4.3**). It is composed of a **phospholipid bilayer** with multiple components interspersed throughout, including **integral proteins, peripheral proteins, cholesterol, glycoproteins,** and **glycolipids.** It is a dynamic, fluid structure that acts as a selectively permeable barrier. In parts of the body where rapid absorption is necessary, the plasma membrane is folded into projections called **microvilli,** which increase its surface area.

2. **Cytoplasm.** The cytoplasm is the material inside the cell. It consists of three parts: **cytosol,** the **cytoskeleton,** and **organelles** (**Figure 4.4**). Cytosol is the fluid portion of the cytoplasm and contains water, solutes, enzymes, and other proteins. The cytoskeleton is a collection of protein filaments including **microtubules, actin filaments,** and **intermediate filaments.** Together, these filaments support the cell, function in cell movement, and move substances within the cell.

   a. In addition, microtubules form the core of motile extensions from the cell called **cilia** and **flagella.** Cilia are small, hairlike extensions that beat rhythmically together to propel substances past the cell. Flagella are single extensions that propel the cell itself (sperm cells are the only flagellated cells in the human body).

   b. Organelles are specialized cellular compartments that carry out a variety of functions. The organelles we cover in this unit include the following:

   ▌ **Ribosomes.** The small, granular ribosomes are not surrounded by a membrane and are composed of two subunits. Some ribosomes float freely in the cytosol, whereas others are bound to the membrane of another organelle or the nucleus. Ribosomes are the sites of protein synthesis in the cell.

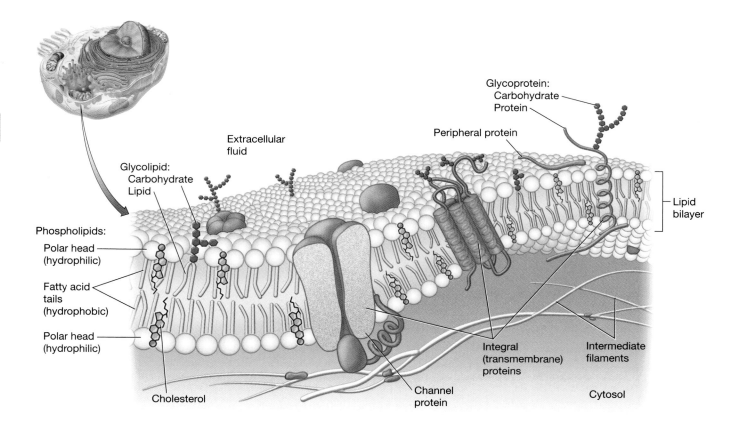

Glycoprotein:
Carbohydrate
Protein

Peripheral protein

Extracellular
fluid

Glycolipid:
Carbohydrate
Lipid

Lipid
bilayer

Phospholipids:

Polar head
(hydrophilic)

Fatty acid
tails
(hydrophobic)

Polar head
(hydrophilic)

Integral
(transmembrane)
proteins

Intermediate
filaments

Cholesterol

Channel
protein

Cytosol

FIGURE **4.3** The plasma membrane

▌ **Peroxisomes.** The small, vesicular organelles known as peroxisomes contain enzymes that detoxify chemicals produced by cellular reactions, metabolize fatty acids, and synthesize certain phospholipids.

▌ **Mitochondria.** The bean-shaped mitochondria are double-membrane-bounded organelles that produce the bulk of the cell's ATP (energy).

▌ **Endoplasmic reticulum.** The series of membrane-enclosed sacs known as the endoplasmic reticulum may be of two types: **rough endoplasmic reticulum,** or **RER,** which has ribosomes on its surface, and **smooth endoplasmic reticulum,** or **SER,** which lacks ribosomes. RER functions in protein synthesis and modifies proteins that the ribosomes have made. SER has multiple functions, including lipid synthesis and detoxification reactions.

▌ **Golgi apparatus.** The Golgi apparatus is a stack of flattened sacs near the RER. Its membrane-enclosed sacs receive vesicles from the RER and other places in the cell and process, modify, and sort the products within the vesicles.

▌ **Lysosomes.** Lysosomes are sacs filled with digestive enzymes that digest particles brought into the cell, old and worn-out organelles, and even the cell itself.

▌ **Centrioles.** Centrioles are paired organelles composed primarily of microtubules that are located in the central area of the cell called the **centrosome.** They appear to be microtubule organizing centers and are important in facilitating the assembly and disassembly of microtubules.

3. **Nucleus.** The third component in nearly all cells is a specialized organelle called the nucleus. The nucleus is the cell's biosynthetic center, which directs the synthesis of all of the body's proteins, as well as certain nucleic acids. The nucleus is surrounded by a double membrane called the **nuclear envelope,** which contains holes called **nuclear pores.** Within the nucleus we find **chromatin,** a ball-like mass of tightly coiled DNA and proteins; RNA; and a dark-staining region called the **nucleolus.** The nucleolus contains a type of RNA called **ribosomal RNA** and is the "birthplace" of ribosomes.

Note that the cell shown in **Figure 4.4** is a **generalized cell** that contains each organelle. Most cells in the body don't look like this and instead are specialized so their forms follow their functions. For example, the cells of the liver contain a large amount of smooth endoplasmic reticulum, and immune cells (phagocytes) house a large number of lysosomes.

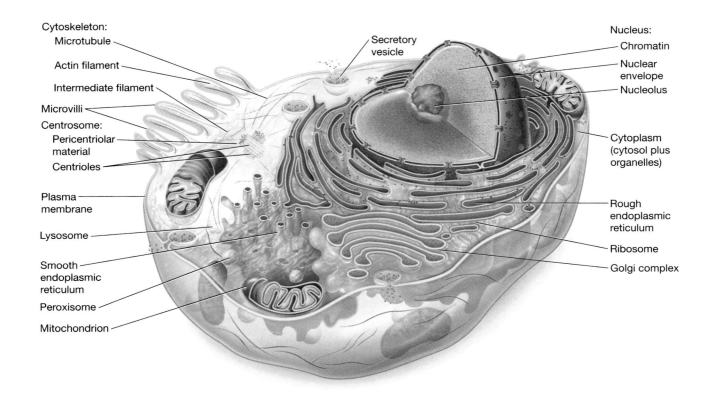

Cytoskeleton:
 Microtubule
 Actin filament
 Intermediate filament
Microvilli
Centrosome:
 Pericentriolar material
 Centrioles
Plasma membrane
Lysosome
Smooth endoplasmic reticulum
Peroxisome
Mitochondrion

Secretory vesicle

Nucleus:
 Chromatin
 Nuclear envelope
 Nucleolus

Cytoplasm (cytosol plus organelles)

Rough endoplasmic reticulum
Ribosome
Golgi complex

FIGURE **4.4** **A generalized cell: sectional view**

## **Procedure** Model Inventory for the Cell

Identify the following structures of the cell on models and diagrams, using your textbook and this unit for reference. As you examine the anatomical models and diagrams, record on the model inventory in **Table 4.2** the name of the model and the structures that you were able to identify. When you have completed this activity, answer Check Your Understanding questions 1 and 2 (p. 83).

1. Plasma membrane
   a. Phospholipid bilayer
   b. Integral proteins
   c. Peripheral proteins
   d. Carbohydrates
2. Nucleus
   a. Nuclear membrane
   b. Nuclear pores
   c. Chromatin and chromosomes
   d. Nucleolus

3. Cytoplasm
4. Ribosomes
5. Smooth endoplasmic reticulum
6. Rough endoplasmic reticulum
7. Golgi apparatus
8. Lysosomes
9. Peroxisomes
10. Mitochondria
11. Centrioles

12. Microtubules, actin filaments, and intermediate filaments
13. Vesicle
14. Microvilli
15. Cilia
16. Flagella

**TABLE 4.2** Cellular Structures Model Inventory

| Model/Diagram | Structures Identified |
|---|---|
|  |  |
|  |  |
|  |  |
|  |  |

## Procedure  Building a Plasma Membrane

In this exercise, you will build a model of the plasma membrane with modeling clay. Refer to **Figure 4.3** (p. 66) for assistance. Use the following color code to build your model:

☐ Phosphate heads: blue
☐ Fatty acid tails: yellow
☐ Integral proteins: red

☐ Peripheral proteins: green
☐ Carbohydrates: orange

## Procedure  Preparing a Cell Sample

Now let's attempt to identify some organelles and cell structures on an actual cell. You will obtain a sample of cells from the mouth (either your own or from a preserved small mammal), and then stain them with methylene blue so that the structures become visible (**Figure 4.5**).

**Safety Note**

⚠ Safety glasses and gloves are required! *Methylene blue stains hands and clothes readily.*

1 Obtain a blank slide and a coverslip.

2 Clean the slide with lens paper.

3 Swab the inside of your cheek with a sterile wooden applicator stick. As an alternative, swab

FIGURE **4.5** Cell sample stained with methylene blue

the inside cheek of a fetal pig or other preserved small mammal. Do not get large chunks of tissue on the swab, or individual cells will not be visible.

4   Wipe the swab with the cheek cells on the blank slide. If you are using your own cheek cells, dispose of the swab in a biohazard bag.

5   Place one drop of methylene blue dye onto the slide. Wait 1 minute.

6   Rinse the dye off the slide with distilled water and pat dry. The blue dye should be barely visible on the slide. If you see large areas of blue, rinse the slide again, or get a new sample of cheek cells.

7   Place a coverslip over the stained area, and place the slide on the stage of a microscope. Focus the image grossly on low power, then switch to the high power objective lens to find individual cells (use oil immersion if available).

8   Draw an individual cell, and label all cell structures and organelles that you can identify.

## HINTS & TIPS

### How to Draw Useful Micrograph Diagrams

You don't need to be a great artist to draw useful micrograph diagrams. Following are some tips for producing effective drawings that can help you learn the material and study for practical exams.

ⓘ  First look at a diagram of the cell or tissue in question in this book or your textbook to get oriented and get an idea of what you should look for in the field of view.

ⓘ  Make note of the image's magnification. As always, start on low power, then advance to higher-powered objectives as needed. You should aim for a magnification about the same as the magnification in the diagram.

ⓘ  Reproduce as closely as possible what you see in the field of view using a pencil. This doesn't have to be a work of art, but you should take care to draw the shape of the cells, the way the cells are organized, and the components of the extracellular matrix. You may draw in the circles in this manual or on a piece of white paper for larger diagrams.

ⓘ  Add color to your drawing with colored pencils using the slide as a guide. Color is a critical component, particularly for study purposes. Some slides will have uniquely colored stains or staining patterns, and studying the colored images can prove quite helpful. In addition, the slides that your lab uses might have different stains than the images in your book. In these cases, it is useful to have a drawing with those specific stain colors.

ⓘ  The final step is to label your drawing. Use your lab manual, textbook, and other resources as a guide to ensure that your labels are accurate.

## Procedure Examining Cellular Diversity with Microscopy

The structure of different cell types can vary drastically. Cells differ not only in size and shape but also in the types and prevalence of organelles in the cell. In this activity you will examine prepared microscope slides of red blood cells, sperm cells, and skeletal muscle cells. Use the techniques you learned in Unit 3: Begin your observation on low power, and advance to high power for each slide. Note that sperm cells can be difficult to find, so an oil-immersion lens is helpful to find the tiny cells. Draw, color, and label the cellular structures and organelles that you see on each slide. You may wish to look at **Figures 5.3K** (p. 101) for red blood cells, **5.4A** (p. 107) for skeletal muscle, and **27.10** (p. 667) for sperm cells. When you have completed this activity, answer Check Your Understanding question 3 (p. 83).

**4**

**1** Red blood cells

**2** Sperm cells

**3** Skeletal muscle cells

# Exercise **4-2**

## Diffusion

### MATERIALS

- ❏ 2 100 ml beakers
- ❏ Food coloring
- ❏ Hot water
- ❏ Ice water
- ❏ 4 agar plates
- ❏ 0.1M potassium permanganate
- ❏ 0.1M methylene blue
- ❏ 1.0M methylene blue
- ❏ 0.1M Congo red
- ❏ Ruler

**Diffusion** is defined as the movement of solute from a high concentration to a low concentration (**Figure 4.6**). Diffusion is a **passive process**—one that requires no net input of energy by a cell—because the energy for diffusion comes from a **concentration gradient**. A concentration gradient is defined as a substance present in different concentrations at two different points. A concentration gradient is illustrated in **Figure 4.6** in the panels "Time 1" and "Time 2."

The rate at which diffusion takes place depends upon several factors, including the steepness of the concentration gradient, the temperature, and the size of the particles. Generally, smaller particle size, steeper concentration gradients, and higher temperatures will increase the rate of diffusion. Diffusion will continue until the concentration of the substance is distributed evenly, a condition known as **equilibrium**. At this point, net diffusion ceases.

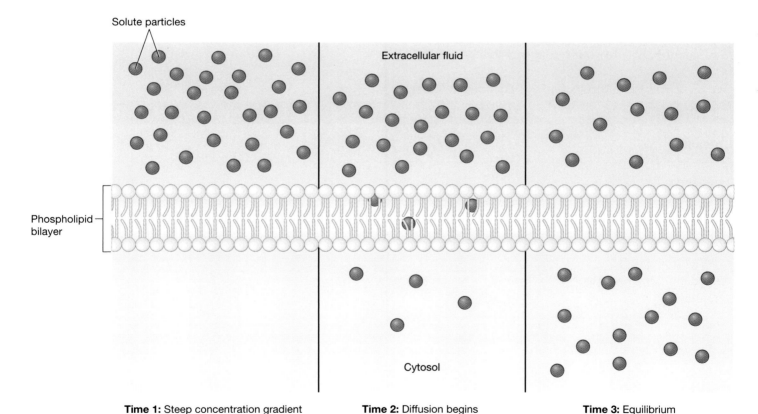

**Time 1:** Steep concentration gradient    **Time 2:** Diffusion begins    **Time 3:** Equilibrium

FIGURE **4.6** Diffusion across a plasma membrane

## Procedure Measuring Rates of Diffusion

Diffusion is a process that occurs all around us and therefore is easy to witness in action (**Figure 4.7**). In this experiment you will examine the effects of temperature, concentration, and particle size on the rate of diffusion by placing dyes in water and in agar. When you have completed the activities, answer Check Your Understanding question 4 (p. 83).

### Part 1

**1** Obtain two 100 ml glass beakers. Label one beaker "cold" and the other "hot." Fill the "cold" beaker with ice water (take care not to get ice in the beaker, though), and fill the "hot" beaker with water that has been heated.

**2** Add two drops of food coloring to each beaker.

**3** Observe the beakers from the side, and measure with a ruler the distance that the dye spreads every minute for 5 minutes. Record your results in **Table 4.3**.

FIGURE **4.7** An example of diffusion is seen with potassium permanganate dye in agar. The dye diffuses outward over time from a high concentration to a low concentration.

TABLE **4.3** Diffusion Results for Food Coloring in Water

| Time | Distance of Dye Diffusion: Cold Water | Distance of Dye Diffusion: Hot Water |
|------|----------------------------------------|---------------------------------------|
| 1 minute | | |
| 2 minutes | | |
| 3 minutes | | |
| 4 minutes | | |
| 5 minutes | | |

**4** Interpret your results. What effect does temperature have on the rate of diffusion?

### Part 2

**1** Obtain 4 agar plates with 1 cm wells scooped out of the middle. Label the plates 1, 2, 3, and 4.

**2** Fill the well of plate 1 with 0.1M potassium permanganate (molecular weight = 158 g/mol).

**3** Fill the well of plate 2 with 0.1M methylene blue (molecular weight = 374 g/mol).

**4** Fill the well of plate 3 with 1.0M methylene blue (molecular weight = 374 g/mol).

**5** Fill the well of plate 4 with 0.1M Congo red (molecular weight = 697 g/mol).

**6** Measure with a ruler the distance that each dye has diffused in 10-minute intervals for 50 minutes. Record your results in Table 4.4.

TABLE **4.4** Diffusion Results for Dyes in Agar

| Time | Distance of Diffusion: Plate 1, 0.1M Potassium Permanganate | Distance of Diffusion: Plate 2, 0.1M Methylene Blue | Distance of Diffusion: Plate 3, 1.0M Methylene Blue | Distance of Diffusion: Plate 4, 0.1M Congo Red |
|---|---|---|---|---|
| 10 minutes | | | | |
| 20 minutes | | | | |
| 30 minutes | | | | |
| 40 minutes | | | | |
| 50 minutes | | | | |

**7** Interpret your results:

  **a** Which dye diffused the fastest?

  **b** Which dye diffused the slowest?

  **c** What effect does particle size have on the rate of diffusion?

  **d** What effect does concentration have on the rate of diffusion?

# Exercise 4-3

## Osmosis and Tonicity

### MATERIALS

- ❑ Animal or human blood cells
- ❑ 3 glass slides (blank)
- ❑ Coverslips
- ❑ Dropper
- ❑ Wooden applicator stick
- ❑ 5% dextrose in water solution
- ❑ Deionized water
- ❑ 25% NaCl in water solution
- ❑ Lens paper/paper towel
- ❑ Light microscope with 40X objective

Whereas diffusion refers to the movement of solute, a separate passive process called **osmosis** refers to the movement of *solvent* (**Figure 4.8**). Specifically, osmosis is the movement of solvent (usually water) from a solution with a lower solute concentration to a solution with a higher solute concentration through a selectively permeable membrane. Notice in **Figure 4.8** that the solvent (water) moves to the more concentrated solution and dilutes it until the concentrations are approximately equal on each side, because the solute particles cannot pass through the membrane.

In discussing biological solutions, it is useful to describe the concentration of a solution outside a cell compared with the concentration of the cytosol. This is called **tonicity**, and it determines how water crosses the plasma membrane. The three variations of tonicity are:

- ▌ **Hypotonic.** A hypotonic solution has a lower solute concentration than the cytosol. A cell placed in a hypotonic solution will gain water by osmosis and may swell and burst.

- ▌ **Isotonic.** An isotonic solution has the same solute concentration as the cytosol. There is no concentration gradient to drive osmosis, so there is no net movement of water into or out of a cell in an isotonic solution.

▌ **Hypertonic.** A hypertonic solution has a higher solute concentration than the cytosol. A cell in a hypertonic environment will lose water by osmosis and shrivel or **crenate**, because water will be drawn toward the more concentrated solution.

FIGURE **4.8** Osmosis

## **Procedure** Watching Osmosis in Action

The following experiment will allow you to watch osmosis in action by placing cells—in this case, red blood cells—in solutions of different tonicity. When you have completed the activities, answer Check Your Understanding questions 5 and 6 (pp. 83–84).

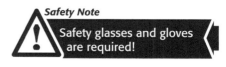

**Safety Note**

Safety glasses and gloves are required!

**1** Obtain three blank slides, and number them 1, 2, and 3.

**2** Place one small drop of animal blood on each slide with a dropper. Take care to keep your droplet fairly small; otherwise you won't be able to see individual cells.

**3** Gently spread the droplet around the center of the slide with a wooden applicator stick, and place a coverslip on it.

**4** On slide 1, place a drop of 5% dextrose solution on one side of the coverslip. On the other side of the coverslip, hold a piece of lens paper or a paper towel. The lens paper will draw the fluid under the coverslip.

**5** Observe the cells under the microscope on high power.

**6** Repeat the procedure by placing 25% NaCl solution on slide 2 and distilled water on slide 3.

**7** Draw and describe what you see on each slide.

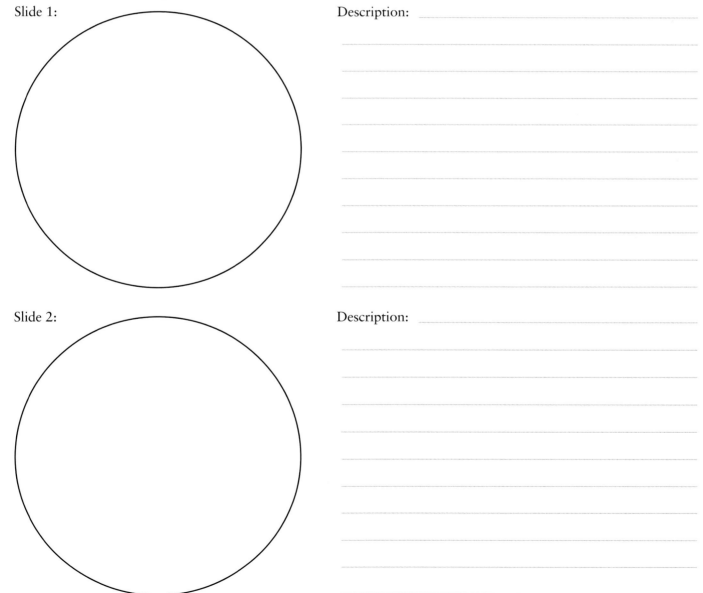

Slide 1:

Description: _____

_____

_____

_____

_____

_____

_____

_____

_____

Slide 2:

Description: _____

_____

_____

_____

_____

_____

_____

_____

_____

_____

_____

_____

Slide 3:

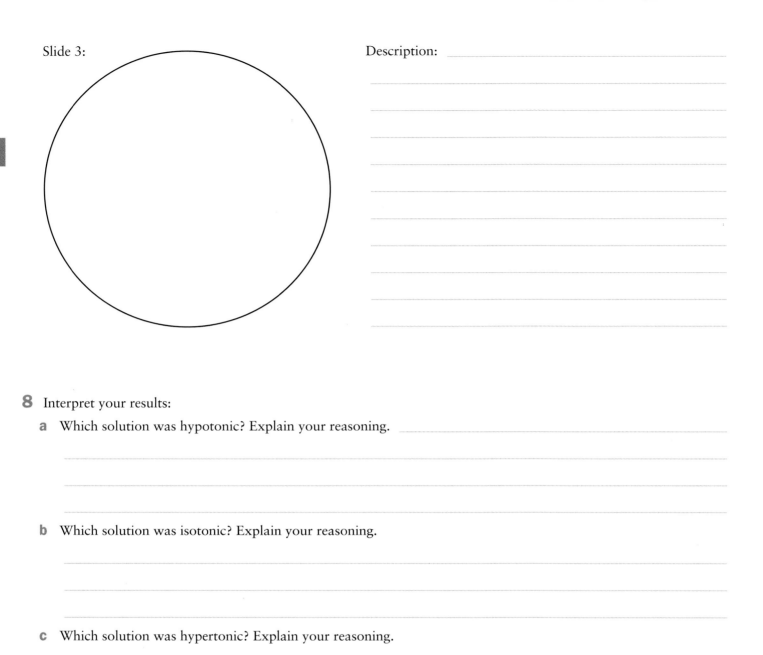

Description: _____

_____

_____

_____

_____

_____

_____

_____

_____

**8** Interpret your results:

**a** Which solution was hypotonic? Explain your reasoning. _____

_____

_____

_____

**b** Which solution was isotonic? Explain your reasoning.

_____

_____

_____

**c** Which solution was hypertonic? Explain your reasoning.

_____

_____

_____

# Exercise 4-4

## Mitosis and the Cell Cycle

### MATERIALS

- ❑ Cell models or diagrams
- ❑ Mitosis models
- ❑ Mitosis slides
- ❑ Light microscope
- ❑ Colored pencils

Most cells go through a continual cycle of growth and replication called the **cell cycle**. The cell cycle consists of four phases:

1. **G1**, or the initial growth phase,
2. **S phase**, during which the DNA is replicated,
3. **G2**, the second growth phase, and
4. **M phase** or **mitosis**, during which the cell divides its organelles, cytosol, and replicated DNA among two identical **daughter cells**.

Each daughter cell has the same exact genetic and structural characteristics as the original cell. The portions of the cycle from G1–G2, when the cell is not dividing, are collectively called **interphase**.

4

INTERPHASE

MITOSIS

Mitotic spindle

Centrioles

Nucleolus

Nucleus

Nuclear membrane

Sister chromatids

Degenerating nuclear membrane

**Prophase**

Spindle fibers

Sister chromatids on the cell's equator

**Metaphase**

Chromatids pulled to opposite poles of cell

**Anaphase**

Chromosomes

Re-forming nuclear membrane

Cleavage furrow

**Telophase**

FIGURE **4.9** Interphase and the stages of mitosis

Mitosis proceeds in the four general stages shown in **Figure 4.9.**

1. **Prophase.** During prophase, the nuclear membrane starts to degenerate, and the DNA condenses so individual **chromosomes** are visible. Also during this stage, we see the formation of a structure called the **mitotic spindle.**

2. **Metaphase.** In metaphase we see the chromosomes line up along the central portion of the cell. Microtubules called **spindle fibers** emanate from the mitotic spindle and attach to the center of each pair of chromosomes.

3. **Anaphase.** During anaphase we see the chromosomes start to move toward opposite poles of the cell as the spindle fibers shorten. In addition, a process called **cytokinesis** begins, during which the cytoplasm is divided up among the two cells.

4. **Telophase.** In the final phase of mitosis, a divot forms between the two cells, called a **cleavage furrow**, that will pinch the cell into two identical daughter cells. In addition, during this stage the nuclear membranes begin to reassemble, the mitotic spindle becomes less visible, and cytokinesis is completed.

**Procedure** Identify Structures of Cell Division

Identify the following structures associated with mitosis on cell models and diagrams.

1. Mitotic spindle
   a. Centrioles
   b. Microtubules (spindle fibers)

2. Nucleus
   a. Nucleolus
   b. Nuclear membrane

3. Chromosomes
   a. Chromatids
   b. Centromere

**Procedure** Model Mitosis

Arrange models of the cell cycle and mitosis in the proper order. As an alternative, build a set of cell cycle models with modeling clay, and arrange them in the proper order of the cycle.

1. Interphase
2. Mitosis
   a. Prophase
   b. Metaphase
   c. Anaphase
   d. Telophase

## Procedure Microscopy of the Cell Cycle

Examine the five phases of the cell cycle on prepared whitefish mitosis slides using the highest-power objective. Note that every stage of the cell cycle may not be visible on one single slide, so you may have to use more than one slide. Note also that most of the cells you see will be in interphase.

Draw and describe what the cell looks like during each phase of the cell cycle, and label your drawing with as many of the structures of cell division (see p. 77) as you can see in each cell. Use your Pre-Lab Exercises and **Figure** 4.9 for reference. When you have completed the activities, answer Check Your Understanding questions 7 and 8 (p. 84).

**1** Interphase

Description: _____

**2** Mitosis

  **a** Prophase

Description: _____

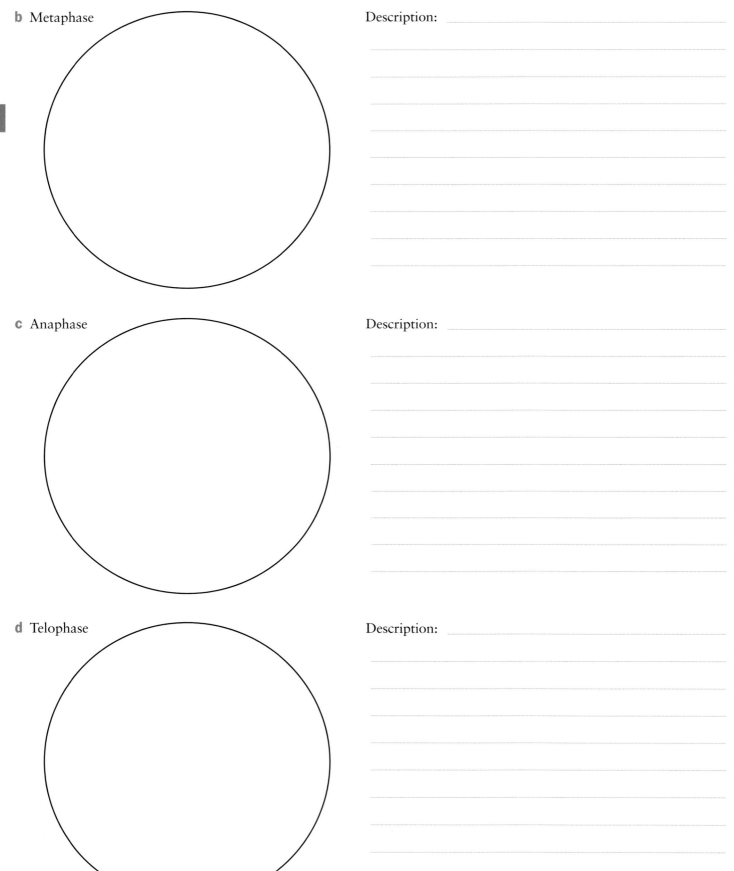

**b** Metaphase

Description: _____
_____
_____
_____
_____
_____
_____
_____

**c** Anaphase

Description: _____
_____
_____
_____
_____
_____
_____
_____

**d** Telophase

Description: _____
_____
_____
_____
_____
_____
_____
_____

## Check Your Recall

**1** Label the following parts of the cell on **Figure 4.10.**

- ☐ Chromatin
- ☐ Golgi apparatus
- ☐ Microvillus
- ☐ Mitochondrion
- ☐ Nuclear envelope
- ☐ Nucleolus
- ☐ Plasma membrane
- ☐ Rough endoplasmic reticulum
- ☐ Smooth endoplasmic reticulum

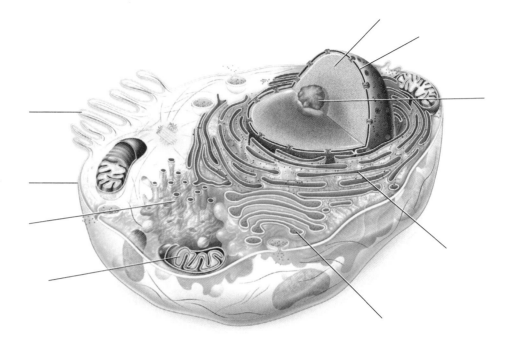

FIGURE **4.10** A generalized cell

**2** Which of the following is not a basic component of most cells?

a. Microvilli

b. Plasma membrane

c. Nucleus

d. Cytoplasm

**3** *Matching:* Match the following organelles and cell structures with the correct definitions.

_____ Plasma membrane

_____ Smooth ER

_____ Mitochondria

_____ Ribosomes

_____ Rough ER

_____ Nucleus

_____ Nucleolus

_____ Lysosome

_____ Microtubule

_____ Golgi apparatus

A. Biosynthetic center of the cell

B. Produce(s) the bulk of the cell's ATP

C. Contain(s) digestive enzymes

D. Stack of flattened sacs that modify and sort proteins

E. Composed of a phospholipid bilayer

F. Series of membrane-enclosed sacs with ribosomes on the surface

G. Cytoskeletal filament found in cilia and flagella

H. Series of membrane-enclosed sacs that detoxify substances and synthesize lipids

I. The cell's "ribosome factory"

J. Granular organelles that are the sites of protein synthesis

**4** *Fill in the blanks:* Diffusion is a _____ process where a solute moves from a _____ concentration to a _____ concentration.

**5** Which of the following factors influence the rate at which diffusion takes place?
a. Size of the particles
b. Temperature
c. Steepness of the concentration gradient
d. All of the above

**6** *Fill in the blanks:* Osmosis is the movement of _____ from a solution with a _____ solute concentration to a solution with a _____ solute concentration.

**7** How do isotonic, hypertonic, and hypotonic solutions differ?

_____

_____

_____

**8** Label the stages of mitosis and the cell cycle on **Figure 4.11**.

FIGURE **4.11** The stages of the cell cycle and mitosis

**9** Which of the following is *not* a phase of mitosis?
a. Prophase
b. Telophase
c. Interphase
d. Anaphase

**10** What is the function of the mitotic spindle and spindle fibers during mitosis?

_____

_____

# Check Your Understanding

### Critical Thinking and Application Questions

**1** Mitochondria contain their own DNA, which encodes 13 proteins. Defects on the mitochondrial DNA can be passed down maternally, leading to a group of diseases called *mitochondrial cytopathies*. Explain which cell populations you think would be most affected by this disease.

_____

_____

_____

**2** Another group of diseases that affect a specific organelle are the *lysosomal storage diseases*. What potential problems could be caused by defective lysosomes?

_____

_____

_____

**3** Which cell type that you observed lacks a nucleus? What functions would this cell be unable to carry out?

_____

_____

_____

**4** Two drugs are injected into the same thigh muscle. Drug A is body temperature (37°C) and has a molecular weight of 150 g/mol. Drug B has been in the refrigerator (5°C) and has a molecular weight of 600 g/mol. Predict which drug will diffuse into the blood more rapidly. Explain your reasoning.

_____

_____

_____

**5** Isotonic saline and 5% dextrose in water are solutions considered isotonic to human blood. What effect on red blood cells would you expect if a patient were given these fluids intravenously? A solution of 10% dextrose in water is hypertonic to human blood. What would happen if you were to infuse your patient with this solution?

_____

_____

**6** Gatorade and other sports drinks are actually hypotonic solutions. How do these drinks rehydrate their user's cells?

**7** Predict locations in the body where cell populations would undergo rapid mitosis. Why would you expect frequent cell division in these locations?

**8** Many anticancer drugs inhibit the formation of the mitotic spindle. What impact will this have on cell division? Why?

# Histology

**5**

85

## OBJECTIVES

*Once you have completed this unit, you should be able to:*

- Identify epithelial tissues by number of layers, cell shape, and specializations.

- Identify and describe connective tissues.

- Identify and describe muscle and nervous tissues.

- Relate tissue structure to tissue function, and describe how organs are formed from two or more tissue types.

- Give examples of organs where each tissue type is found.

Name _____ Section _____ Date _____

Complete the following exercises prior to coming to lab, using your textbook and lab manual for reference.

## Pre-Lab Exercise **5-1**

5

### ✎ Key Terms

You should be familiar with the following terms before coming to lab.

| Term | Definition |
|---|---|
| **Epithelial Tissue** | |
| Simple epithelial tissue | |
| Stratified epithelial tissue | |
| Pseudostratified epithelial tissue | |
| Squamous cell | |
| Cuboidal cell | |
| Columnar cell | |
| Basement membrane | |
| **Connective Tissue** | |
| Ground substance | |
| Collagen fiber | |
| Elastic fiber | |
| Reticular fiber | |

Loose connective tissue

Dense connective tissue

Cartilage

Bone

Blood

**Muscle Tissue**

Striated

Skeletal muscle tissue

Cardiac muscle tissue

Smooth muscle tissue

**Nervous Tissue**

Neuron

Neuroglial cell

# EXERCISES

The histology labs can be some of the more intimidating and frustrating labs for beginning anatomy and physiology students. The subjects are somewhat abstract and unfamiliar and require use of a complicated tool—the microscope. The best way to approach this subject is to be systematic, and let your lab manual walk you through it step-by-step. If you get confused, don't despair. With the help of this book, your lab instructor, and a little patience, you can do it! Before you begin, you may wish to review the "Hints and Tips" box on p. 56.

The exercises in this unit introduce you to the four basic types of tissue: **epithelial tissue**, **connective tissue**, **muscle tissue**, and **nervous tissue** (**Figure 5.1**). All four of these tissue types have two main components:

1. cells, which are unique for each tissue type, and

2. the **extracellular matrix (ECM)**, which is produced largely by the tissue's cells.

ECM consists of two components: ground substance and protein fibers. **Ground substance** is a gelatinous substance that contains water, ions, nutrients, large polysaccharides, and glycoproteins. It enables the tissue to resist compression. **Protein fibers** are found within the ground substance and give the tissue distensibility and tension resistance. Let's now begin our exploration of this fascinating level of organization.

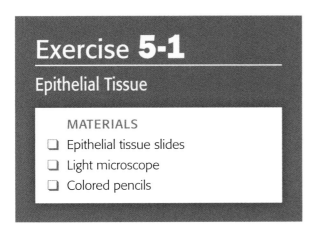

## Exercise 5-1

### Epithelial Tissue

**MATERIALS**
- ❏ Epithelial tissue slides
- ❏ Light microscope
- ❏ Colored pencils

**Epithelial tissues** are our covering and lining tissues. They are found covering body surfaces, lining body passageways, lining body cavities, and forming glands. Epithelia predominantly contain cells called **epithelial cells**, and their ECM is limited mostly to the space underneath the cells in a layer called the **basal lamina**. The basal lamina adheres to another layer of ECM produced by the connective tissues deep to the epithelium (known as the *lamina reticularis*). Together, these two structures are called the **basement membrane**.

Epithelial tissues are all **avascular**; they have no blood vessels to supply them directly and rely on oxygen and nutrients diffusing up from deeper tissues. For this reason, epithelial tissues can be only a certain number of cell layers in thickness. If they are too thick, oxygen and nutrients will not reach the more superficial cells. and they will die.

The many types of epithelia are classified according to the number of cell layers and the shape of the epithelial cells. The classes include the following (see **Figure 5.2**):

1. **Simple epithelia** have only one layer of cells and include:
   a. **Simple squamous epithelium.** This type of epithelium, shown in **Figure 5.2A**, consists of a single layer of flat cells with a flattened nucleus. We often find simple squamous epithelium in places where substances have to cross the epithelium quickly, such as the air sacs of the lungs.
   b. **Simple cuboidal epithelium.** Note in **Figure 5.2B** that the cells of simple cuboidal epithelium are about as wide as they are tall, with a round, central nucleus. Simple cuboidal epithelium is found lining glands such as the thyroid gland, certain respiratory passages, and in the kidneys.
   c. **Simple columnar epithelium.** The cells of simple columnar epithelium, shown in **Figure 5.2C**, are taller than they are wide, with round nuclei located near the base of the cell. These cells line certain respiratory passages, much of the digestive tract, and the genitourinary tract. The surfaces of simple columnar epithelial cells often contain cilia or are folded into microvilli.

2. **Stratified epithelia** have two or more layers of cells and include the following:
   a. **Stratified squamous epithelium.** This type of epithelium consists of many layers of flattened cells. There are two variants of stratified squamous epithelium. The first, shown in **Figure 5.2D**, is **stratified squamous keratinized epithelium,** which consists of epithelial cells called **keratinocytes** that produce the hard protein keratin. Note in **Figure 5.2D** that the more superficial cells are dead and are filled with keratin. This is because they are too far away from the blood supply in the deeper tissues and, as a result, they harden and

## Epithelial Tissue

Epithelial tissue covers the outside of the body and lines all body cavities. Its primary function is to provide protection.

**Simple squamous epithelium**

Nucleus

Plasma membrane

Basal lamina

Lamina reticularis

Basement membrane

**Simple cuboidal epithelium**

Goblet cell

**Simple columnar epithelium**

## Connective Tissue

Connective tissue functions as a binding and supportive tissue for all other tissues in the body.

Collagen fibers

Fibroblasts

**Dense regular connective tissue**

Nucleus

Fat droplet

**Adipose tissue**

Osteocyte

Matrix

**Bone tissue**

## Muscle Tissue

Muscle tissue is a tissue adapted to contract.

Striation

Nucleus

**Skeletal muscle**

Intercalated disk

Striation

Nucleus

**Cardiac muscle**

Nucleus

**Smooth muscle**

## Nervous Tissue

Nervous tissue functions to receive stimuli and transmit signals from one part of the body to another.

Neuroglial cell

Axon

Dendrite

**Neuron**

**Neuroglial cell**

FIGURE **5.1** The four different types of tissue

5

die. Stratified squamous keratinized epithelial cells are well-structured to resist mechanical stresses and, accordingly, they are found in the outer layer of the skin, where they protect the deeper tissues.

Stratified squamous nonkeratinized epithelium (Figure 5.2E) contains no keratin and is located in places subject to lesser degrees of mechanical stress, such as the oral cavity, the pharynx (throat), the anus, and the vagina. This type of epithelium is not as thick and, as a result, the superficial cells are alive and much different in appearance than those of keratinized epithelium.

b. **Stratified cuboidal epithelium** and **stratified columnar epithelium**. Both of these types of epithelium are rare in the human body (**Figure 5.2F** shows stratified cuboidal epithelium only) and are found lining the ducts of certain glands.

Not all epithelial tissues fit this classification scheme. One such epithelial tissue is called **transitional epithelium** (**Figure 5.2G**). This type of epithelial tissue is stratified but is not classified by its shape because its cells can change shape. Typically, the apical or surface cells are dome-shaped, but when the tissue is stretched, they flatten and are squamous in appearance. Transitional epithelium is found lining the urinary bladder and ureters.

Another type of epithelial tissue that does not fit neatly into this classification system is **pseudostratified ciliated columnar epithelium** (**Figure 5.2H**). This epithelium has the appearance of having many layers but actually has only one layer of cells (*pseudo* = false). Note that this type of epithelium usually has cilia, and the cell shape is always columnar. It is found lining the nasal cavity and much of the respiratory tract.

FIGURE **5.2** Epithelial tissues: (**A**) simple squamous epithelium from the lungs; (**B**) simple cuboidal epithelium from the kidney *(continues)*

FIGURE **5.2** Epithelial tissues *(cont.)*: (**C**) simple ciliated columnar epithelium from the uterine tube; (**D**) stratified squamous keratinized epithelium from the skin; (**E**) stratified squamous nonkeratinized epithelium from the vagina *(continues)*

**F**

Loose connective tissue

Stratified cuboidal
epithelium

Nuclei

Basement membrane

**G**

Transitional
epithelial cells

Basement membrane

Nuclei

Loose connective tissue

**H**

Pseudostratified
epithelial cells

Nuclei

Loose connective tissue

Cilia

Goblet cell

Simple cuboidal
epithelium (from a gland)

FIGURE **5.2** Epithelial tissues *(cont.)*: (**F**) stratified cuboidal epithelium from a sweat gland; (**G**) transitional epithelium from the urinary bladder; (**H**) pseudostratified ciliated columnar epithelium from the trachea

## HINTS & TIPS

### How to Approach Epithelium

Before you start, remember how to approach any slide: First examine the slide with the naked eye, then begin on low power and scan the slide, and advance progressively to higher power to see details. Use Figure 5.2 as a guide—if your slide looks nothing like the tissue in Figure 5.2, scroll around the slide, and keep looking. But do not rely completely on Figure 5.2, because the slides that your lab uses may be prepared with different stains or from different tissues. A few simple things will help you distinguish epithelial tissues from other tissues, including:

ⓘ Epithelial tissues are all avascular, so you won't see any blood vessels in epithelium.

ⓘ Epithelial tissues are typically on the outer edge of the slide. Keep in mind that most slides have several tissues in each section. To find the epithelial tissue, scroll to one end of the slide or the other.

ⓘ Epithelial tissues consist mostly of cells. If you aren't sure if something is a cell, look for a nucleus. If you can see a nucleus, you typically will be able to see the plasma membrane surrounding the cell as well, which will help you define borders between cells.

## Procedure Microscopy of Epithelial Tissues

Examine prepared slides of the following epithelial tissues. Use colored pencils to draw what you see under the microscope, and label your drawings with the terms from **Figure 5.2**. Then (a) describe what you see, and (b) give examples of locations in the body where this tissue is found.

**1** Simple squamous epithelium

a _____

b _____

**2** Simple cuboidal epithelium

a _____

b _____

**3** Simple columnar epithelium

a _____

b _____

**4** Stratified squamous keratinized epithelium

a _____

b _____

**5** Stratified squamous nonkeratinized epithelium

a _____

b _____

**6** Stratified cuboidal epithelium

a _____

b _____

**7**  Transitional epithelium

**8**  Pseudostratified ciliated columnar epithelium

5

a _____

b _____

a _____

b _____

# Exercise 5-2

## Connective Tissue

**MATERIALS**

❑ Connective tissue slides
❑ Light microscope
❑ Colored pencils

Connective tissues are found throughout the body. They have a variety of functions, most of which serve to *connect*, as their name implies (blood is an exception). All connective tissues stem from a common embryonic tissue called **mesenchyme**. Connective tissues are distinguished easily from epithelial tissues by the prominence of their extracellular matrices. Typically, connective tissues contain few cells and have an extensive ECM consisting of ground substance and three main types of protein fibers:

1. **collagen fibers**, composed of the thick protein **collagen**, which gives a tissue tensile strength,

2. **elastic fibers**, composed of the protein **elastin**, which makes a tissue distensible, and

3. thin **reticular fibers** that interweave to form networks that support blood vessels, nerves, and other structures.

The four general types of connective tissue (CT) are as follows (**Figure 5.3**):

1. **Connective tissue proper.** Connective tissue proper, the most widely distributed class of connective tissue in the body, consists of scattered cells called **fibroblasts** that secrete an extensive ECM filled with many types of protein fibers. This tissue is highly vascular with an extensive blood supply. The subclasses of CT proper include the following:

   a. **Loose (areolar) CT.** You can see in **Figure 5.3A** that the primary element in loose CT is ground substance, which gives it a "loose" appearance on a slide. All three types of protein fibers are scattered in loose CT ground substance. Loose CT is found as part of the basement membrane and in the walls of hollow organs.

   b. **Reticular CT.** As you can guess by the name, reticular CT consists of many reticular fibers produced by cells called **reticular cells** (**Figure 5.3B**). It is located in the spleen and lymph nodes, where the fine reticular fibers interweave to form "nets" that trap pathogens and foreign cells. Reticular CT also is located around blood vessels and nerves, where it forms supportive networks.

   c. **Adipose tissue.** Notice in **Figure 5.3C** that adipose tissue (fat tissue) has a much different appearance than the other types of CT proper. It consists of mostly cells, with little visible ECM, and the huge cells are called **adipocytes**. Each adipocyte contains a large lipid droplet that occupies most of its cytoplasm. The nucleus and other organelles are barely visible, because they are pushed to the periphery of the cell against the plasma membrane. Adipose tissue is distributed widely throughout the body under the skin and around organs.

   d. **Dense regular collagenous CT.** The difference between loose and dense CT is obvious in **Figure 5.3D**. Dense regular collagenous CT consists primarily of collagen fibers arranged in parallel bundles with little ground substance and few cells. It is exceptionally strong and makes up structures that require tensile strength in a single plane, such as tendons and ligaments.

   e. **Dense irregular collagenous CT.** Like dense regular collagenous CT, dense *irregular* collagenous CT consists of bundles of collagen fibers. Note in **Figure 5.3E**, however, that these collagen bundles are arranged in an irregular, haphazard fashion without a consistent pattern. Dense irregular collagenous CT also is quite strong and is located in places that require tensile strength in multiple planes, such as the dermis and joint and organ capsules.

   f. **Dense elastic CT.** This tissue type contains elastic fibers arranged in parallel bundles (**Figure 5.3F**). We find dense elastic CT lining large blood vessels and in certain ligaments.

2. **Cartilage.** Cartilage is a tough but flexible tissue resistant to tension, twisting, and compressive forces. It consists of cells called **chondrocytes** located in cavities called **lacunae** embedded in the ECM. Cartilage is notable among the connective tissues for being avascular. Each of the three types of cartilage has a different ECM composition.

   a. **Hyaline cartilage**. Notice in **Figure 5.3G** that hyaline cartilage contains mostly chondrocytes scattered in ground substance with few visible protein fibers. This lack of protein fibers gives hyaline cartilage a smooth, glassy appearance and makes it an ideal tissue to cover the ends of bones where they form joints with another bone. The smooth texture of hyaline cartilage provides a nearly frictionless surface on which bones can articulate. Hyaline cartilage also is found connecting the ribs to the sternum, in the nose, and forming the framework for certain respiratory passages.

b. **Fibrocartilage**. As you can see in **Figure 5.3H**, fibrocartilage is named appropriately, because it is full of protein fibers (mostly collagen). This makes fibrocartilage tough and extremely strong but not at all smooth (think of the surface of fibrocartilage like a flannel sheet, with the cotton fibers representing the protein fibers). For this reason, fibrocartilage does not cover the ends of bones, but it does reinforce ligaments and form *articular disks*, tough structures that improve the fit of two bones. In addition, fibrocartilage is found in joints where hyaline cartilage has been damaged.

c. **Elastic cartilage**. The final type of cartilage, elastic cartilage, is shown in **Figure 5.3I**. It is filled with elastic fibers that allow it to stretch and recoil. Elastic cartilage is found in the ear and in the epiglottis.

3. **Bone**. Bone tissue, also called **osseous tissue**, consists of bone cells called **osteocytes** encased in an ECM that contains collagen fibers and calcium hydroxyapatite crystals. Note in **Figure 5.3J** that the ECM is arranged in concentric layers called **lamellae**, with the osteocytes sandwiched between them. This structure makes bone the hardest tissue in the body and the most resistant to mechanical stresses.

4. **Blood**. Blood (**Figure 5.3K**) is unique in that it is the only connective tissue that doesn't actually connect anything physically. It consists of cells called **erythrocytes** (red blood cells) and **leukocytes** (white blood cells), cellular fragments called platelets, and an ECM called **plasma**.

FIGURE **5.3** Connective tissues: (**A**) loose (areolar) CT; (**B**) reticular CT from the spleen *(continues)*

Adipocyte nucleus

Lipid (fat) droplet

Adipocyte plasma membrane

Adipocyte

Bundles of collagen fibers

Fibroblast nuclei

Ground substance

Bundles of collagen fibers

Fibroblast nuclei

Ground substance

FIGURE **5.3** Connective tissues *(cont.)*: (**C**) adipose tissue; (**D**) dense regular collagenous CT from a tendon; (**E**) dense irregular collagenous CT from the dermis *(continues)*

Ground substance

Elastic fibers

Chondrocytes in lacunae

Extracellular matrix

Chondrocytes in lacunae

Bundles of collagen fibers

FIGURE **5.3** Connective tissues *(cont.)*: (**F**) dense elastic CT from the aorta; (**G**) hyaline cartilage from a joint; (**H**) fibrocartilage from an articular disk *(continues)*

Chondrocytes in lacunae

Bundles of elastic fibers

Osteocytes in lacunae

Lamellae containing collagen fibers and calcium salts in the extracellular matrix

Leukocytes (white blood cells)

Erythrocytes (red blood cells)

FIGURE **5.3** Connective tissues *(cont.)*: (**I**) elastic cartilage from the ear; (**J**) bone tissue; (**K**) blood

### How to Approach Connective Tissue

Students typically find connective tissue the most difficult of the tissues to identify. Don't let this scare you away. Simply take care to approach these slides systematically in the same manner as you do the epithelial slides, and use Figure 5.3 as a guide. The following points will help you to identify the various connective tissues and differentiate them from other tissue types:

ⓘ The cells typically are not densely packed together, and you usually will see a large amount of space between connective tissue cells. Remember to look for the nucleus and plasma membrane to discern the borders of a cell.

ⓘ Generally connective tissue, particularly connective tissue proper, contains a lot of ground substance and protein fibers in the extracellular matrix. One exception is adipose tissue, which consists of densely packed adipocytes filled with a large lipid droplet.

ⓘ Many types of CT can be distinguished by the types of fibers they contain:

1. Reticular fibers are the thinnest fibers, which typically stain brown or black. Look for reticular fibers in loose and reticular CT.

2. Collagen fibers are thick fibers that often stain pink. Look for collagen fibers in fibrocartilage, dense regular collagenous CT, dense irregular collagenous CT, and loose CT.

3. Elastic fibers are fairly thick and may have a wavy appearance. They are thinner than collagen fibers. Their color ranges from purple-black to blue, depending on the stain used. Look for them in elastic tissue, elastic cartilage, and loose CT.

ⓘ Cartilage is easy to discern from CT proper by looking at the shape of the cells. Fibroblasts are generally small and flat, whereas chondrocytes are much larger and round. In addition, chondrocytes sit in lacunae.

ⓘ Blood and bone are perhaps the two easiest tissues you will examine in this lab. They should look much like they do in Figure 5.3, and no other tissues resemble them.

## Procedure Microscopy of Connective Tissue Proper

View prepared slides of each type of connective tissue proper. Use colored pencils to draw pictures of what you see under the microscope, and label your drawings with the terms from **Figure 5.3**. Then (a) describe what you see, and (b) give examples of locations in the body where this tissue is found. When you have completed the exercise, answer Check Your Understanding questions 1–3 (p. 117).

**1** Loose (areolar) CT

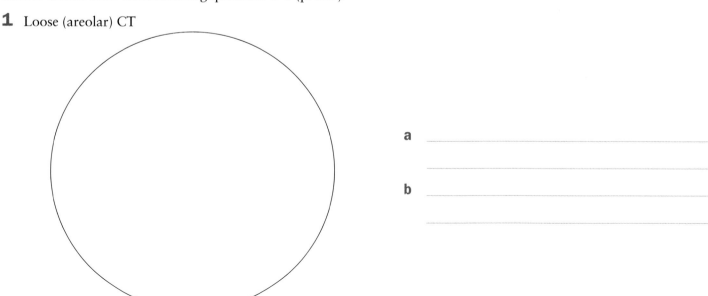

a _____

b _____

**2** Reticular CT

a _____

_____

b _____

_____

**3** Adipose tissue

a _____

_____

b _____

_____

**4** Dense regular collagenous CT

a _____

_____

b _____

_____

**5** Dense irregular collagenous CT

a _____

_____

b _____

_____

**6** Dense elastic CT

a _____

_____

b _____

_____

### Procedure Microscopy of Cartilage

View prepared slides of the three types of cartilage. Use colored pencils to draw pictures of what you see under the microscope, and label your drawings with the terms from **Figure 5.3**. Then (a) describe what you see, and (b) give examples of locations in the body where this tissue is found. When you have completed this activity, answer Check Your Understanding question 4 (p. 117).

**7** Hyaline cartilage

a _____

_____

b _____

_____

**8** Fibrocartilage

a _____

_____

b _____

_____

**9** Elastic cartilage

a _____
_____
b _____
_____

**Procedure** Microscopy of Bone and Blood

View prepared slides of bone and blood. Use colored pencils to draw pictures of what you see under the microscope, and label your drawings with the terms in **Figure 5.3**. Then (a) describe what you see, and (b) give examples of locations in the body where this tissue is found. When you have completed all of the connective tissue slides, answer Check Your Understanding question 5 (p. 118).

**10** Bone

a _____
_____
b _____
_____

**11** Blood

a _____
_____
b _____
_____

# Exercise 5-3

## Muscle Tissue

### MATERIALS

❑ Muscle tissue slides

❑ Light microscope

❑ Colored pencils

Muscle tissue is located in skeletal muscles, in the walls of hollow organs, in the heart, and in other locations such as the iris of the eye. It consists of muscle cells, sometimes called **myocytes** or **muscle fibers**, and a small amount of ECM called the **endomysium**. Notice in **Figures 5.4A–C** that myocytes aren't shaped like the cells you are accustomed to seeing. For this reason, muscle tissue is easy to discern from the other tissue types.

There are three types of muscle tissue:

1. **Skeletal muscle tissue**. The myocytes of skeletal muscle tissue are long, tubular, and **striated** (striped) in appearance (**Figure 5.4A**). The striations result from the arrangement of proteins within the muscle fiber called **myofilaments**. Skeletal muscle fibers are formed from the fusion of cells called **myoblasts** and for this reason have multiple nuclei.

2. **Cardiac muscle tissue**. The myocytes of cardiac muscle tissue, located in the heart, are short, fat, striated, and tend to be branching (**Figure 5.4B**). Adjacent myocytes are linked by specialized junctions called **intercalated disks** that contain desmosomes and gap junctions. Cardiac myocytes typically have only one nucleus, but some may have two or more.

3. **Smooth muscle tissue**. The myocytes of smooth muscle tissue are flat with one nucleus in the center of the spindle (**Figure 5.4C**). The arrangement of myofilaments within smooth muscle fibers differs from that of skeletal and cardiac muscle fibers, and as a result, these cells lack noticeable striations (hence the name *smooth* muscle). It lines all hollow organs and is found in the skin, the eyes, and surrounding many glands.

We revisit each type of muscle tissue in several units in this book.

FIGURE **5.4** Muscle tissue: (**A**) skeletal muscle tissue; (**B**) cardiac muscle tissue; (**C**) teased muscle cells; (**D**) smooth muscle cells in tissue

## Procedure Microscopy of Muscle Tissue

View prepared slides of skeletal, smooth, and cardiac muscle tissue. Use colored pencils to draw what you see under the microscope, and label your drawing with the terms from **Figure 5.4**. Record your observations of each slide in **Table 5.1**. When you have completed the microscope slides, answer Check Your Understanding question 6 (p. 118).

**5**

**1** Skeletal muscle

a _____

_____

b _____

_____

**2** Cardiac muscle

a _____

_____

b _____

_____

**3** Smooth muscle:

a _____

_____

b _____

_____

TABLE **5.1** Characteristics of Muscle Tissues

| Muscle Tissue Type | Striated or Nonstriated | One or Multiple Nuclei | Size and Shape of Cells | Special Features |
|---|---|---|---|---|
| Cardiac muscle | | | | |
| Skeletal muscle | | | | |
| Smooth muscle | | | | |

5

# Exercise 5-4

## Nervous Tissue

### MATERIALS
☐ Nervous tissue slide
☐ Light microscope
☐ Colored pencils

Nervous tissue (**Figure 5.5**) is the primary component of the brain, the spinal cord, and the peripheral nerves. It consists of a unique ECM and two main cell types:

1. **Neurons**. The neurons are responsible for sending and receiving messages within the nervous system. On your slide they are the larger of the two cell types. The large, central portion of the neuron is called the **cell body**. Within the cell body we find the nucleus and most of the neuron's organelles, including clusters of rough endoplasmic reticulum called **Nissl bodies**. Most neurons contain two types of long armlike processes extending from the cell body—the **dendrites**, which receive messages from other neurons, and the **axon**, which sends messages to other neurons, muscle cells, or gland cells.

2. **Neuroglial cells**. The smaller and more numerous cells around the neurons are the neuroglial cells. The six different types of neuroglial cells vary significantly in shape and appearance. Neuroglial cells in general perform functions that support the neurons in some way.

## Procedure  Microscopy of Nervous Tissue

View a prepared slide of nervous tissue (the slide might be called a "motor neuron smear"). Use colored pencils to draw a picture of what you see under the microscope, and label your drawing with the terms from **Figure 5.5**. Then (a) describe what you see, and (b) give examples of locations in the body where this tissue is found.

Nervous tissue

FIGURE **5.5** Nervous tissue

a _____

_____

b _____

_____

# Exercise 5-5

## Organology

### MATERIALS
☐ 5 colors of modeling clay

All organs consist of two or more tissues that must work together to enable the organ to function properly. The study of the tissues that make up the body's organs is called **organology**. Most organs are made of layers of tissues stacked upon one another and "glued" together by proteins and other molecules in the ground substance. This exercise introduces you to organology, a topic we will explore repeatedly in the remainder of this lab manual.

**Procedure** Determine Major Tissue Types of Organs

Use your textbook to determine the main tissue types that constitute each of the following organs, and record this information in **Table 5.2**. Be specific about which type of muscle, epithelial tissue, and connective tissues are found in the organ.

TABLE **5.2** Organs and Their Component Tissues

| Organ | Major Tissue Types |
|---|---|
| Urinary bladder | |
| Blood vessel | |
| Skin | |
| Lymph node | |
| Knee joint and capsule | |
| Heart | |

*(continues)*

TABLE **5.2** Organs and Their Component Tissues *(cont.)*

| Organ | Major Tissue Types |
|---|---|
| Trachea | |
| Esophagus | |
| Auricle (external ear) | |
| Brain (and brain coverings) | |

## Procedure  Build an Organ

For this exercise, choose one or more of the organs from **Table 5.2,** and build it with modeling clay. Use the following color code when building your organ:

▌ Yellow  =  epithelial tissue (be sure to use more than one layer if the tissue is stratified!)

▌ Blue  =  connective tissue proper

▌ Orange  =  cartilage

▌ Red  =  muscle tissue

▌ Green  =  nervous tissue (don't forget that nearly every organ in the body has nerves that supply it. To place the nerves properly, find out which tissue layer of the organ is innervated.)

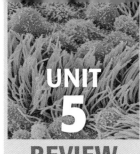

## Check Your Recall

**1** Identify each of the following tissues in **Figure 5.6.**

a _____     e _____

b _____     f _____

c _____     g _____

d _____     h _____

FIGURE **5.6** Unknown tissues for question 1

**2** List the four basic tissue types:

_____          _____

_____          _____

**3** _Fill in the blanks:_ All tissues consist of two main components: _____ and _____, which

consists of a gelatinous substance called _____ and numerous different _____.

**4** Which of the following statements about epithelial tissue is _false_?
   a.  Epithelial tissues are avascular.
   b.  Epithelial tissues consist of few cells and an extensive ECM.
   c.  The ECM of epithelial tissues is located in the basal lamina.
   d.  Epithelial tissues are our covering and lining tissues.

**5** How do simple and stratified epithelial tissues differ?

_____

_____

_____

_____

**6** Which of the following statements about connective tissue is _false_?
   a.  All connective tissues stem from a common embryonic tissue called mesenchyme.
   b.  Connective tissues may contain three types of protein fibers: collagen, elastic, and reticular fibers.
   c.  Most connective tissues are highly vascular, with the exception of cartilage.
   d.  Most connective tissues consist largely of cells with little ECM.

**7** How do loose and dense connective tissues differ?

_____

_____

_____

**8** Which of the following statements about muscle tissue is _true_?
   a.  Skeletal muscle and cardiac muscle tissues have no striations.
   b.  Smooth muscle tissue is found in the heart.
   c.  The cells of skeletal muscle tissue are long, tubular, and multinucleated.
   d.  Smooth muscle cells are joined by intercalated disks.

**9** _Fill in the blanks:_ Nervous tissue is composed of _____ and _____.

**10** *Matching:* Match the tissue type with its location in the body.

_____ Simple squamous epithelium

_____ Transitional epithelium

_____ Loose CT

_____ Cardiac muscle

_____ Nervous tissue

_____ Stratified squamous epithelium

_____ Hyaline cartilage

_____ Smooth muscle

_____ Dense regular collagenous CT

_____ Reticular CT

A. Part of the basement membrane, walls of hollow organs

B. Spleen and lymph nodes

C. Air sacs of the lungs

D. Oral cavity, pharynx, vagina, anus

E. Lining hollow organs, in the skin, in the eye, and surrounding many glands

F. Brain, spinal cord, peripheral nerves

G. Urinary bladder

H. Tendons and ligaments

I. Joints, connecting the ribs to the sternum, nose

J. In the heart

5

# Check Your Understanding

### Critical Thinking and Application Questions

5

**1** The condition *Marfan syndrome* results from the production of abnormal elastin, the main protein component of elastic fibers. Predict the potential consequences of defective elastic fibers. (*Hint:* Consider the organs and tissues in which elastic fibers are found. How would defective elastic fibers impair the functions of these organs and tissues?)

_____

_____

_____

_____

**2** Other diseases, called collagen vascular diseases, affect the protein collagen, the main component of collagen fibers. Predict the possible consequences of defective collagen fibers on specific organs and tissues.

_____

_____

_____

_____

**3** How would the effects of a collagen vascular disease that affected only the collagen in reticular fibers differ from the effects seen with question 2? Why?

_____

_____

_____

_____

**4** The formation of fibrocartilage is a common response to injury of hyaline cartilage. Do you think fibrocartilage would provide an articular surface (i.e., the cartilage in joints) as smooth as the original hyaline cartilage? Why or why not?

_____

_____

_____

_____

**5** Explain how the structure of each of the following tissues follows its function:

**a** Stratified squamous keratinized epithelium:

_____

_____

_____

**b** Simple squamous epithelium:

_____

_____

_____

**c** Hyaline cartilage:

_____

_____

_____

**d** Bone:

_____

_____

_____

**6** When muscle tissue dies, it usually is replaced with dense irregular collagenous connective tissue. How do these tissues differ in structure? Will the muscle be able to function normally? Why or why not?

_____

_____

_____

_____

# Integumentary System

## OBJECTIVES

*Once you have completed this unit, you should be able to:*

- Identify structures of the integumentary system.

- Describe the gross and microscopic structure of thick and thin skin.

- Map the distribution of touch receptors on different areas of the body.

Name _____ Section _____ Date _____

Complete the following exercises prior to coming to lab, using your textbook and lab manual for reference.

## Pre-Lab Exercise **6-1**

### ✎ Key Terms

You should be familiar with the following terms before coming to lab.

| Term | Definition |
|------|------------|

**Epidermal Structures**

Epidermis _____

_____

Keratinocyte _____

_____

Melanocyte _____

_____

Stratum corneum _____

_____

Stratum lucidum _____

_____

Stratum granulosum _____

_____

Stratum spinosum _____

_____

Stratum basale _____

_____

**Structures of the Dermis**

Dermis _____

_____

Dermal papillae _____

_____

Pacinian (lamellated) corpuscle _____

_____

Meissner's (tactile) corpuscle _____

_____

**Other Structures**

Hypodermis _____

_____

Sweat gland _____

_____

Sebaceous gland _____

_____

Hair follicle _____

_____

Nails _____

_____

Arrector pili muscle _____

_____

## Pre-Lab Exercise **6-2**

### Skin Anatomy

Label and color the structures of the skin depicted in **Figure 6.1** with the terms from Exercise 6-1 (p. 123). Use your text and Exercise 6-1 in this unit for reference.

FIGURE **6.1** Skin section

Label and color the structures of the hair and nail depicted in **Figure 6.2** with the terms from Exercise 6-1 (p. 123). Use your text and Exercise 6-1 in this unit for reference.

6

A

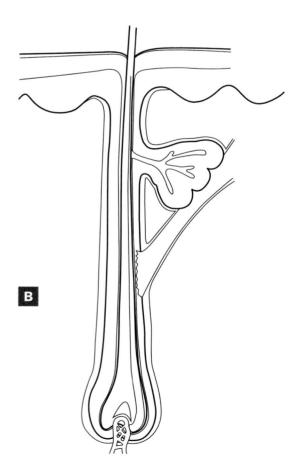

B

FIGURE **6.2** (**A**) Nail anatomy; (**B**) Hair structure

# EXERCISES

Although the skin is the largest organ in the body, most people don't realize that it is actually an organ. Like all organs, the skin is composed of several tissue types, including several layers of epithelial tissue, connective tissue, muscle tissue, and nervous tissue. In the following exercises you will examine the tissues of this organ, along with the other structures of the integumentary system, and apply your knowledge of this system to perform fingerprinting procedures.

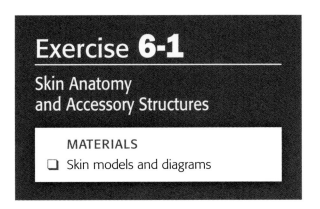

## Exercise **6-1**

### Skin Anatomy and Accessory Structures

#### MATERIALS
☐ Skin models and diagrams

The **integumentary system** is composed of the skin (the **integument**) and its **accessory structures**: the **hair**, **glands**, and **nails**. The skin is composed of two layers, the **epidermis** and the **dermis** (Figure 6.3). The tissue beneath the dermis, sometimes called the **hypodermis** or the *subcutaneous tissue*, connects the skin to the underlying tissues and is not considered part of the integument.

The epidermis contains layers (or *strata*) of stratified squamous keratinized epithelium. From superficial to deep, the layers are as follows:

1. **Stratum corneum**. This superficial layer is composed of dead cells called **keratinocytes**. Under microscopic examination,

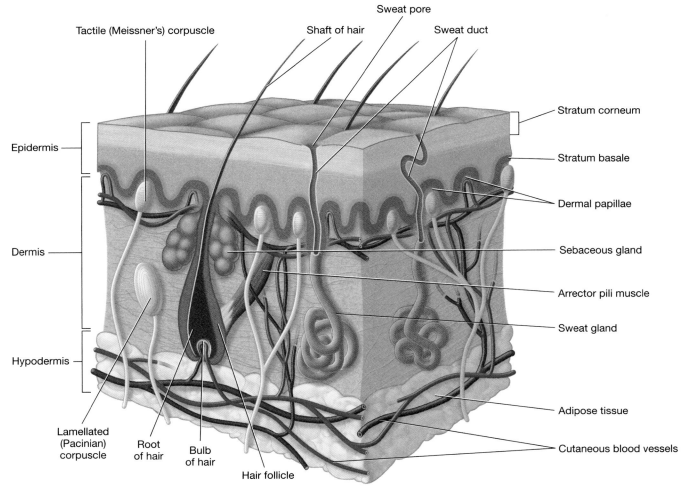

**FIGURE 6.3 Skin section**

the cells of the stratum corneum bear little resemblance to living cells and have a dry, flaky appearance. These cells contain a hard protein called **keratin** that protects the underlying layers of cells.

2. **Stratum lucidum.** This is a single layer of translucent, dead cells found only in the skin of the palms and the soles of the feet.

3. **Stratum granulosum.** The superficial cells of the stratum granulosum are dead, but the deeper cells are alive. This layer is named for the cells' cytoplasmic granules, which contain the protein keratin and an oily waterproofing substance.

4. **Stratum spinosum.** The first actively metabolizing cells are encountered in the stratum spinosum. The pigment **melanin** is found in this layer, which provides protection from UV light and also decreases production of vitamin D, so the body does not overproduce it.

5. **Stratum basale.** The stratum basale is the deepest layer and contains a single row of actively dividing cells. It is often combined with the stratum spinosum and called the **stratum germinativum**.

Why does the epidermis have so many dead cells? Recall that the epidermis is composed of epithelial tissue, and epithelial tissue is avascular (has no blood supply). All epithelial tissues require oxygen and nutrients to be delivered from the deeper tissues. In the case of the epidermis, this deeper tissue is the dermis. Only the cells of the deeper parts of the stratum granulosum, the stratum spinosum, and the stratum basale are close enough to the blood supply in the dermis to get adequate oxygen and nutrients for survival. As the cells migrate farther away from the blood supply, they begin to die.

Immediately deep to the stratum basale of the epidermis is the dermis. The dermis is composed of highly vascular connective tissue and contains two layers:

1. **Papillary layer.** The superficial papillary layer is composed of loose connective tissue. It contains fingerlike projections called the **dermal papillae** that project into the epidermis. These dermal papillae contain touch receptors called **Meissner's** (or tactile) **corpuscles** and capillary loops that provide blood supply to the avascular epidermis.

2. **Reticular layer.** The thick reticular layer is composed of dense irregular collagenous connective tissue. It houses structures such as **sweat glands**, oil-producing **sebaceous glands**, blood vessels, and pressure receptors called **Pacinian corpuscles** (also known as *lamellated corpuscles*).

Both hair and nails are considered accessory structures of the integumentary system (**Figure 6.4**). A hair consists of two basic parts: (a) the long, slender **shaft** composed of dead keratinized cells that projects from the skin's surface, and

FIGURE **6.4** (A) Hair structure; (B) nail anatomy

(b) the hair **root** or **bulb** embedded in the dermis (**Figure 6.4A**). The structure surrounding the hair bulb is an epithelial tissue-lined sheath known as the **hair follicle**.

Like hairs, nails are composed primarily of dead, keratinized cells (**Figure 6.4B**). A nail consists of a **nail plate** surrounded by folds of skin on all three sides, known as **nail folds**. The nail plate is formed by a group of dividing cells proximal to the nail fold called the **nail matrix**.

Other accessory structures of the integumentary system are its **sebaceous glands** and the **sweat glands**, both of which are exocrine glands that secrete their products onto the skin's surface. Sebaceous glands secrete sebum (oil) into a hair follicle, and sweat glands secrete sweat through a small pore.

## Procedure  Model Inventory for the Integumentary System

Identify the following structures of the integumentary system on models and diagrams, using your textbook and this unit for reference. As you examine the anatomical models and diagrams, record on the model inventory in **Table 6.1** the name of the model and the structures you were able to identify. When you have completed this activity, answer Check Your Understanding questions 1–4 (p. 133).

### Structures of the Skin
1. Epidermal layers
   a. Stratum corneum
   b. Stratum lucidum
   c. Stratum granulosum
   d. Stratum spinosum
   e. Stratum basale

### Other Structures
1. Sweat gland
   a. Sweat duct
   b. Sweat pore
2. Hair
   a. Hair follicle
   b. Hair shaft
   c. Arrector pili muscle
   d. Hair root
   e. Hair bulb
   f. Connective tissue sheath

2. Dermal layers
   a. Papillary layer
   b. Reticular layer
3. Dermal papillae
4. Blood vessels
5. Nerves:
   a. Pacinian (lamellated) corpuscle
   b. Meissner's (tactile) corpuscles

3. Nail
   a. Eponychium
   b. Nail matrix
   c. Nail fold (proximal and lateral)
   d. Nail plate
   e. Lunula
4. Sebaceous gland
5. Hypodermis
   a. Subcutaneous fat
   b. Subcutaneous blood vessels

TABLE **6.1** Model Inventory for the Integumentary System

| Model/Diagram | Structures Identified |
|---|---|
|  |  |
|  |  |
|  |  |

# Exercise 6-2

## Histology of Integument

### MATERIALS

- ☐ Slide of thick skin
- ☐ Slide of thin skin
- ☐ Colored pencils
- ☐ Light microscope

6

In this exercise we will examine prepared slides of skin from different regions of the body so we can compare and contrast two types of skin: (a) **thin skin**, found on the palms and soles of the feet, and (b) **thick skin**, found everywhere else (**Figure 6.5**).

Before moving on, you may wish to review the basics of microscopy from Unit 3. Remember to follow a step-by-step approach when examining the slides: Look at the slide with the naked eye first, then begin your examination on low power, and advance to higher power to see more details.

FIGURE **6.5** Photomicrographs of (**A**) thick skin, and (**B**) thin skin

## Procedure Microscopy of Thick Skin

Obtain a prepared slide of thick skin (which may be labeled "Palmar Skin"), and examine it with the naked eye to get oriented. Once you are oriented, place the slide on the stage of the microscope, and scan it on low power. You should be able to see the epidermis with its superficial layers of dead cells and the dermis with its pink clusters of collagen bundles that make up the dense irregular collagenous connective tissue. Advance to higher power to see the cells and associated structures in greater detail.

Use your colored pencils to draw what you see in the field of view (you will be able to see the most structures on low power). Label your drawing with the following terms, using **Figure 6.5** for reference. When you have completed your drawing, fill in **Table 6.2**.

1. Epidermis
   a. Stratum corneum
   b. Stratum lucidum
   c. Stratum granulosum
   d. Stratum spinosum
   e. Stratum basale
2. Dermis
   a. Dermal papillae
   b. Collagen bundles
   c. Sweat gland

## Procedure Microscopy of Thin Skin

Obtain a prepared slide of thin skin (which may be called "Scalp Skin"). As before, examine the slide with the naked eye, then scan the slide on low power, advancing to higher power as needed to see the structures more clearly.

Use your colored pencils to draw what you see in the field of view (you will be able to see the most structures on low power). Label your drawing using the terms below, using **Figure 6.5** for reference. When you have completed your drawing, fill in the remainder of **Table 6.2**.

1. Epidermis
   a. Stratum corneum
   b. Stratum granulosum
   c. Stratum spinosum
   d. Stratum basale

2. Dermis
   a. Dermal papillae
   b. Collagen bundles

3. Hair follicle

4. Sebaceous gland

5. Sweat gland

6. Arrector pili muscle

TABLE **6.2** Characteristics of Thick and Thin Skin

| Characteristic | Thick Skin | Thin Skin |
|---|---|---|
| Thickness of stratum corneum | | |
| Hair follicles present? | | |
| Sebaceous glands present? | | |
| Stratum lucidum present? | | |
| Arrector pili muscles present? | | |

# Exercise 6-3

## Touch Receptor Distribution

### MATERIALS
- ❏ Water-soluble marking pen
- ❏ Ruler
- ❏ Monofilament

The two types of sensory receptors in the skin that detect fine touch are called **Meissner's** (or tactile) **corpuscles** and **Merkel disks**. The distribution of these receptors can be mapped using an instrument called a monofilament (**Figure 6.6**) that applies 10 grams of force as it is pressed on the skin (10 grams of force is generally accepted as the maximum amount of force required to activate the Meissner's corpuscles and Merkel disks). If a monofilament is not available, a similar instrument may be made using a stiff-bristle hair glued to a toothpick. When you have completed the activity, answer Check Your Understanding questions 5 and 6 (pp. 133–134).

## Procedure  Mapping Touch Receptor Distribution

**1**  Use a centimeter ruler to measure a 2-centimeter square on your partner's palm. Mark this square with a water-soluble marking pen.

**2**  Have your partner close his or her eyes.

**3**  Apply the monofilament to one of the corners of the square. Press only until the monofilament bends slightly.

**4**  If your partner can perceive the sensation of the monofilament, mark this spot with a different-colored water-soluble marker.

**5**  Repeat this process for the remainder of the square, advancing the monofilament 1–2 millimeters each time. Be certain to apply the same amount of pressure with each application.

**6**  When you have completed the square, record the number of receptors found on the palm.

Number of receptors on palm: _____

**7**  Repeat this process on the posterior shoulder. Record the number of receptors found on this location.

Number of receptors on the posterior shoulder: _____

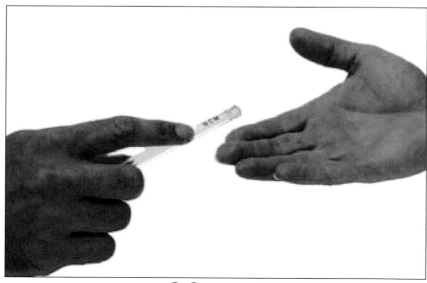

FIGURE **6.6** A monofilament

# Exercise 6-4

## Fingerprinting

### MATERIALS

- ❏ Blank glass slide (one per person)
- ❏ Ink pad
- ❏ Fingerprint card (one per person)
- ❏ Dusting powder
- ❏ Dusting brush
- ❏ Fingerprint lifting tape
- ❏ Latex gloves
- ❏ Sharpie® marker

Specific patterns in the epidermis are created in thick skin, where the dermal papillae are folded into dermal ridges. These dermal ridges indent the overlying epidermis to create **epidermal ridges**, which increase gripping ability. They also lead to the characteristic patterns of finger, toe, palm, and footprints. The patterns of the epidermal ridges are unique to each individual, even identical twins.

Sweat pores open at the top of the epidermal ridges, which causes us to leave behind a thin film when we touch surfaces. This film can be detected in a variety of ways, most commonly using a powder that binds to it.

Fingerprinting was not a widely accepted means of identification in the United States until 1902, when the New York Civil Service began collecting and using fingerprints. Prior to this, fingerprints had been studied widely and even were used to solve a murder case in Argentina in 1892.

Fingerprints can be broadly classified into three ridge flow patterns: loop, arch, and whorl (**Figure 6.7**). Once the ridge flow pattern has been identified, the characteristics of the ridges, also called *Galton points*, are examined. The three basic ridge characteristics are the ridge ending, the island (or the dot), and the bifurcation (**Figure 6.8**). In professional laboratories the fingerprint is analyzed further on the basis of minute details, or "points," in the fingerprint. Much of this is done currently with the help of computers.

In this exercise you will be playing the role of detective, searching for a "thief" among your group members (which may end up being you!). Your group will determine the thief's identity by analyzing the ridge flow patterns and ridge characteristics of the fingerprints of your group members. When you have completed the activity, answer Check Your Understanding question 7 (p. 134).

Loop    Whorl    Arch

FIGURE **6.7** The three basic patterns of fingerprints

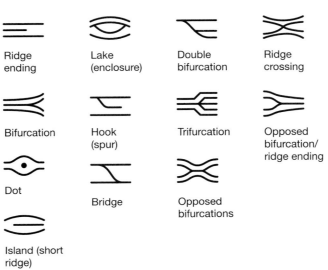

Ridge ending

Lake (enclosure)

Double bifurcation

Ridge crossing

Bifurcation

Hook (spur)

Trifurcation

Opposed bifurcation/ ridge ending

Dot

Bridge

Opposed bifurcations

Island (short ridge)

FIGURE **6.8** The Galton points

## Procedure Fingerprint Analysis

**1** Assemble into groups of a minimum of four people.

**2** Obtain one blank glass slide, and handle it by the edges as you clean it off with a paper towel and water.

**3** Once the slide is clean and dry, place two or three large fingerprints (use different fingers) somewhere on one side of the slide.

**4** After placing one or more fingerprints on the slide, put on gloves, and use the Sharpie at your table to write your initials on the upper right-hand corner of the slide.

**5** Bring the slides for the entire table to your instructor, and your instructor will pick one slide from the group, cover the initials in the corner, and designate that person as the "thief." The thief is one of your group members, so you are all suspects.

**6** Fingerprint each member of the group (using the ink and cards provided):

   **a** Obtain a blank fingerprinting card or a blank sheet of paper.

   **b** Roll each finger individually on the ink pad from right to left. Be certain not to get excessive ink on your fingers.

   **c** Roll each finger individually on the fingerprinting card slowly from left to right.

**7** When all of the fingerprinting is completed, don a pair of latex gloves.

**8** Use a dusting brush to lightly sweep some dusting powder onto the slide containing the thief's fingerprints. Use the dusting powder *sparingly*, because too much powder will obscure your prints. Brush away the excess dusting powder. (You can tap the slide on its side to help with this.)

**9** Place a piece of fingerprint lifting tape on the fingerprints that have appeared. Pat the tape into place firmly, and remove it slowly. Place the tape on a blank sheet of white paper for comparison.

**10** With the help of **Figures 6.7** and **6.8**, identify the thief by comparing the fingerprints lifted from the slide with the fingerprints taken from the suspects. Be careful not to tell your group mates which fingers you used. Remember— you don't want to be caught!

Who is the thief from your table? _____

How did you arrive at this conclusion? _____

_____

_____

_____

**UNIT 6 REVIEW**

## Check Your Recall

**1** Label the following parts of the skin on **Figure 6.9**.

☐ Arrector pili muscle
☐ Cutaneous blood vessels
☐ Dermal papillae
☐ Dermis
☐ Epidermis
☐ Hair bulb (root)
☐ Hair follicle
☐ Hair shaft
☐ Hypodermis
☐ Sebaceous gland
☐ Sweat gland
☐ Sweat pore

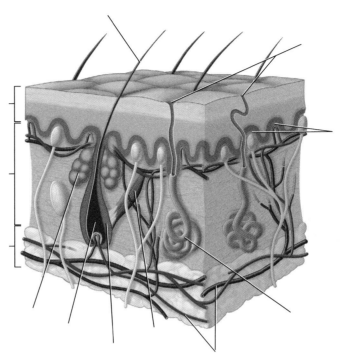

FIGURE **6.9** Skin section

**2** The main cell type in skin is the
  a. melanocyte.
  b. reticulocyte.
  c. monocyte.
  d. keratinocyte.

**3** Number the layers of the epidermis, with 1 being the most superficial layer and 5 being the deepest layer.

_____ Stratum lucidum

_____ Stratum basale

_____ Stratum corneum

_____ Stratum spinosum

_____ Stratum granulosum

**4** Which layers of the epidermis contain living cells?
  a. Stratum granulosum only
  b. Stratum corneum, stratum granulosum, stratum lucidum
  c. Stratum basale, stratum spinosum, stratum granulosum
  d. All of the layers of the epidermis contain living cells.
  e. None of the layers of the epidermis contain living cells.

**5** From where do the cells of the epidermis obtain oxygen and nutrients?

   a. From blood vessels in the epidermis

   b. Diffusion from blood vessels in the dermis

   c. Diffusion from the air

   d. From blood vessels in other epithelial tissues

**6** Match the following terms with the correct description.

| | |
|---|---|
| _____ Papillary layer | A. Secrete product through a pore |
| _____ Sebaceous gland | B. Pressure receptor in the dermis |
| _____ Pacinian corpuscle | C. Projections of the dermis that indent the epidermis |
| _____ Hair follicle | D. Superficial layer of the dermis; loose connective tissue |
| _____ Dermal papillae | E. Sheath of epithelial and connective tissue around a hair |
| _____ Hair shaft | F. Deep layer of the dermis; dense irregular collagenous connective tissue |
| _____ Reticular layer | G. Secretes sebum (oil) |
| _____ Sweat gland | H. Portion of the hair that projects from the skin's surface |

**7** Which of the following are characteristics of thick skin? (*Circle all that apply.*)

   a. Located over the palms and the soles of the feet

   b. Contains hair and arrector pili muscles

   c. Contains sweat glands

   d. Very thick stratum corneum

   e. Contains sebaceous glands

   f. Contains a stratum lucidum

**8** The dividing cells of a nail are located in the

   a. nail fold.

   b. nail plate.

   c. nail bed.

   d. nail matrix.

**9** The two types of receptors in the skin that detect fine touch are the

   a. Merkel disks.

   b. Pacinian corpuscles.

   c. Meissner's corpuscles.

   d. Both a and b are correct.

   e. Both a and c are correct.

**10** Fingerprints are the result of

   a. epidermal ridges.

   b. projections from the hypodermis.

   c. epidermal papillae that create dermal ridges.

   d. dermal papillae that house sebaceous glands.

# Check Your Understanding

### Critical Thinking and Application Questions

**1**  Explain why a superficial skin scrape (such as a paper cut) doesn't bleed. Why don't you bleed when a hair is pulled out?

**2**  Shampoos and hair conditioners often claim to have nutrients and vitamins your hair must have to grow and be healthy. Taking into account the composition of hair, do you think these vitamins and nutrients will be beneficial? Why or why not?

**3**  Many drugs can be applied to the skin without entering the blood. How is this fact explained by the structure of the skin?

**4**  The disease bullous pemphigoid results in the destruction of proteins within the basement membrane that holds the epidermis and dermis together. How would this likely affect the epidermis? Would the skin be an effective barrier to this disease?

**5**  Would you have expected to have found more Meissner's corpuscles and Merkel disks on the palm or on the posterior shoulder? Why?

**6** In which areas of the body would you expect to find the most Merkel disks and Meissner's corpuscles? Why?

_____

_____

_____

_____

**6** **7** Why do you have unique markings on your fingers, toes, palms, and soles but not on any other part of the body? Why do you leave behind fingerprints when you touch certain surfaces?

_____

_____

_____

_____

# Introduction to the Skeletal System

**7**

Name _____ Section _____ Date _____

# PRE-LAB EXERCISES

Complete the following exercises prior to coming to lab, using your textbook and lab manual for reference.

## Pre-Lab Exercise **7-1**

 Key Terms

You should be familiar with the following terms before coming to lab.

| Term | Definition |
| --- | --- |

**Types of Bone Tissue**

Compact (cortical) bone _____

_____

Spongy (cancellous) bone _____

_____

**Structures of the Osteon**

Osteon _____

_____

Lamellae _____

_____

Central (Haversian) canal _____

_____

Perforating (Volkmann's) canal _____

_____

Lacunae _____

_____

Osteocytes _____

_____

Canaliculi _____

_____

Periosteum _____

_____

Endosteum _____

_____

**Bone Shapes**

Long bone _____

_____

Short bone _____

_____

Flat bone _____

_____

Irregular bone _____

_____

Sesamoid bone _____

_____

Sutural bone _____

_____

**Structures of Long Bones**

Diaphysis _____

_____

Epiphysis _____

_____

Epiphyseal plate/line _____

_____

Medullary cavity _____

_____

Bone marrow _____

_____

## Pre-Lab Exercise **7-2**

Microscopic Anatomy of Compact Bone

Label and color the microscopic anatomy of compact bone tissue depicted in **Figure 7.1** with the terms from Exercise 7-1 (p. 141). Use your text and Exercise 7-1 in this unit for reference.

**7**

FIGURE **7.1** Microscopic anatomy of compact bone

## Structure of a Long Bone

Label and color the diagram of a long bone (the femur) depicted in **Figure 7.2** with the terms from Exercise 7-4 (p. 148).
Use your text and Exercise 7-4 in this unit for reference.

FIGURE **7.2** A long bone (the femur)

# EXERCISES

The **skeletal system** consists of the bones and their cartilages. At first it might seem odd that a set of bones makes up an organ system, but remember that each bone is considered an organ. A bone consists of many tissue types, including osseous tissue, epithelial tissue, dense irregular collagenous connective tissue, and adipose tissue. The following exercises will introduce you to these complex organs and the histology of osseous tissue.

## Exercise **7-1**

### Histology of Osseous Tissue

**MATERIALS**
❑ Osteon model
❑ Slide of compact bone
❑ Light microscope
❑ Colored pencils

The most superficial tissue of a bone is called the **periosteum**. The periosteum is composed of dense irregular collagenous connective tissue richly supplied with blood vessels. The innermost layer of the periosteum contains osteogenic cells that become cells called **osteoblasts,** which secrete bone matrix and build new bone, and **osteoclasts,** which secrete enzymes that catalyze the breakdown of bone matrix. The periosteum is anchored to the bone by bundles of collagen called **perforating** or **Sharpey's fibers.**

Deep to the periosteum we find osseous tissue. The two general types of **osseous tissue** are

1. **compact** (or *cortical*) **bone,** and
2. **spongy** (or *cancellous*) **bone.**

Compact bone is hard, dense bone tissue found immediately deep to the periosteum. Its hardness comes from its structure, which consists of repeating, densely packed subunits called **osteons (Figures 7.3 and 7.4).** Osteons contain several features, including the following:

1. **Lamellae.** Lamellae are concentric rings of bone matrix. The lamellae give compact bone a great deal of strength, much like a tree's rings.

Lacunae with osteocytes

Interstitial lamellae

Concentric lamellae

Central canal

FIGURE **7.3** Compact bone, light micrograph

FIGURE **7.4** Compact bone

2. **Central (Haversian) canal.** Running down the center of each osteon is a central or Haversian canal. Each central canal contains blood vessels and nerves and is lined with a connective tissue membrane called the **endosteum**. Like the periosteum, the endosteum has an inner layer of osteoblasts, which secrete bone matrix, and osteoclasts, which degrade bone.

3. **Lacunae.** Situated between the lamellae are small cavities called lacunae. Lacunae contain mature osteoblasts called **osteocytes** that monitor and maintain the bone matrix. Neighboring lacunae and osteocytes are connected to each other by tiny canals called **canaliculi**.

4. **Perforating (Volkmann's) canals.** The perforating or Volkmann's canals lie perpendicular to the osteon and carry blood vessels into the bone from the periosteum. Like the central canals, perforating canals are lined by the endosteum.

Spongy bone is found on the inside of a bone deep to compact bone. As its name implies, it somewhat resembles a sponge and consists of a latticework-type structure with tiny bone spicules called **trabeculae** (Figure 7.5). The latticework structure of spongy bone allows it to house another important tissue, the **bone marrow**. The two types of bone marrow are: (a) **red bone marrow**, which produces blood cells, and (b) **yellow bone marrow**, which is composed

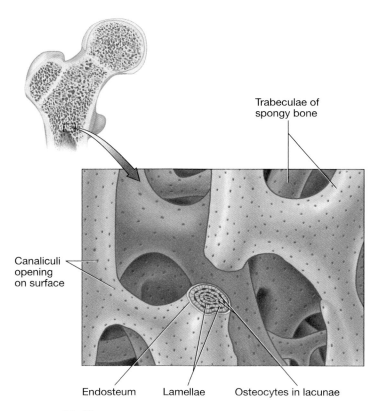

FIGURE **7.5** Microscopic anatomy of spongy bone tissue

primarily of adipose tissue. As you can see in **Figure 7.5**, trabeculae are composed of lamellae but are not organized into osteons. For this reason, spongy bone lacks the hardness of compact bone.

## **Procedure** Model Inventory for Compact and Spongy Bone

Identify the following structures of compact and spongy bone on models and diagrams, using your textbook and this unit for reference. As you examine the anatomical models and diagrams, record on the model inventory in **Table 7.1** the name of the model and the structures you were able to identify. When you have completed the activity, answer Check Your Understanding questions 1–3 (p. 151).

**Compact Bone Structures**

Osteon
Central (Haversian) canal
Perforating (Volkmann's) canal
Lamellae
Lacunae
  ▮ Osteocytes
Canaliculi
Periosteum
  ▮ Sharpey's fibers

**Spongy Bone Structures**

Trabeculae (bone spicules)
Red bone marrow
Endosteum
Lacunae
Osteocytes

TABLE **7.1** Model Inventory for Osseous Tissue

| Model | Bone Structures Identified |
|---|---|
|  |  |
|  |  |
|  |  |
|  |  |

## **Procedure** Microscopy

View a prepared slide of compact bone. The structures on the slide should look similar to the osteon models you viewed in lab earlier and **Figure 7.3**.

Use colored pencils to draw a picture of what you see under the microscope, and label your drawing with the compact bone structures from Exercise 7-1 (p. 141).

# Exercise 7-2

## Chemical Components of Bone Tissue

### MATERIALS

- ❏ 1 piece of normal bone
- ❏ 1 piece of oven-heated bone
- ❏ 1 piece of bone soaked in nitric acid

Osseous tissue is a type of connective tissue composed of two primary chemical components:

1. The **organic component** consists mostly of protein fibers such as **collagen**. Collagen provides a supportive network that gives the bone tensile strength (the ability to withstand stretching and twisting forces).

2. The **inorganic component** consists mostly of calcium ions in the form of calcium hydroxyapatite crystals. The inorganic component provides the bone compressional strength (the ability to withstand compressive forces).

As you will see in this exercise, both chemical components of bone are required for bone to remain strong and perform its functions.

## Procedure  Examining the Chemical Components of Bone

In this exercise you will compare the effects of removing different chemical components of bone. One sample has been heated in an oven to destroy the organic components of the bone, and another sample has been treated with nitric acid to destroy the inorganic components of the bone (**Figure 7.6**). When you have completed the activity, answer Check Your Understanding questions 4–6 (p. 152).

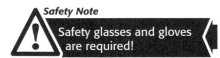

**Safety Note**

⚠️ Safety glasses and gloves are required!

**1** Obtain three pieces of bone:
   a. one piece of untreated bone,
   b. one piece of bone that has been baked in an oven for a minimum of two hours, and
   c. one piece of bone that has been soaked in nitric acid.

**2** Place the three pieces of bone side by side. Describe their appearance. Do they look different from one another?

_____

_____

_____

_____

_____

_____

_____

_____

_____

_____

_____

_____

_____

_____

_____

FIGURE **7.6** Three samples of bone: (**A**) untreated; (**B**) heated; (**C**) soaked in nitric acid

**3** Squeeze each bone between your fingers, and twist each sample with your hands. What happens to:

**a** the untreated bone? _____

_____

_____

**b** the heated bone? _____

_____

_____

**c** the bone treated with nitric acid? _____

_____

_____

# Exercise 7-3

## Classification of Bones by Shape

### MATERIALS
- ❏ Disarticulated bones
- ❏ Articulated skeleton

One way in which bones are classified is by their shape. Note in **Figure 7.7** the four general shapes of bones:

1. **Long bones** are longer than they are wide and include the bones of the upper and lower extremities excluding the ankle and wrist bones.

2. **Short bones** are about as long as they are wide. The bones of the wrist and the ankle are short bones.

3. **Flat bones** are shaped exactly as they're named—they are flat. Flat bones include the ribs, the sternum, the clavicle, certain skull bones, and the bones of the pelvis.

4. **Irregular bones** are those whose shape doesn't fit into any of the other classes. Irregular bones include the vertebrae and certain bones of the skull, such as the mandible (lower jaw bone).

Note that there are two other, less common bone shapes: sesamoid bones and sutural bones. **Sesamoid bones** are roughly oval-shaped bones located within tendons. The sesamoid bone with which you are most likely familiar is the patella, or the kneecap. **Sutural bones** are generally small bones located between the flat bones of the skull. You will likely see sutural bones when you study the skull in the next unit.

## HINTS & TIPS

### Bone Shape, Not Length

Note that long bones are not named for their length but rather for their shape. Many long bones actually are quite short in length, including the bones of the fingers and the toes. Be careful when identifying bone shapes to look at the overall shape of the bone rather than its size.

## Procedure Identifying Bone Shapes

Obtain a set of disarticulated bones, and classify them according to their shape. Identify examples of long bones, short bones, flat bones, and irregular bones, and record them in **Table 7.2**. You may wish to use the figures in your text or Unit 8 (p. 153) to help you identify the disarticulated bones.

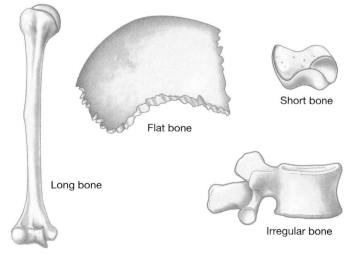

Short bone

Flat bone

Long bone

Irregular bone

FIGURE **7.7** The four shapes of bones

TABLE **7.2** Examples of Bone Shapes

| Bone | Shape |
|---|---|
| | |
| | |
| | |
| | |
| | |
| | |
| | |
| | |
| | |

# Exercise 7-4

## Anatomy of Long Bones

### MATERIALS
❑ Long bone, sectioned

All long bones share common structures and parts, illustrated in **Figure 7.8** with the example of the femur. The **diaphysis** is the shaft of the long bone. As you can see in the figure, it consists of a thick collar of compact bone surrounding a hollow area called the **medullary cavity**. This collar of compact bone makes long bones quite strong and able to support the body's weight. The medullary cavity has sparse trabeculae and generally is filled with yellow bone marrow in living bone.

The ends of a long bone are called the **epiphyses**. Each epiphysis contains a shell of compact bone surrounding the inner spongy bone. The spongy bone within the epiphyses contains either red or yellow bone marrow. The end of each epiphysis is covered with hyaline cartilage (sometimes called **articular cartilage**), which allows two bones to articulate during movement with minimal friction. At certain epiphysis-diaphysis junctions you will note a thin, calcified line called the **epiphyseal line**. This structure is the remnant of the **epiphyseal plate**, a band of hyaline cartilage from which long bones grow in length. When longitudinal growth ceases, the chondrocytes of the epiphyseal plate die and are replaced by calcified bone tissue.

## Procedure Identification of Long Bone Structures

Identify the following structures of long bones on specimens and X-rays (if available). Check off each structure as you identify it.

☐ Compact bone
☐ Diaphysis
☐ Epiphyseal line
☐ Epiphyseal plate
  (may be visible
  only on X-ray)
☐ Epiphysis
☐ Hyaline (articular)
  cartilage
☐ Medullary cavity
☐ Red bone marrow
☐ Spongy bone
☐ Yellow bone marrow

FIGURE **7.8** A long bone: the femur

## Check Your Recall

**1** Label the following parts of compact bone on **Figure 7.9**.

- ☐ Blood vessels
- ☐ Canaliculi
- ☐ Central canal
- ☐ Concentric lamellae
- ☐ Lacunae
- ☐ Osteon
- ☐ Perforating canal
- ☐ Trabeculae of spongy bone

FIGURE **7.9** Microscopic anatomy of compact bone tissue

**2** *Mark the following statements as true (T) or false (F).* If the statement is false, correct it to make it a true statement.

_____ The periosteum contains osteocytes and osteoclasts.

_____ Osteoblasts secrete bone matrix.

_____ Compact bone is the hard, outer bone.

_____ Spongy bone is composed of spicules called osteons.

_____ Spongy bone houses red and yellow bone marrow.

_____ Perforating fibers anchor the endosteum to the superficial surface of a bone.

**3** The organic component of bone consists of _____ and functions to _____.
   a. collagen; provide compressional strength
   b. calcium hydroxyapatite; provide compressional strength
   c. collagen; provide tensile strength
   d. calcium hydroxyapatite; provide tensile strength

**4** The inorganic component of bone consists of _____ and functions to _____ .

    a. Collagen; provide compressional strength

    b. Calcium hydroxyapatite; provide compressional strength

    c. Collagen; provide tensile strength

    d. Calcium hydroxyapatite; provide tensile strength

**5** Which of the following statements about long bones is/are false? *Mark all that apply.*

    a. A long bone is longer than it is wide.

    b. They include the bones of the upper and lower extremities except the finger and toes.

    c. They include the bones of the upper extremities except the bones of the ankle and wrist.

    d. A long bone is named for its length.

**6** Examples of flat bones include

    a. the clavicle.

    b. the mandible.

    c. ankle bones.

    d. Both a and b are correct.

    e. Both a and c are correct.

**7** Short bones are

    a. short in length.

    b. about as long as they are wide.

    c. irregular in shape.

    d. flat.

**8** Label the following parts of a long bone on **Figure 7.10**.

☐ Articular cartilage
☐ Diaphysis
☐ Epiphyseal line
☐ Epiphysis
☐ Medullary cavity
☐ Periosteum
☐ Yellow bone marrow

**9** The epiphyseal plate is

    a. the structure from which long bones grow in length.

    b. a remnant of the structure from which long bones grow in length.

    c. composed of osseous tissue.

    d. found lining the surface of the epiphysis.

FIGURE **7.10** Structure of a long bone

**10** The medullary cavity is found in the _____ and usually contains the _____ .

    a. diaphysis; red bone marrow

    b. epiphysis; red bone marrow

    c. diaphysis; yellow bone marrow

    d. epiphysis; yellow bone marrow

## Check Your Understanding

### Critical Thinking and Application Questions

**1** Explain how each of the following structures follows its function.

**a** Compact bone _____

_____

_____

**b** Spongy bone _____

_____

_____

**c** A long bone _____

_____

_____

**2** A bone tumor disrupts the normal structure of osteons, replacing the organized rings with disorganized, irregular masses of bone. How will this affect the ability of a bone to perform its functions?

_____

_____

_____

_____

**3** When a surgeon is operating on a bone, great care is taken to preserve the periosteum. Why is it important to bone healing to take care of this delicate structure?

_____

_____

_____

_____

_____

**4** *Osteogenesis imperfecta* is a congenital condition in which collagen synthesis is defective. Of the three samples of bone you tested in Exercise 7-2, which is most similar to the bones in osteogenesis imperfecta? What symptoms would you expect to find in this disease?

_____

_____

_____

_____

_____

**5** Would it benefit a patient with osteogenesis imperfecta to take supplemental calcium salts? Why or why not?

_____

_____

_____

_____

_____

**6** The diseases *rickets* and *osteomalacia* result from insufficient vitamin D intake, which decreases the amount of calcium ions available for synthesis of the inorganic component of bone. Of the three samples of bone you tested in Exercise 7-2, which is the most similar to the bones in rickets and osteomalacia? What symptoms would you expect to find in these diseases?

_____

_____

_____

_____

_____

# Skeletal System

**8**

## OBJECTIVES

*Once you have completed this unit, you should be able to:*

- Identify bones and markings of the axial skeleton.
- Identify bones and markings of the appendicular skeleton.
- Build a skeleton using disarticulated bones.

Name _____ Section _____ Date _____

# PRE-LAB EXERCISES

Complete the following exercises prior to coming to lab, using your textbook and lab manual for reference.

## Pre-Lab Exercise **8-1**

 Key Terms

You should be familiar with the following terms before coming to lab.

| Term | Definition |
|------|------------|
| **Axial Skeleton** | |
| Cranial bones | |
| Facial bones | |
| Suture | |
| Hyoid bone | |
| Vertebrae | |
| Ribs | |
| Sternum | |
| **Appendicular Skeleton—Upper Limb** | |
| Pectoral girdle | |
| Clavicle | |
| Scapula | |
| Humerus | |
| Radius | |
| Ulna | |
| Carpals | |
| Metacarpals | |
| Phalanges | |
| **Appendicular Skeleton—Lower Limb** | |
| Pelvic girdle | |
| Femur | |
| Tibia | |
| Fibula | |
| Tarsals | |
| Metatarsals | |

## Pre-Lab Exercise **8-2**

### Bones of the Skull

Label and color the structures of the skull depicted in **Figure 8.1** with the terms from Exercise 8-1 (p. 159). Use your text and Exercise 8-1 in this unit for reference.

**A**

**B**

FIGURE **8.1** Skull: (**A**) anterior view; (**B**) lateral view *(continues)*

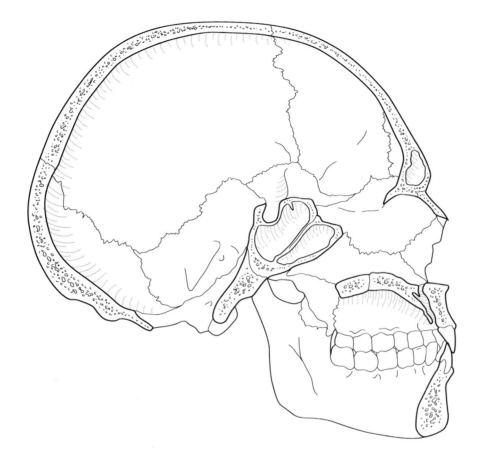

FIGURE **8.1** Skull *(cont.)*: (**C**) midsagittal section

# Pre-Lab Exercise **8-3**

Whole Skeleton

Label and color the structures of the skeleton depicted in **Figure 8.2** with the terms from Exercises 8-1 to 8-3 (pp. 159–182). Use your text and Exercises 8-1 to 8-3 in this unit for reference.

FIGURE **8.2** Skeleton:
(**A**) anterior view *(continues)*

8

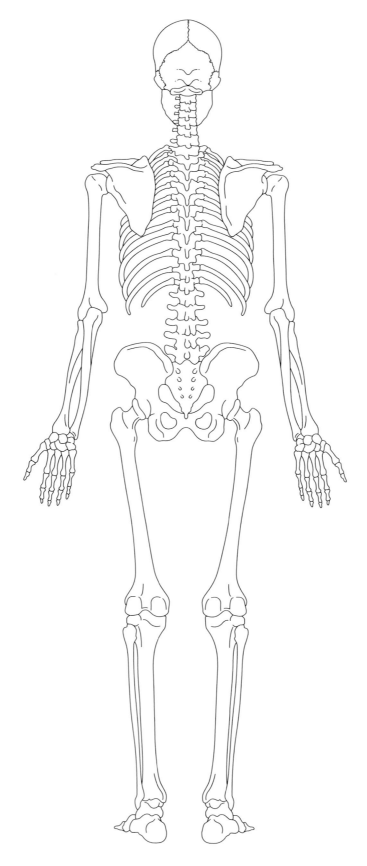

FIGURE **8.2** Skeleton *(cont.)*: (**B**) posterior view

# EXERCISES

The **skeletal system** consists of the bones, its associated cartilages, and the joints. Its two divisions are the **axial skeleton** and the **appendicular skeleton**. The axial skeleton is composed of the bones of the head, neck, and trunk—specifically, the cranial bones, the facial bones, the vertebral column, the hyoid bone, the sternum, and the ribs. The appendicular skeleton consists of the bones of the upper limbs, the lower limbs, the pectoral girdle (the bones forming the shoulder joint), and the pelvic girdle (the bones forming the pelvis and hip joint).

In this unit we explore the anatomy of the bones and bone markings of the skeletal system, which will serve as a foundation for later chapters. For example, the radial and ulnar arteries parallel the radius and the ulna, and the frontal, parietal, temporal, and occipital lobes of the brain are named for the cranial bones under which they are located.

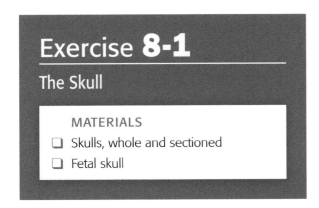

## Exercise **8-1**

### The Skull

#### MATERIALS
☐ Skulls, whole and sectioned
☐ Fetal skull

The skull is composed of two general types of bones—the **cranial bones** and the **facial bones** (**Figures 8.3–8.7**). The cranial bones encase the brain and together form the **calvaria** (also known as the "skullcap"), which consists of several of the cranial bones joined at immovable joints called **sutures**. These bones also form the cranial base, which contains indentations that accommodate the brain, called the **anterior**, **middle**, and **posterior cranial fossae** (visible in **Figure 8.6**). The eight cranial bones include the following:

1. **Frontal bone**. The frontal bone forms the anterior portion of the cranium. Internally it contains hollow spaces called the **frontal sinuses** (visible in **Figure 8.3**). The frontal sinuses are part of a group of cavities that surround the nasal cavity, called the **paranasal sinuses**. Air from the nasal cavity enters the paranasal sinuses via small openings in the bones, and in the sinuses the air gets filtered, warmed, and humidified.

2. **Parietal bones**. The paired parietal bones form the superior and part of the lateral walls of the cranium. Note in **Figures 8.3** and **8.4** that they articulate with one another and many other cranial bones at sutures: They meet one another at the **sagittal suture** (not visible in the figures), they meet the frontal bone at the **coronal suture**, they meet the temporal bones at the **squamous sutures**, and they meet the occipital bone at the **lambdoid suture**.

3. **Temporal bones**. The paired temporal bones form the lateral walls of the cranium (best seen in **Figure 8.4**). Each temporal bone has a complex shape with four general regions.
   a. The flat **squamous region** is the temporal bone's most lateral aspect. It contains a projection called the **zygomatic process** that forms part of the cheekbone.
   b. The temporal bone's inferior region is called the **tympanic region**. It houses the **external acoustic meatus** and the needlelike **styloid process**.
   c. Posterior to the tympanic region is the **mastoid region**, which contains the large **mastoid process**.
   d. Internally, the temporal bone is shaped like a mountain ridge and is accordingly called the **petrous region**.

4. **Occipital bone**. The posterior cranial bone is the occipital bone (best seen in **Figures 8.5** and **8.6**). Its most conspicuous feature is found in its base—a large hole called the **foramen magnum** through which the spinal cord passes.

5. **Sphenoid bone**. The butterfly-shaped sphenoid bone is posterior to the frontal bone on the interior part of the skull (best seen in **Figures 8.5–8.7**). Its superior surface contains a saddle-like formation called the **sella turcica** that normally houses the pituitary gland. Inferior to the sella turcica is the body of the sphenoid, which contains the second paranasal sinus—the **sphenoid sinus**. Laterally it consists of two sets of "wings"—the small **lesser wings** and the larger **greater wings**. Inferiorly, it has another set of "wings" called the **pterygoid processes**.

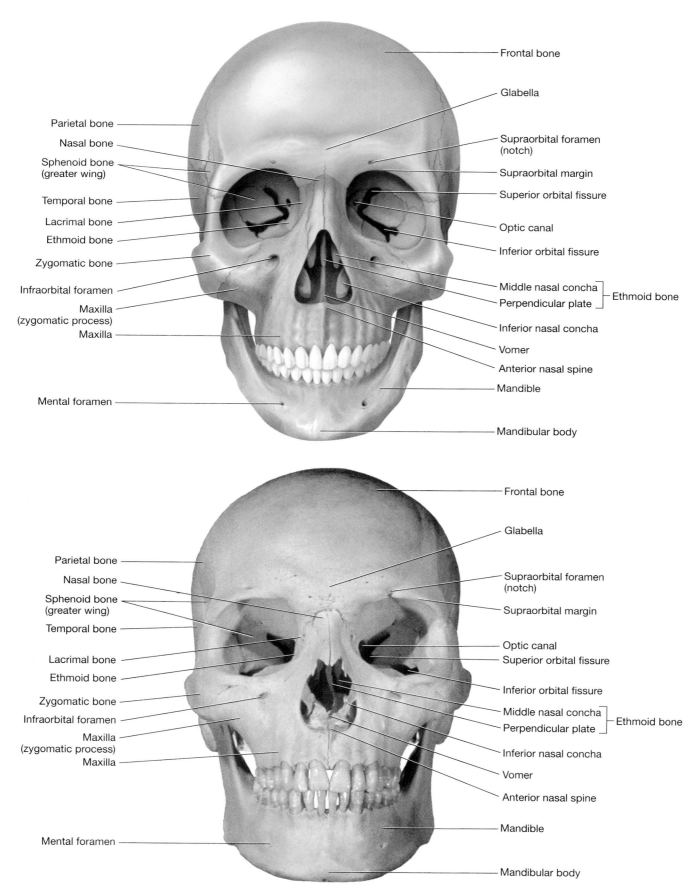

FIGURE **8.3** Anterior view of the skull

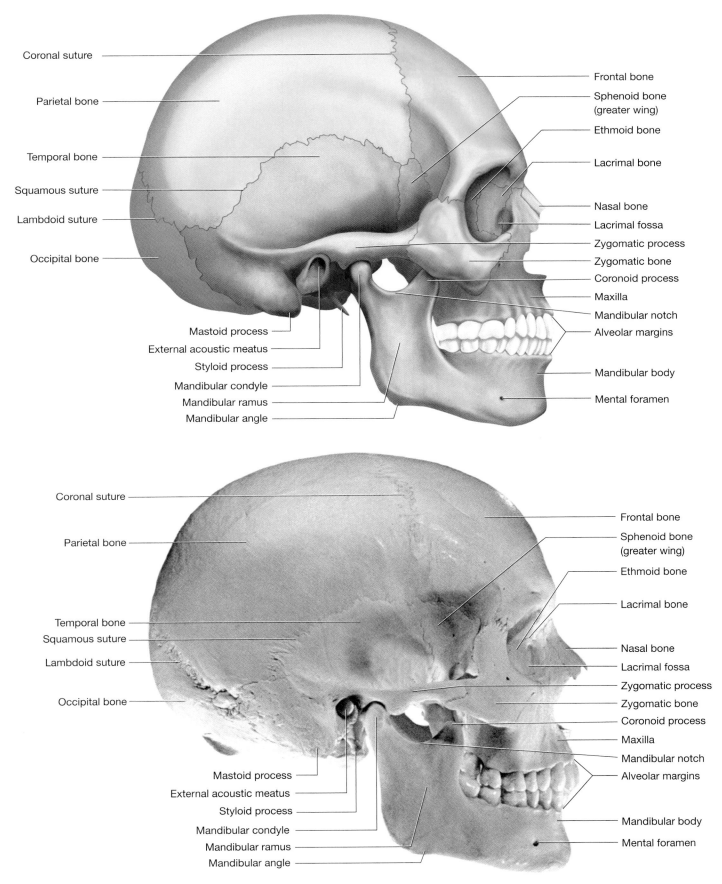

FIGURE **8.4** Lateral view of the skull

Hard palate
Maxilla (palatine process)
Palatine bone (horizontal plate)
Sphenoid bone (greater wing)
Temporal bone (zygomatic process)
Foramen lacerum
Mandibular fossa
Styloid process
Mastoid process
Parietal bone
Inferior nuchal line
External occipital protuberance

Incisive fossa
Maxilla (zygomatic process)
Infraorbital foramen
Maxilla
Zygomatic bone
Vomer
Foramen ovale
Foramen spinosum
Carotid canal
External acoustic meatus
Jugular foramen
Occipital condyle
Foramen magnum
Superior nuchal line

Hard palate
Maxilla (palatine process)
Palatine bone (horizontal plate)
Sphenoid bone (greater wing)
Temporal bone (zygomatic process)
Foramen lacerum
Mandibular fossa
Styloid process
Mastoid process
Parietal bone
External occipital protuberance

Incisive fossa
Maxilla (zygomatic process)
Maxilla
Zygomatic bone
Vomer
Foramen ovale
Foramen spinosum
Carotid canal
External acoustic meatus
Jugular foramen
Occipital condyle
Foramen magnum
Inferior nuchal line
Superior nuchal line

FIGURE **8.5** Inferior view of the skull (mandible removed)

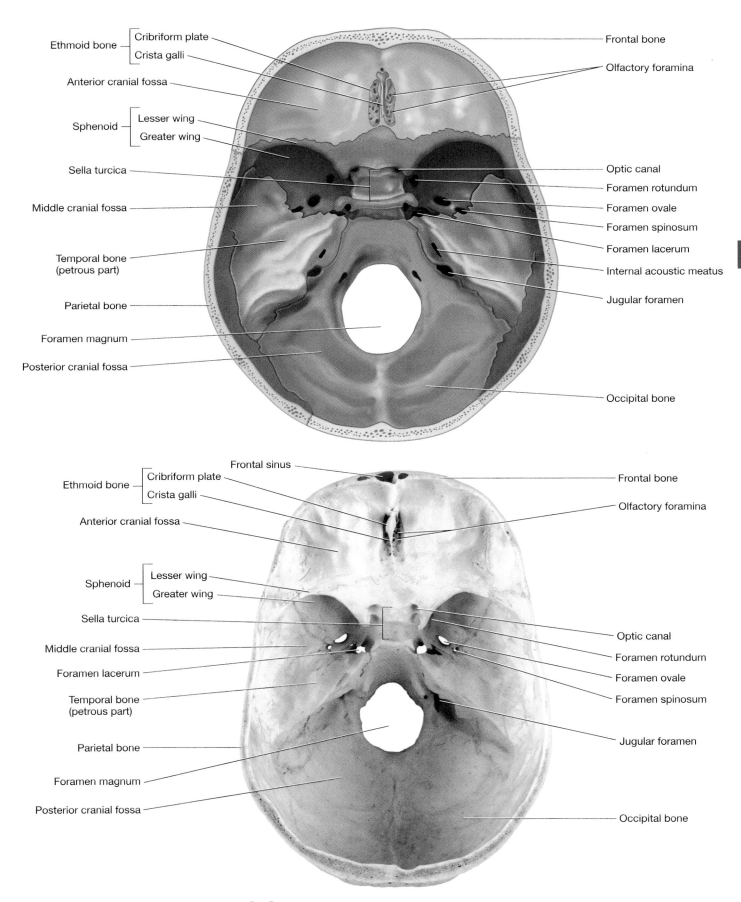

Ethmoid bone — Cribriform plate
— Crista galli

Frontal bone

Olfactory foramina

Anterior cranial fossa

Sphenoid — Lesser wing
— Greater wing

Sella turcica

Optic canal

Foramen rotundum

Middle cranial fossa

Foramen ovale

Foramen spinosum

Foramen lacerum

Temporal bone
(petrous part)

Internal acoustic meatus

Jugular foramen

Parietal bone

Foramen magnum

Posterior cranial fossa

Occipital bone

Frontal sinus

Ethmoid bone — Cribriform plate
— Crista galli

Frontal bone

Olfactory foramina

Anterior cranial fossa

Sphenoid — Lesser wing
— Greater wing

Sella turcica

Optic canal

Middle cranial fossa

Foramen rotundum

Foramen lacerum

Foramen ovale

Temporal bone
(petrous part)

Foramen spinosum

Parietal bone

Jugular foramen

Foramen magnum

Posterior cranial fossa

Occipital bone

FIGURE **8.6** Interior view of the skull (calvaria removed)

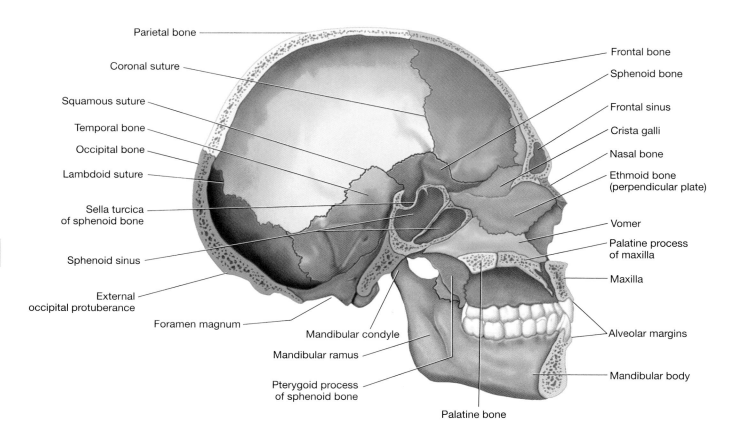

Parietal bone

Coronal suture

Squamous suture

Temporal bone

Occipital bone

Lambdoid suture

Sella turcica
of sphenoid bone

Sphenoid sinus

External
occipital protuberance

Foramen magnum

Mandibular condyle

Mandibular ramus

Pterygoid process
of sphenoid bone

Palatine bone

Frontal bone

Sphenoid bone

Frontal sinus

Crista galli

Nasal bone

Ethmoid bone
(perpendicular plate)

Vomer

Palatine process
of maxilla

Maxilla

Alveolar margins

Mandibular body

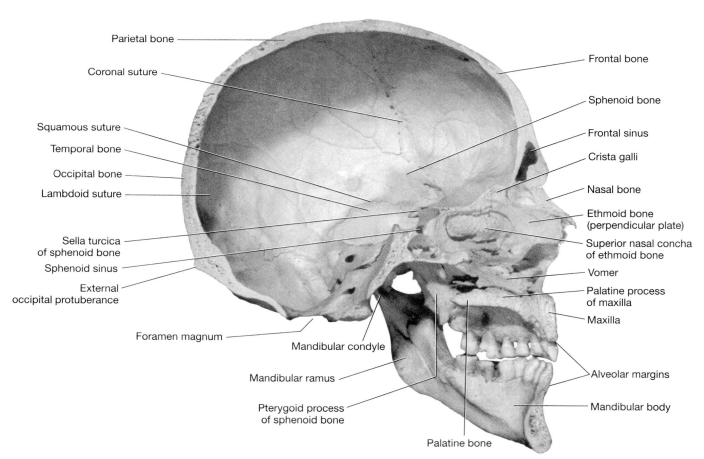

Parietal bone

Coronal suture

Squamous suture

Temporal bone

Occipital bone

Lambdoid suture

Sella turcica
of sphenoid bone

Sphenoid sinus

External
occipital protuberance

Foramen magnum

Mandibular condyle

Mandibular ramus

Pterygoid process
of sphenoid bone

Palatine bone

Frontal bone

Sphenoid bone

Frontal sinus

Crista galli

Nasal bone

Ethmoid bone
(perpendicular plate)

Superior nasal concha
of ethmoid bone

Vomer

Palatine process
of maxilla

Maxilla

Alveolar margins

Mandibular body

FIGURE **8.7** Midsagittal section of the skull

6. **Ethmoid bone.** The complex ethmoid bone is the deepest cranial bone and the most difficult to see from standard views of the skull. It can be best seen in **Figure 8.7**. Located anterior to the sphenoid bone and posterior to the nasal bones of the face, it contains a number of features, including the following:

   a. Its superior surface, called the **cribriform plate**, forms the roof of the nasal cavity. It has small holes, called **olfactory foramina**, through which olfactory nerves pass, and a projection called the **crista galli**.

   b. The lateral bodies of the ethmoid bone form part of the orbit and the walls of the nasal cavity. Internally, the bone contains numerous cavities called the **ethmoid sinuses** that constitute the third set of paranasal sinuses. Extending medially from the lateral bodies are two projections into the nasal cavity—the **superior nasal conchae** and **middle nasal conchae**.

   c. The middle portion of the ethmoid bone, called the **perpendicular plate**, forms the superior part of the bony nasal septum.

The 14 facial bones form the framework for the face, provide openings for ventilation and eating, and form cavities for the sense organs. Several of the facial bones are located deeper in the skull, and you will want to refer to several different figures (noted with each bone) to best locate them and appreciate their structure. These bones are the following:

1. **Mandible.** The mandible, or the lower jaw bone, consists of a central **body** and two "arms" called the **mandibular rami**. The mandibular rami turn superiorly at the **mandibular angle** and have two processes—an anterior process called the **coronoid process** and a posterior process called the **mandibular condyle**. The mandibular condyle fits into a depression in the temporal bone to form the **temporomandibular joint**. As you can see in the figures, the upper teeth are attached to the mandible along a border called the **alveolar margin**.

2. **Maxillae.** The two fused maxillae are the upper jaw bones. They form part of the orbit and the anterior portion of the hard palate (via their **palatine processes**). Within their walls are cavities called the **maxillary sinuses**, the fourth and final set of paranasal sinuses. Laterally the maxillae have projections called **zygomatic processes** that form part of the cheekbone. Like the mandible, the maxillae also have an alveolar margin along which the lower teeth are attached.

3. **Lacrimal bones.** The tiny lacrimal bones are located in the medial part of the orbit, where they form part of the structure that drains tears produced by the lacrimal gland of the eye. They are best seen in **Figure 8.4**.

4. **Nasal bones.** The two nasal bones form the anterior framework of the bridge of the nose.

5. **Vomer.** The single vomer forms the inferior portion of the bony nasal septum (best seen in **Figure 8.7**).

6. **Inferior nasal conchae.** The small inferior nasal conchae form part of the lateral walls of the nasal cavity. As their name implies, they are located inferior to the middle nasal conchae (**Figure 8.3**).

7. **Palatine bones.** The two palatine bones form the posterior part of the hard palate (seen in **Figure 8.5**) and the posterolateral walls of the nasal cavity.

8. **Zygomatic bones.** The two zygomatic bones form the bulk of the cheek and a significant portion of the "cheekbone" or zygomatic arch.

The **orbit** and the **nasal cavity** are complicated structures with contributions from several bones. The orbit is formed by parts of seven bones: the frontal bone, the maxilla, the sphenoid bone, the ethmoid bone, the lacrimal bone, the zygomatic bone, and a tiny piece of the palatine bone. The nasal cavity is formed by parts of six bones: the ethmoid bone, the maxilla, the palatine bone, the inferior nasal concha, the sphenoid bone, and the vomer.

As you can see in **Figure 8.8**, the fetal skull contains notable differences from the adult skulls you have seen so far. In adults, the sutures are fused, but in the fetus, the sutures have not yet fused and are instead joined by fibrous membranes. This can be seen with the **frontal suture** (also known as the **metopic suture**), seen where the two fetal frontal bones fuse. Where several sutures meet, we find large, membrane-covered areas called the **fontanels**, known to many as "soft spots." The two main fontanels are the **anterior fontanel**, where the sagittal and coronal sutures meet, and the **posterior fontanel**, where the sagittal and lambdoid sutures meet.

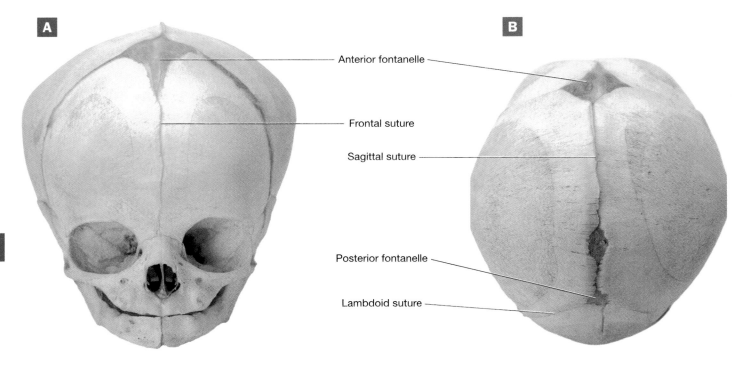

FIGURE **8.8** Fetal skull: (**A**) anterior view; (**B**) posterior view

## Procedure Cooperative Learning for the Skull

This exercise takes an approach called *cooperative learning*, in which you work with your lab partners to teach one another the bones and bone markings. The process may seem a bit confusing at first, but by the end of the first couple of rotations, it should move more quickly and smoothly. After you have completed the activity, answer Check Your Understanding questions 1–3 (p. 199).

**1** Assemble into groups of a minimum of four students; five is optimum.

**2** Distribute a skull to each member of the group, and assign each student one of the following five groups of bones and bone markings (the specific structures of each bone are listed on p. 167):

*Group A:* calvaria, base, and structures of the frontal bone and parietal bones

*Group B:* temporal bone and occipital bone structures

*Group C:* ethmoid bone and sphenoid bone structures

*Group D:* mandible and maxillary bone structures

*Group E:* remainder of the facial bones, orbit, sutural bones, and anterior and posterior fontanels.

**3** Spend approximately 3 minutes learning the assigned structures on your own.

**4** Then have each student spend approximately 1 to 2 minutes teaching the group his or her assigned structures.

**5** Rotate the assigned structures clockwise so each student has a new set of structures to learn (the student that was assigned group A will take group B, and so on).

**6** Spend approximately 2 to 3 minutes learning your newly assigned structures.

**7** Have each student spend approximately 1 to 2 minutes teaching the group his or her assigned structures.

**8** Repeat this process (it begins to speed up significantly at this point) until each student has taught each group of structures once.

By the end of this activity, each group member will have learned and presented each group of structures.

The following is a list of bones and bone markings of the skull that will be covered in this exercise.

## Cranial Bones

### Group A Structures
1. Calvaria
2. Base of cranial cavity
   a. Anterior cranial fossa
   b. Middle cranial fossa
   c. Posterior cranial fossa
3. Frontal bone
   a. Frontal sinuses
   b. Glabella
   c. Supraorbital foramen
4. Parietal bones
   a. Coronal suture
   b. Sagittal suture
   c. Squamous suture
   d. Lambdoid suture

### Group B Structures
1. Temporal bones
   a. Zygomatic process
   b. Mandibular fossa
   c. External acoustic (auditory) meatus
   d. Styloid process
   e. Mastoid process with mastoid air cells
   f. Carotid canal
   g. Jugular foramen
2. Occipital bone
   a. Occipital condyles
   b. Foramen magnum
   c. External occipital protuberance
   d. Superior nuchal line

### Group C Structures
1. Sphenoid bone
   a. Body
   b. Greater and lesser wings
   c. Optic foramen (canal)
   d. Superior orbital fissure
   e. Sella turcica
   f. Sphenoid sinus
2. Ethmoid bone
   a. Perpendicular plate
   b. Superior nasal conchae
   c. Middle nasal conchae
   d. Crista galli
   e. Cribriform plate
   f. Ethmoid sinuses
   g. Olfactory foramina

## Facial Bones and Other Structures

### Group D Structures
1. Mandible
   a. Mandibular condyle
   b. Coronoid process
   c. Mandibular notch
   d. Angle of mandible
   e. Mandibular ramus
   f. Body of mandible
   g. Mental foramen
   h. Alveolar margin
   i. Mandibular foramen
2. Maxilla
   a. Palatine processes
   b. Infraorbital foramen
   c. Maxillary sinuses
   d. Alveolar margin

### Group E Structures
1. Palatine bones
2. Zygomatic bones and zygomatic arch
3. Lacrimal bones and lacrimal fossa
4. Nasal bones
5. Vomer
6. Inferior nasal conchae
7. Sutural (Wormian) bones
8. Auditory ossicles
9. Orbit
10. Anterior fontanel
11. Posterior fontanel
12. Frontal (metopic) suture

Your instructor may wish to omit certain structures included above or add structures not included in these lists. List any additional structures below:

# Exercise 8-2

## Remainder of the Axial Skeleton

### MATERIALS

❏ Vertebral column, articulated
❏ Disarticulated vertebrae, sacrum, sternum, ribs, hyoid bone
❏ Skeleton, articulated

Another key component of the axial skeleton is the **vertebral column,** which consists of 24 vertebrae, the **sacrum,** and the **coccyx** (Figures 8.9–8.10). The vertebrae are divided into 7 **cervical vertebrae,** 12 **thoracic vertebrae,** and 5 **lumbar vertebrae.** Nearly all vertebrae share certain general features, including a posterior **spinous process,** an anterior **vertebral body,** two lateral **transverse processes,** and a central **vertebral foramen.** Extending posteriorly from the vertebral body to the transverse processes are two short extensions known as **pedicles.** Posterior to the pedicles are the **laminae,** which connect the pedicles to the spinous process. Between each vertebral body is a fibrocartilage pad called an **intervertebral disk.**

The basic properties of each region of the vertebral column are as follows:

1. The seven cervical vertebrae are located in the neck (**Figure 8.10A–C**). All cervical vertebrae have holes in their transverse processes called **transverse foramina.** These foramina permit the passage of blood vessels called the

Cervical vertebrae: $C_1 – C_7$

Thoracic vertebrae: $T_1 – T_{12}$

Intervertebral discs

Intervertebral foramen

Lumbar vertebrae: $L_1 – L_5$

Sacral vertebrae: $S_1 – S_1$ (fused)

Coccyx $C_1 – C_4$ (fused)

*Right lateral view*        *Anterior view*

Occipital bone

Cervical vertebrae (C1–C7)

Thoracic vertebrae (T1–T12)

Lumbar vertebrae (L1–L5)

Ilium

Sacrum

Pubic bone

Coccyx

Ischium

FIGURE **8.9** Vertebral column

*vertebral artery* and *vein*. In addition, the spinous processes of cervical vertebrae are often bifid or forked. Two cervical vertebrae are named differently than the others because of their unique features:

a. **Atlas** (C1): The atlas is the first cervical vertebra that articulates with the occipital bone. It is easily identified because it has a large vertebral foramen, no body, and no spinous process (**Figure 8.10B**).

b. **Axis** (C2): The axis is the second cervical vertebra. It is also easily identified by a superior projection called the **dens** (or the **odontoid process; Figure 8.10C**). The dens fits up inside the atlas to form the *atlanto-axial joint*, which allows rotation of the head.

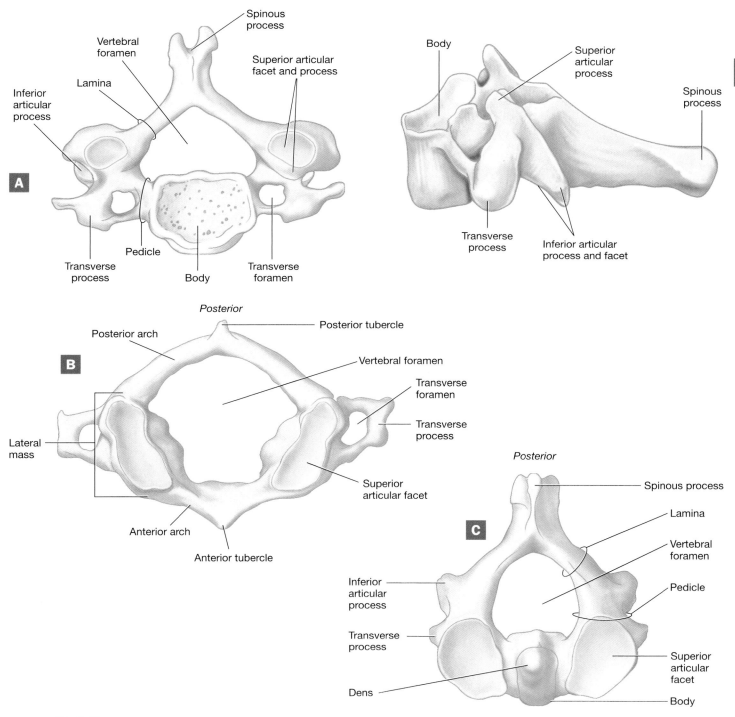

FIGURE **8.10** Vertebrae: (**A**) typical cervical vertebra, superior and lateral views; (**B**) atlas (C1), (**C**) axis (C2) *(continues)*

2. The 12 thoracic vertebrae share the following common features (**Figure 8.10D**):
   a. The spinous processes are thin and point inferiorly.
   b. All have two *costal facets* that articulate with the ribs (there are 12 pairs of ribs).
   c. All have triangular vertebral foramina.
   d. If you look at a thoracic vertebra from the posterior side, it looks like a giraffe (seriously!).

3. The five lumbar vertebrae share the following common features (**Figure 8.10E**):
   a. All have a large, block-like body.
   b. The spinous processes are thick and point posteriorly.
   c. If you look at a lumbar vertebra from the posterior side, it looks like a moose (really!).

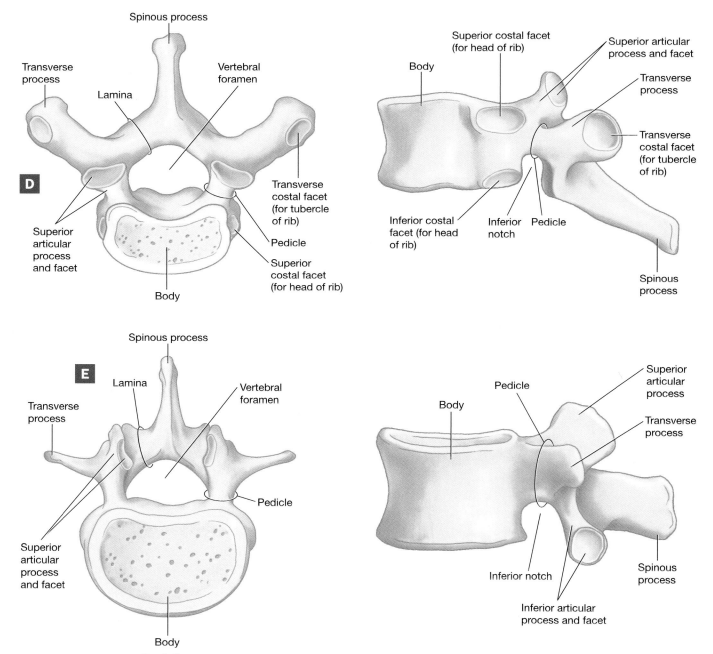

FIGURE **8.10** Vertebrae *(cont.)*: (**D**) typical thoracic vertebra, superior and lateral views; (**E**) typical lumbar vertebra, superior and lateral views *(continues)*

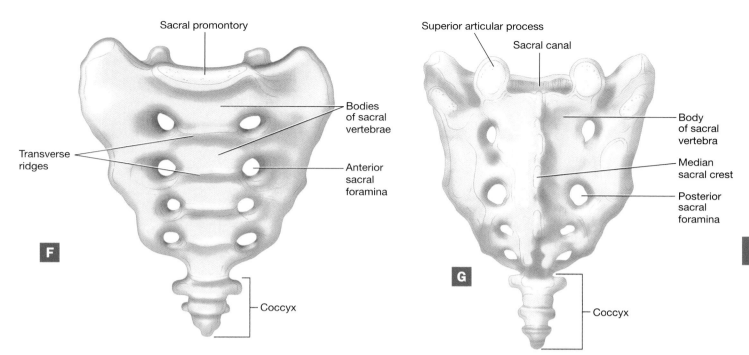

FIGURE **8.10** Vertebrae *(cont.)*: (**F**) sacrum, anterior view; (**G**) sacrum, posterior view

4. The sacrum consists of five fused vertebrae (**Figures 8.10F** and **8.10G**). You can see the fused spinous processes of the vertebrae on the sacrum's posterior side as the **median sacral crest**. Spinal nerves pass through holes called **sacral foramina** that flank both sides of the sacral bodies. The lateral surfaces of the sacrum articulate with the hip bones to form the **sacroiliac joints**.

5. The coccyx consists of three to five small, fused vertebrae that articulate superiorly with the sacrum.

The remainder of the axial skeleton consists of the bones of the thoracic cavity (the **sternum** and the **ribs, Figure 8.11**) and the **hyoid bone** in the neck (**Figure 8.12**). Note that the photo (Figure 8.11B) also shows the bones of the pectoral girdle. The sternum is divided into three parts—the upper **manubrium**, the middle **body**, and the lower **xiphoid process**. The ribs are classified according to how they attach to the sternum: Ribs 1–7 are considered **true ribs,** because they attach directly to the sternum by their own cartilage, ribs 8–10 are classified as **false ribs,** because they attach to the cartilage of the true ribs rather than directly to the sternum, and ribs 11–12 are called **floating ribs,** because they lack an attachment to the sternum.

The hyoid bone often is classified as a skull bone, although it does not articulate with any skull bone or any other bone. It is held in place in the superior neck by muscles and ligaments, and it helps to form part of the framework for the larynx (voice box). It also serves as an attachment site for the muscles of the tongue and aids in swallowing. When a person is choked manually, the hyoid bone is often broken. (I would not recommend testing this on your lab partner!)

The bones and bone markings of the axial skeleton are listed with Exercise 8-3 (p. 173), where they will be included with another cooperative learning exercise.

FIGURE **8.11** Thoracic cage

FIGURE **8.12** Hyoid bone

# Exercise 8-3

## The Appendicular Skeleton

### MATERIALS
- ❏ Disarticulated bones
- ❏ Skeleton, articulated
- ❏ Male and female pelvises

The bones of the appendicular skeleton can be divided into those of the **pectoral girdle**, the upper limb, the **pelvic girdle**, and the lower limb.

## Pectoral Girdle

The pectoral girdle consists of the two bones that frame the shoulder—the **scapula** and the **clavicle**. The scapula, shown in **Figure 8.13**, has a posterior ridge called the **spine** and a lateral depression called the **glenoid cavity** that forms the shoulder joint with the humerus. Superiorly it has two projections—the anterior **coracoid process** and the posterior **acromion**, which forms a joint with the clavicle called the **acromioclavicular** (AC) joint.

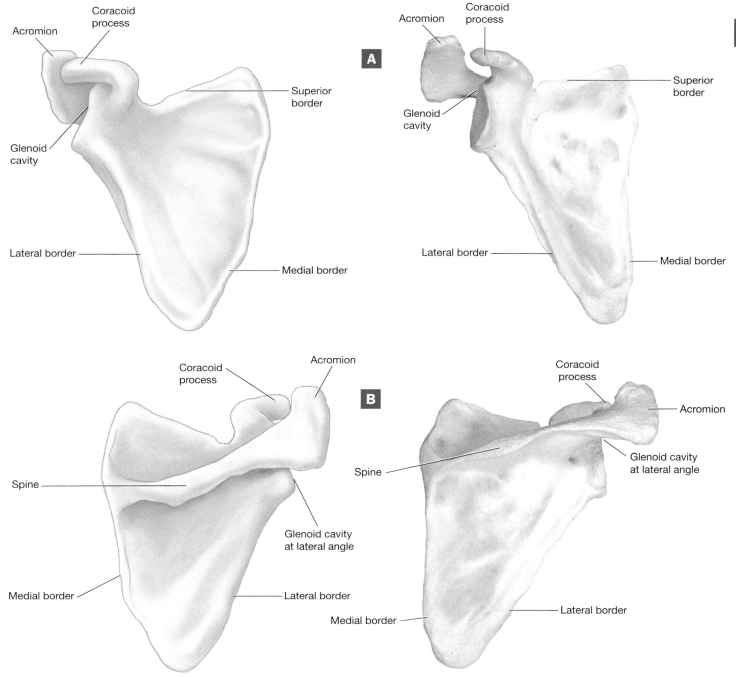

FIGURE **8.13** Right scapula: (**A**) anterior view; (**B**) posterior view

# Upper Limb

The upper limb consists of the arm, the forearm, the wrist, and the hand. The only bone within the arm is the **humerus** (**Figures 8.14** and **8.16**). The humerus has a number of features, some of which include the following:

- At the proximal end of the humerus is a rounded **head** that fits into the glenoid cavity. Just lateral to the head is the **lesser tubercle**, separated from the larger **greater tubercle** by the **intertubercular sulcus**.
- The middle of the humerus features a projection called the **deltoid tuberosity**, where the deltoid muscle attaches.
- At the humerus' distal end we find two condyles—the medial **trochlea**, which is shaped like a spool of thread, and the lateral **capitulum**, which is ball-shaped. Just proximal to these condyles are indentations in the humerus where the bones of the forearm articulate—the anterior **coronoid fossa** and the posterior **olecranon fossa**.

The two forearm bones, shown in **Figures 8.15** and **8.16**, are the lateral **radius** and medial **ulna** (if you have a hard time remembering which is which, stand in anatomical position and take your radial pulse; it is on the lateral side of the forearm, just like the radius). The ulna is wide proximally where it articulates with the humerus and thin distally where it articulates with the bones of the wrist. Its proximal end has two processes—the large, posterior **olecranon process** and the smaller, anterior **coronoid process**—separated by a deep curve called the **trochlear notch**.

FIGURE **8.14** Right humerus: (**A**) anterior view; (**B**) posterior view

FIGURE **8.15** Right radius and ulna, anterior view

As its name implies, the trochlear notch fits around the trochlea of the humerus to form the elbow joint. The olecranon process is the actual "elbow bone," which you can feel on your posterior arm. The trochlear notch and the ulna's two processes form a "U" shape when the ulna is held on its side. This makes the ulna easy to differentiate from the radius.

The radius has a width distribution opposite from that of the ulna. It is skinny proximally and wide distally. Proximally it consists of a **radial head** that articulates with the ulna at the **proximal radioulnar joint**. Note in **Figure 8.15** that the two bones also articulate at their distal ends at the **distal radioulnar joint**. Both the radius and the ulna have projections on their distal ends called **styloid processes**.

The wrist is composed of eight short bones called **carpals**, labeled individually in **Figure 8.17**. The carpals articulate with the radius and the ulna, as well as the five long bones in the hand called **metacarpals**. The metacarpals articulate distally with the fingers, which are formed from 14 long bones called **phalanges**. The second through fifth digits have three phalanges each (the *proximal*, *intermediate* [or *middle*], and *distal phalanges*); the thumb has only two (a *proximal* and a *distal phalanx*).

FIGURE **8.16** Bones of the upper limb: (**A**) humerus, radius, and ulna, anterior view; (**B**) humerus, radius, and ulna, posterior view

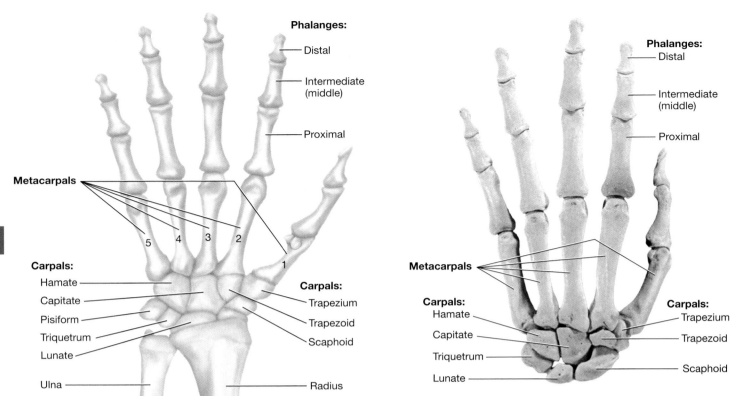

FIGURE **8.17** Right wrist and hand, anterior view.

## Pelvic Girdle

The pelvic girdle connects the lower limbs to the trunk, supports the pelvic organs, and transmits the weight of the trunk to the legs. It is formed by two **coxal bones**. Each coxal bone is also known as a **hemipelvis**. Three fused bones—the **ilium**, the **ischium**, and the **pubis**—make up each coxal bone (**Figures 8.18–8.19**). Notice that the lateral side of the hip (**Figure 8.19**) has a place where all three bones come together to form a deep socket. This socket, called the **acetabulum**, forms the hip joint with the femur. Notice also that where the ischium and pubis meet there is a large hole called the **obturator foramen**. In a living person this hole is covered with a membrane and allows only small blood vessels and nerves to pass through.

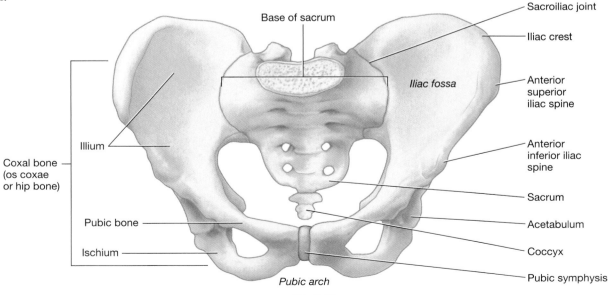

FIGURE **8.18** Pelvis

The three bones of the hip have the following features:

1. **Ilium.** The ilium is the largest of the three bones. Its main portion is called the **body,** and its superior "wing" is called the **ala.** The ridge of the ala, called the **iliac crest,** is where you rest your hands when your hands are on your hips. At the anterior end of the crest we find a projection called the **anterior superior iliac spine.** A similar

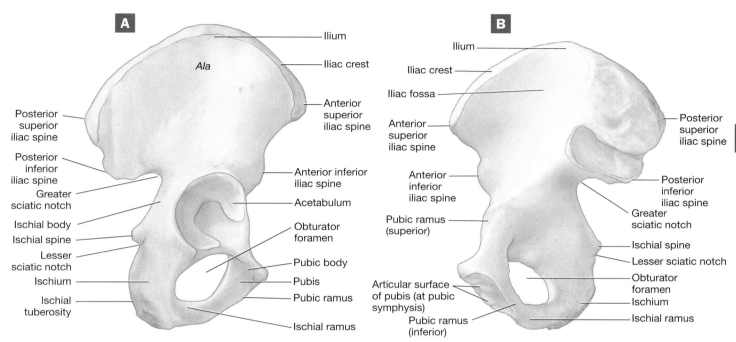

FIGURE **8.19** Pelvis: (**A**) right coxal bone, lateral view; (**B**) right coxal bone, medial view

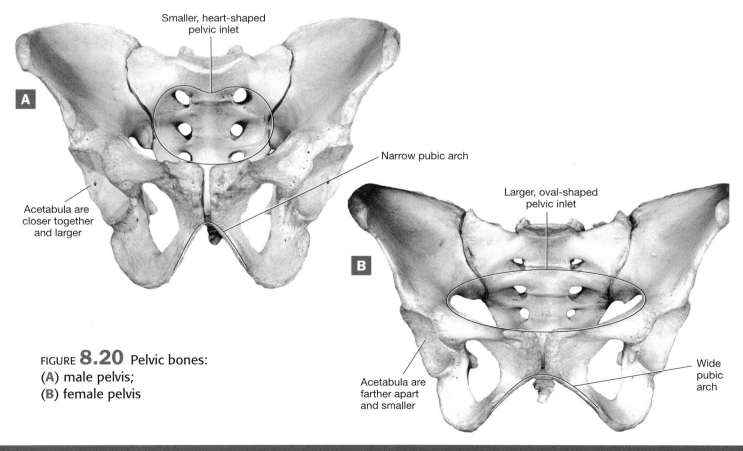

FIGURE **8.20** Pelvic bones:
(**A**) male pelvis;
(**B**) female pelvis

## How to Determine Gender Based upon Pelvic Features

A skeleton's gender can be determined based upon the differences between the male and female pelvises. In general, the male and female pelvis can be distinguished by the features listed below, shown in Figure 8.20:

| Feature | Female Pelvis | Male Pelvis |
| --- | --- | --- |
| Pelvic inlet shape | Wider and oval-shaped | Narrower and heart-shaped |
| Pubic arch | Wide angle | Narrow angle |
| Acetabulae | Farther apart | Closer together |
| Ischial tuberosities | Everted | Inverted |
| Coccyx | Straighter, more movable | Curved anteriorly, less movable |

but smaller **posterior superior iliac spine** is located posteriorly. The posterior ilium also features a notch called the **greater sciatic notch**, which allows the large greater sciatic nerve to pass between the pelvis and the thigh.

2. **Ischium.** The ischium makes up the posteroinferior pelvis. It contains three features on its posterior side: the superior **ischial spine**, the middle **lesser sciatic notch**, and the thick, inferior **ischial tuberosity**. The ischial tuberosities are the "butt bones," the bones that bear your weight when you sit down.

3. **Pubis.** The pelvis' anterior portion is formed by the pubis or the **pubic bone**. The pubis consists of a **body** and two extensions called the **superior** and **inferior rami**. The bodies of the two pubic bones meet at a fibrocartilage pad called the **pubic symphysis**. The angle that these bones form when they meet, called the **pubic arch**, can help to determine the gender of a skeleton (see the "Hints and Tips" box).

## Lower Limb

The lower limb consists of the thigh, the **patella** (kneecap), the leg, the ankle, and the foot. The thigh contains only a single bone, the large, heavy **femur** (**Figures 8.21** and **8.23**). Proximally, the femur articulates with the acetabulum at its rounded **head**. In the center of the femoral head we find a pit called the **fovea capitis**. This is the point of attachment for a ligament that holds the femur in the acetabulum. Just distal to the femoral head is the **neck** of the femur, the weakest part of the femur and the most common location for it to fracture (when the femoral neck fractures, it is usually called a "broken hip"). Where the femoral neck meets the femoral shaft, we find two large prominences—the anterolateral **greater trochanter** and the posteromedial **lesser trochanter**.

Distally, the femur expands into the **medial** and **lateral condyles**, which form the knee joint with the largest bone of the leg, the **tibia**. The tibia is flattened proximally at its articular surface, and its **medial** and **lateral condyles** fit together with those of the femur (**Figures 8.22** and **8.23**). Distally, the tibia articulates with a tarsal bone called the **talus**, with which it forms the ankle joint. At its terminal end is a projection called the **medial malleolus**, the medial ankle bone.

The other leg bone is the thin, lateral **fibula**. The proximal fibular end is called the **head**, and its distal end is the **lateral malleolus** (*lateral ankle bone*). The fibula doesn't articulate directly with either the femur or the talus. It does, however, articulate with the lateral side of the tibia at *proximal* and *distal tibiofibular joints*.

The ankle is composed of seven short bones called **tarsals**, labeled individually in **Figure 8.24**. The tarsals articulate with the five long bones in the foot called **metatarsals**. Like the bones of the fingers, the bones of the toes also consist of 14 **phalanges**. The second through fifth digits have three phalanges each (the *proximal*, *intermediate* [or *middle*], and *distal phalanges*); and the big toe or *hallux* has only two (a *proximal* and a *distal phalanx*).

FIGURE 8.21 Right femur: (A) anterior view; (B) posterior view

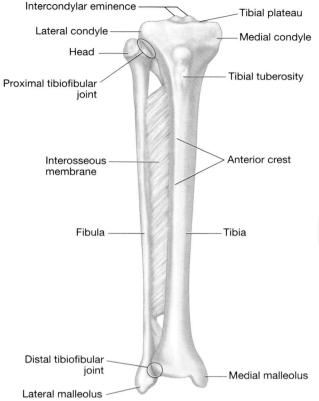

FIGURE 8.22 Right tibia and fibula, anterior view

FIGURE 8.23 Bones of the lower limb: (A) femur, tibia, and fibula, anterior view; (B) femur, tibia, and fibula, posterior view

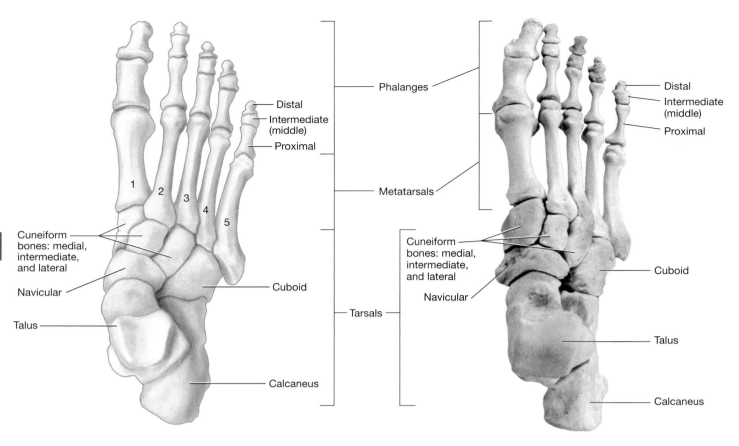

FIGURE **8.24** Right foot and ankle, superior view

**Procedure** Cooperative Learning for the Axial and Appendicular Skeletons

We will follow essentially the same procedure here as we did in Exercise 8-1, but with different groups of bones and bone markings. Note that this cooperative learning exercise includes the structures from Exercises 8-2 and 8-3. When you have completed this activity, answer Check Your Understanding questions 4–6 (p. 200).

**1** Assemble into groups with a minimum of four students; five is optimum.

**2** Distribute a bone or set of bones to each member of the group, and assign each student one of the following five groups of bones and bone markings (the specific structures of each bone are listed on pp. 167 and 168):

*Group A:*   cervical, thoracic, and lumbar vertebrae and vertebral markings; ribs and rib markings; sternum structures and the hyoid bone

*Group B:*   scapula structures, clavicle structures, humerus structures

*Group C:*   radius and ulna markings, carpals, metacarpals, and phalanges

*Group D:*   ilium, ischium, and pubis structures, and the difference between the male and female pelvises

*Group E:*   femur, tibia, fibula structures, tarsals, metatarsals, phalanges.

**3** Spend approximately 3 minutes learning the assigned structures on your own.

**4** Ask each student to spend approximately 1 to 2 minutes teaching the group his or her assigned structures.

**5** Rotate the assigned structures clockwise so each student has a new set of structures to learn (the student that was assigned group A will take group B, and so on).

**6** Spend approximately 2 to 3 minutes learning your newly assigned structures.

**7** Then ask each student to take approximately 1 to 2 minutes to teach the group his or her assigned structures.

**8** Repeat this process (it begins to speed up significantly at this point) until each student has been taught each group of structures once. By the end of this "game," each group member will have learned and presented each group of structures.

The following is a list of bone and bone markings of the appendicular skeleton covered in this exercise.

## Remaining Structures of the Axial Skeleton

### Group A Structures

1. Vertebrae
   a. Body
   b. Spinous process
   c. Vertebral foramen
   d. Transverse processes
   e. Lamina
   f. Pedicle
   g. Intervertebral disks
2. Cervical vertebrae
   a. Atlas
   b. Axis
      ▌ dens
   c. Transverse foramina
3. Thoracic vertebrae
   a. Costal facets
4. Lumbar vertebrae

5. Sacrum
   a. Median sacral crest
   b. Posterior sacral foramina
   c. Anterior sacral foramina
6. Coccyx
7. Ribs
   a. True ribs
   b. False ribs
   c. Floating ribs
8. Sternum
   a. Manubrium
   b. Body
   c. Xiphoid process
9. Hyoid bone

## Pectoral Girdle and Upper Limb

### Group B Structures

1. Clavicle
   a. Acromial end
   b. Sternal end
2. Scapula
   a. Acromion
   b. Spine
   c. Coracoid process
   d. Glenoid cavity (fossa)
   e. Lateral, medial, and superior border
3. Humerus
   a. Head
   b. Greater tubercle
   c. Lesser tubercle
   d. Intertubercular groove
   e. Deltoid tuberosity
   f. Medial and lateral epicondyles
   g. Capitulum
   h. Trochlea
   i. Olecranon fossa
   j. Coronoid fossa

### Group C Structures

1. Radius
   a. Head
   b. Neck
   c. Radial tuberosity
   d. Styloid process
2. Ulna
   a. Olecranon process
   b. Coronoid process
   c. Trochlear notch
   d. Styloid process
   e. Radial notch
3. Carpals
   a. Scaphoid
   b. Lunate
   c. Triquetrum
   d. Pisiform
   e. Trapezium
   f. Trapezoid
   g. Capitate
   h. Hamate
4. Metacarpals
5. Phalanges

## Pelvic Girdle and Lower Limb

Group D Structures

1. Ilium
   a. Iliac crest
   b. Anterior superior iliac spine
   c. Anterior inferior iliac spine
   d. Posterior superior iliac spine
   e. Posterior inferior iliac spine
   f. Greater sciatic notch
2. Ischium
   a. Ischial tuberosity
   b. Lesser sciatic notch
   c. Ischial spine
   d. Obturator foramen
3. Pubis
   a. Superior and inferior rami
   b. Pubic arch
   c. Pubic symphysis
4. Acetabulum
5. Male and female pelvises
6. Sacroiliac joint

Group E Structures

1. Femur
   a. Head
   b. Neck
   c. Greater trochanter
   d. Lesser trochanter
   e. Linea aspera
   f. Lateral and medial epicondyles
   g. Medial and lateral condyles
   h. Fovea capitis
2. Patella
3. Tibia
   a. Medial and lateral condyles
   b. Tibial tuberosity
   c. Medial malleolus
4. Fibula
   a. Head
   b. Lateral malleolus
5. Tarsals
   a. Talus
   b. Calcaneus
   c. Navicular
   d. Cuboid
   e. Cuneiforms
6. Metatarsals
7. Phalanges

Your instructor may wish to omit certain structures included above or add structures not included in these lists. List any additional structures below:

# Exercise **8-4**

## More Practice

### MATERIALS
- ❑ Disarticulated bones in a box
- ❑ Skeleton, articulated
- ❑ Colored pencils

On average, the human body contains 206 bones and thousands of bone markings, processes, fossae, and foraminae. The previous cooperative learning exercises gave you a solid foundation on which to build your knowledge of the skeleton—but a good foundation is just a start. A lot more work is needed to actually build a house on that foundation. This exercise contains activities to help you build that "house" and practice what you learned in Exercises 8-1 through 8-3.

## **Procedure** Building a Skeleton

**1** Obtain a set of disarticulated bones (real bones are best).

**2** Assemble the bones into a full skeleton. (If you have an articulated vertebral column and rib cage, go ahead and use them.)

**3** Be certain to keep your skeleton in anatomical position. **Figure 8.25** gives an overall "big picture" view of the skeleton that you may use for reference.

**4** Assemble the bones into a full skeleton.

FIGURE **8.25** The articulated skeleton: (**A**) anterior view; (**B**) posterior view

## **Procedure** Identifying Bones Blindly

**1** Place your set of disarticulated bones in a box.

**2** Working with your lab partner, close your eyes, reach into the box, grab a bone randomly, and attempt to identify it only by its feel.

**3** If you are unable to identify the bone, have your lab partner give you hints to help you work out the bone's identity.

## **Procedure** Comparing Bones on the Right and Left Sides

Examine the following bones, and figure out how you can determine if the bones are from the left or the right side of the body.

Femur

Tibia

Humerus

Scapula

## **Procedure** Draw a Skeleton

Anatomy and physiology coloring books have always been a popular study tool because they engage several parts of the brain and facilitate more rapid learning and memorization. An even better memorization tool than coloring is drawing. Drawing even the crudest picture of an anatomical structure or physiological pathway engages multiple parts of the brain and greatly enhances memory.

**1** Draw a skull and a skeleton in the spaces provided.

**2** Label and color your drawings.

These drawings are not meant to be works of art, so don't worry if you aren't even remotely skilled as an artist (however, if they turn out well, feel free to take them out of your lab book and put them on your refrigerator).

**1** Skull: anterior view

**8**

**2** Skull: lateral view

8

**UNIT**
**8**
**REVIEW**

## Check Your Recall

**1** Label the following bones in **Figure 8.26**.

- ☐ Clavicle
- ☐ Femur
- ☐ Fibula
- ☐ Humerus
- ☐ Ilium
- ☐ Metacarpals
- ☐ Pubis
- ☐ Radius
- ☐ Sternum
- ☐ Tarsals
- ☐ Tibia
- ☐ Ulna

8

FIGURE **8.26** Anterior view of the skeleton

**2** Label the following bones of the skull in **Figure 8.27.**

- ☐ Ethmoid bone
- ☐ Frontal bone
- ☐ Inferior nasal concha
- ☐ Lacrimal bone
- ☐ Mandible
- ☐ Maxilla
- ☐ Nasal bones
- ☐ Occipital bone
- ☐ Palatine bone
- ☐ Parietal bone
- ☐ Sphenoid bone
- ☐ Temporal bone
- ☐ Vomer
- ☐ Zygomatic bone

FIGURE **8.27** Skull: (**A**) anterior view; (**B**) lateral view; (**C**) midsagittal section

**3** Label the following skull markings
and processes in **Figure 8.28**.

- ☐ Cribriform plate
- ☐ External acoustic meatus
- ☐ Foramen magnum
- ☐ Frontal sinus
- ☐ Mastoid process
- ☐ Palatine process of maxilla
- ☐ Perpendicular plate of ethmoid bone
- ☐ Sella turcica
- ☐ Sphenoid sinus
- ☐ Styloid process
- ☐ Zygomatic arch

**A**

**8**

**B**

**C**

FIGURE **8.28** Skull: (**A**) lateral view; (**B**) midsagittal section; (**C**) internal view

**4** Label the following parts of the vertebral column in **Figure 8.29**.

☐ Atlas
☐ Axis
☐ Cervical vertebrae
☐ Coccyx
☐ Lumbar vertebrae
☐ Sacrum
☐ Thoracic vertebrae

FIGURE **8.29** Vertebral column

**5** Label the following parts of the axial skeleton in **Figure 8.30**.

☐ False ribs
☐ Floating ribs
☐ Manubrium
☐ True ribs
☐ Xiphoid process

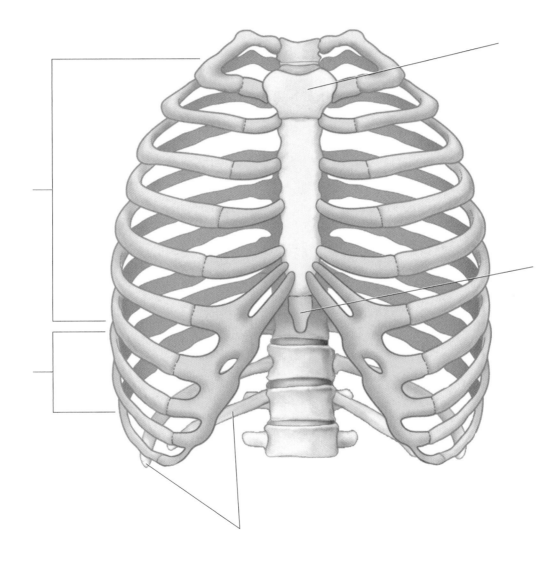

FIGURE **8.30** Thoracic cage

**6** Identify each of the following vertebrae in **Figure 8.31** as being cervical, thoracic, or lumbar. Additionally, explain how you arrived at this conclusion.

a _____

_____

b _____

_____

c _____

_____

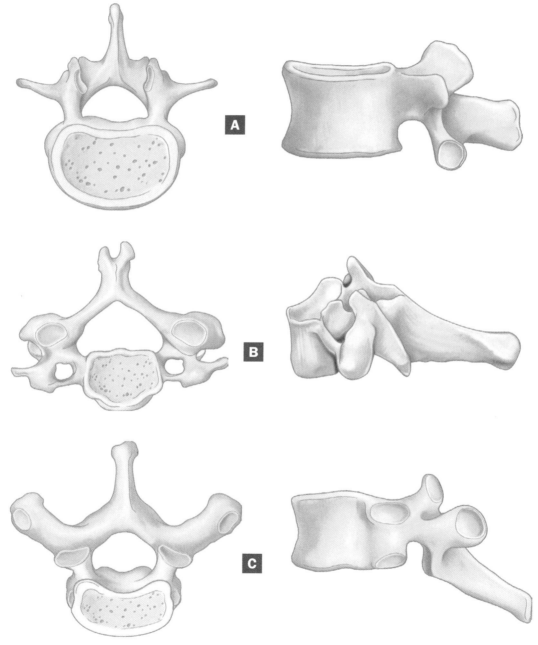

FIGURE **8.31** (**A**) unknown vertebra 1; (**B**) unknown vertebra 2; (**C**) unknown vertebra 3

**7** Label the following parts of the scapula in **Figure 8.32**.

☐ Acromion
☐ Coracoid process
☐ Glenoid cavity
☐ Spine

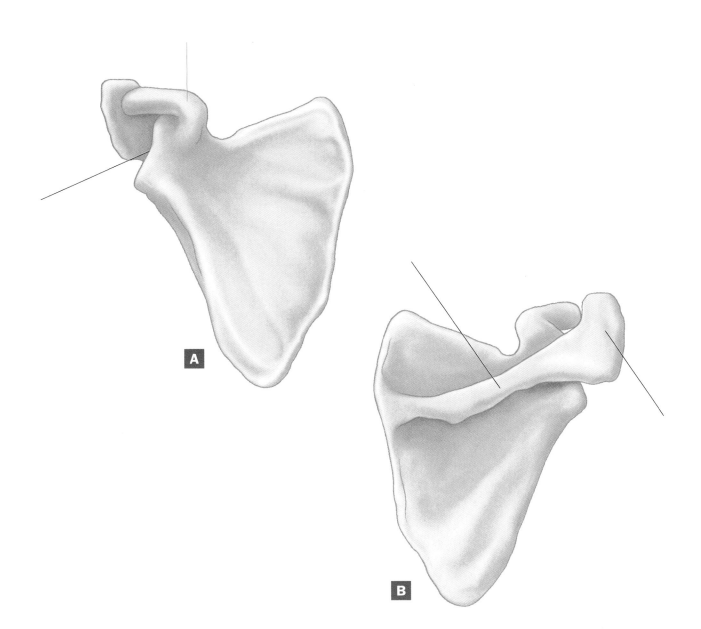

FIGURE **8.32** Scapula: (**A**) anterior view; (**B**) posterior view

**8** Label the following parts of the upper limb in **Figure 8.33.**

- ☐ Capitulum
- ☐ Coronoid process
- ☐ Deltoid tuberosity
- ☐ Greater tubercle
- ☐ Head
- ☐ Intertubercular sulcus

- ☐ Olecranon
- ☐ Radial head
- ☐ Styloid process
- ☐ Trochlea
- ☐ Trochlear notch

FIGURE **8.33** Upper limb: (**A**) arm, anterior view; (**B**) forearm, anterior view

**9** Label the following parts of the coxal bone (hemipelvis) in **Figure 8.34.**

☐ Acetabulum          ☐ Ischial tuberosity
☐ Anterior superior iliac spine    ☐ Ischium
☐ Greater sciatic notch      ☐ Obturator foramen
☐ Iliac crest             ☐ Pubis
☐ Ilium

8

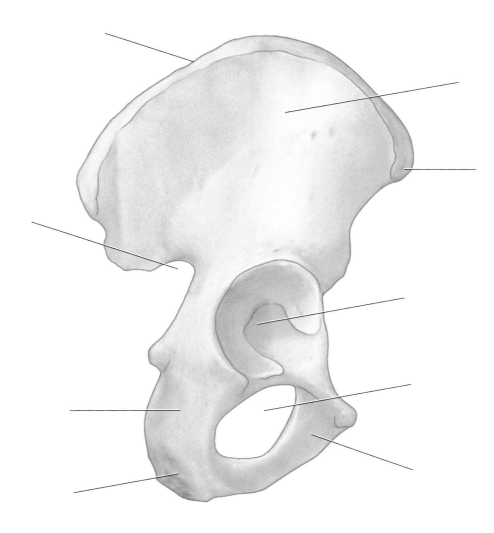

FIGURE **8.34** Right coxal bone, lateral view

**10** Label the following parts of the lower limb in **Figure 8.35.**

☐ Greater trochanter          ☐ Medial condyle
☐ Head                        ☐ Medial malleolus
☐ Lateral condyle             ☐ Neck
☐ Lateral malleolus           ☐ Tibial tuberosity
☐ Lesser trochanter

FIGURE **8.35** Lower limb: (**A**) thigh, anterior view; (**B**) leg, anterior view

# Check Your Understanding

**Critical Thinking and Application Questions**

8

**1** Which bones potentially could be involved in a fracture of the orbit? Why do you think orbital fractures are often difficult to fixate surgically with screws and plates?

_____

_____

_____

_____

_____

**2** Your friend has suffered a "broken nose." Which bones potentially could be involved in a fracture involving the nose and the nasal cavity?

_____

_____

_____

_____

_____

**3** The bones of the fetal skull are not yet fused, giving the skull many "soft spots," or fontanels. Why do you think the bones of the fetal skull are not fused at birth?

_____

_____

_____

_____

_____

**4** Police officers have found human skeletal remains, including a pelvis, and they ask your help in determining the individual's gender. Which features will you examine to help you make this determination?

8

**5** Many people mistakenly believe that the pelvis is fractured when one has a broken hip. What is actually fractured in a broken hip? Examine this structure, and compare it to the bones of the pelvis. Why do you think it is more likely to fracture than the pelvis?

**6** You are presented the following X-ray from a six-year-old child (**Figure 8.36**).

   **a**  Identify the bones in the X-ray.

   **b**  Your colleague thinks that there may be a fracture present at the arrow in **Figure 8.36**. What do you think? Is this a fracture or a normal anatomical feature? Explain.

FIGURE **8.36** X-ray from six-year-old child

# Articulations

**9**

## OBJECTIVES

*Once you have completed this unit, you should be able to:*

- Classify joints based upon structure and function.
- Identify examples of the different types of joints.
- Identify structures associated with synovial joints.
- Classify synovial joints according to range of motion.
- Identify structures of the knee joint.
- Demonstrate and describe motions allowed at synovial joints.

201

# PRE-LAB EXERCISES

Complete the following exercises prior to coming to lab, using your textbook and lab manual for reference.

## Pre-Lab Exercise **9-1**

✎ Key Terms

You should be familiar with the following terms before coming to lab.

| Term | Definition |
|------|------------|

**Classes of Joints by Function**

Synarthrosis _____

_____

Diarthrosis _____

_____

Amphiarthrosis _____

_____

**Classes of Joints by Structure**

Fibrous _____

_____

Cartilaginous _____

_____

Synovial _____

_____

**Classes of Synovial Joints by Structure**

Plane _____

_____

Hinge _____

_____

Pivot _____

_____

Condyloid _____

_____

Saddle _____

_____

Ball and socket _____

**Types of Movement in Synovial Joints**

Flexion _____

Extension _____

Abduction _____

Adduction _____

Circumduction _____

Rotation _____

Inversion _____

Eversion _____

Plantarflexion _____

Dorsiflexion _____

Pronation _____

Supination _____

Elevation _____

Depression _____

Opposition _____

# Pre-Lab Exercise **9-2**

## Structural Classes of Joints

Label and color the structural classes of joints depicted in **Figure 9.1** with the terms from Exercise 9-2 (p. 209). Use your text and Exercise 9-2 in this unit for reference.

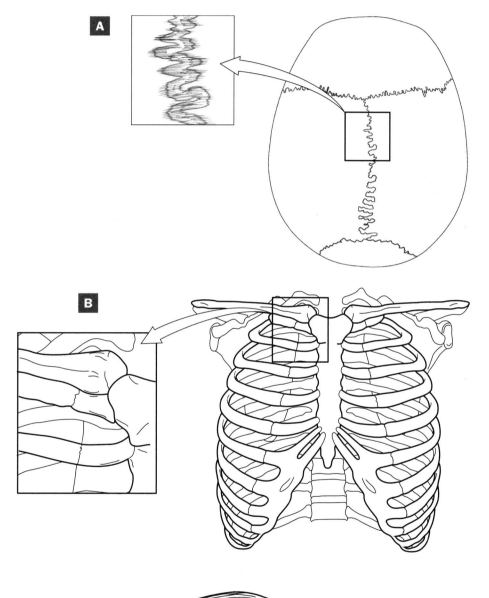

FIGURE **9.1** The three structural classes of joints:
(**A**) fibrous joint;
(**B**) cartilaginous joint;
(**C**) synovial joint

Label and color the knee joint depicted in **Figure 9.2** with the terms from Exercise 9-2 (p. 212). Use your text and Exercise 9-2 in this unit for reference.

**9**

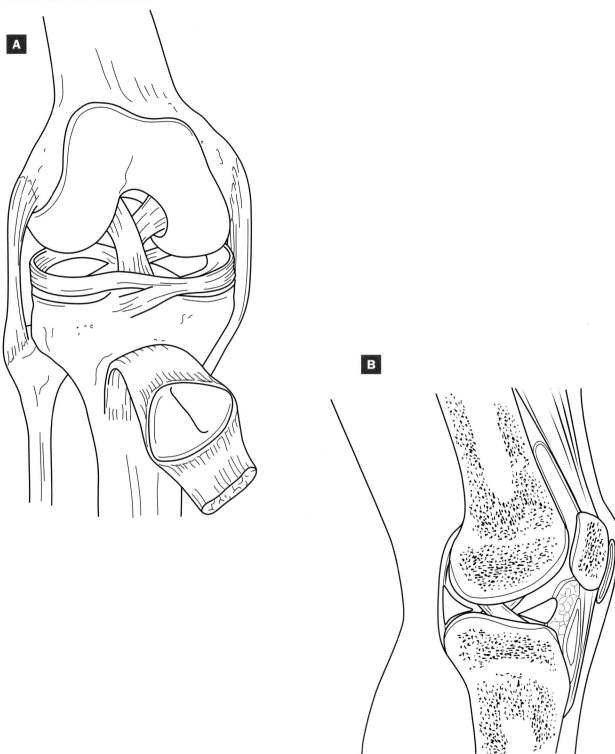

FIGURE **9.2** Knee joint: (**A**) anterior view; (**B**) midsagittal section

# EXERCISES

Most bones in the body form **articulations**, or **joints**, with another bone. These articulations allow you to perform the wide variety of motions involved in everyday movement. You probably don't realize just how many joints are articulating when you perform routine activities such as answering the phone, walking to the door, or holding a model in A&P lab. In the following exercises you will identify and classify joints according to structure, function, and amount of motion; examine the knee joint; and determine which joints are involved in simple everyday movements.

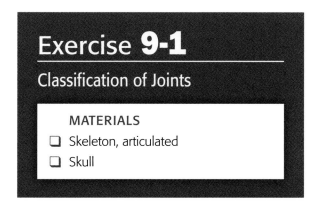

## Exercise 9-1

### Classification of Joints

**MATERIALS**
- ❏ Skeleton, articulated
- ❏ Skull

Structurally, joints are classified as (**Figure 9.3**):

1. **Fibrous joints.** Fibrous joints consist of bones joined by short connective tissue fibers. Most fibrous joints allow no motion and are synarthroses.

2. **Cartilaginous joints.** Cartilaginous joints consist of bones united by cartilage rather than fibrous connective tissue. Most cartilaginous joints allow some motion and are amphiarthroses. However, the epiphyseal plate, a structure composed of hyaline cartilage found in growing bones, is a cartilaginous joint that is a synarthrosis. When the epiphyseal plate ossifies, it becomes a type of immovable joint called a **synostosis**.

3. **Synovial joints.** Synovial joints have a true joint cavity and consist of two bones whose articular ends are covered with hyaline cartilage. The joint cavity is lined by a **synovial membrane** that secretes a watery fluid called **synovial fluid** similar in composition to blood plasma without the proteins. The fluid bathes the joint to permit frictionless motion. We discuss synovial joints further in Exercise 9-2.

Joints are classified according to both their structure and their function. Functionally, joints are classified as:

1. **Synarthroses**—immovable joints,
2. **Amphiarthroses**—joints that allow some motion, or
3. **Diarthroses**—freely moveable joints.

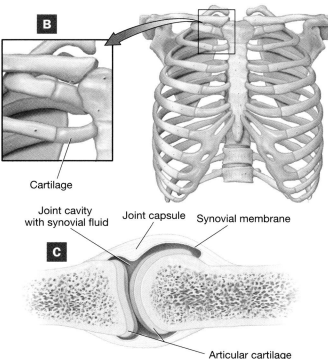

Connective tissue fibers

**A**

**B**

Cartilage

Joint cavity with synovial fluid    Joint capsule    Synovial membrane

**C**

Articular cartilage

FIGURE **9.3** The three classes of joints: (**A**) fibrous joint; (**B**) cartilaginous joint; (**C**) synovial joint

## Procedure Classifying Joints by Structure and Function

Classify each joint listed in **Table 9.1** by its structure. Then examine and manipulate the joint to determine the amount of motion allowed at the joint. When you have determined how much movement is allowed at the joint, classify it functionally. After you have completed the activity, answer Check Your Understanding questions 1 and 2 (p. 221).

TABLE **9.1** Structural and Functional Classification of Joints

| Joint | Structural Classification | Amount of Motion | Functional Classification |
|---|---|---|---|
| Intervertebral joint | | | |
| Shoulder (glenohumeral) joint | | | |
| Intercarpal joint | | | |
| Coronal suture | | | |
| Pubic symphysis | | | |
| Interphalangeal joint | | | |

# Exercise 9-2

## Synovial Joints

### MATERIALS
- ☐ Fresh pig or chicken joints
- ☐ Dissection trays and kits
- ☐ Skeleton, articulated
- ☐ Knee joint model

Synovial joints have a fluid-filled cavity lined by a synovial membrane. Features common to synovial joints are the following:

- **Joint capsule.** The joint capsule is made of dense irregular collagenous connective tissue and provides strength and structural reinforcement for the joint. It is lined by the synovial membrane.

- **Articular cartilage.** The articulating ends of the bones are covered with articular cartilage, usually hyaline cartilage. The cartilage provides a smooth, nearly frictionless surface for articulation.

- **Ligaments.** The bones in a synovial joint are held together by ligaments that further reinforce the joint. Some ligaments (called extrinsic ligaments) are within the joint cavity, whereas others (intrinsic ligaments) are embedded in the capsule (they are called "intrinsic" because they are an intrinsic part of the capsule).

- **Articular discs.** Also known as **menisci,** articular discs are fibrocartilage pads that improve the fit of two bones to prevent dislocation.

Synovial joints typically are surrounded by tendons that move the bones involved in the joint. The tendons generally are wrapped in a sheath of connective tissue in which they can slide with a minimum of friction. Fluid-filled sacs called **bursae** are often located between tendons and joints, and this also reduces friction.

## Procedure Identifying Structures of Synovial Joints

Identify the following structures on fresh specimens such as pigs' feet. If fresh specimens are not available, use anatomical models instead. When you have completed the activity, answer Check Your Understanding question 3 (p. 221).

- ☐ Joint cavity
- ☐ Joint capsule
- ☐ Articular cartilage
- ☐ Articular discs (menisci)
- ☐ Synovial membrane
- ☐ Synovial fluid

- ☐ Bursae
- ☐ Tendon with tendon sheath
- ☐ Ligaments
  - Intrinsic
  - Extrinsic

## Types of Synovial Joints

The range of motion of a synovial joint is described conventionally in terms of an invisible *axis* about which the bone moves. Synovial joints are classified according to the number of planes of motion in which the bones can move around this axis. The classes are as follows:

1. **Nonaxial joints.** As implied by its name, a nonaxial joint does not move around an axis. Instead, the bones in a nonaxial joint simply glide past one another. An example of a nonaxial joint is the vertebrocostal joint.

2. **Uniaxial joints.** Joints that allow motion in one plane or direction only are called uniaxial joints. A classic example is the elbow, which permits only flexion and extension.

3. **Biaxial joints.** Joints that allow motion in two planes are called biaxial joints. An example of a biaxial joint is the wrist, which allows both flexion/extension and abduction/adduction.

4. **Multiaxial joints.** The joints with the greatest range of motion are multiaxial joints, which allow motion in multiple planes. An example of a multiaxial joint is the shoulder joint.

In addition to this classification scheme, synovial joints are classified according to their structure. The structural classes are illustrated in **Figure 9.4** and include the following:

- **Plane.** The bones of plane joints have flat articular surfaces that allow the bones to glide past one another.
- **Condyloid.** Condyloid joints consist of one bone that fits into the concave surface of another bone.
- **Saddle.** Note in **Figure 9.4** that saddle joints somewhat resemble condyloid joints but permit a greater range of motion.
- **Hinge.** In a hinge joint, the bones fit together much like the hinge of a door. Generally, the convex articular surface of one bone fits into a concave articular surface of another bone.
- **Pivot.** In a pivot joint, one bone rotates or "pivots" around another bone. Generally, pivot joints consist of one bone with a rounded projection that fits into a groove of another bone.
- **Ball and socket.** Ball-and-socket joints are named for the rounded, ball-like end of one bone that fits into the concave "socket" of another bone.

9

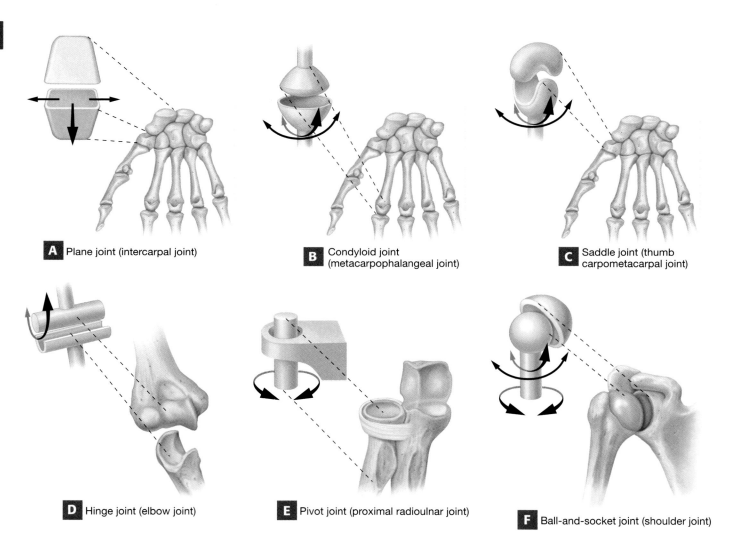

**A** Plane joint (intercarpal joint)

**B** Condyloid joint (metacarpophalangeal joint)

**C** Saddle joint (thumb carpometacarpal joint)

**D** Hinge joint (elbow joint)

**E** Pivot joint (proximal radioulnar joint)

**F** Ball-and-socket joint (shoulder joint)

FIGURE **9.4** The structural classes of synovial joints

## Procedure Classifying Synovial Joints

For each of the following joints, first list the joint's structural classification in **Table 9.2**. Then obtain an articulated skeleton so you can manipulate each joint and determine if the joint is nonaxial, uniaxial, biaxial, or multiaxial. When you have completed this activity, answer Check Your Understanding question 4 (p. 221).

TABLE **9.2** Classification of Synovial Joints

| Joint | Structural Classification | Range of Motion (nonaxial, uniaxial, biaxial, or multiaxial) |
|---|---|---|
| Shoulder joint | | |
| Intercarpal joint | | |
| Proximal radioulnar joint | | |
| Radiocarpal joint | | |
| Thumb carpometacarpal joint | | |
| Interphalangeal joint | | |
| Knee joint | | |
| Atlantoaxial joint | | |
| Hip joint | | |

# Knee Joint

The knee joint, illustrated in **Figure 9.5**, is a modified hinge joint. It is stabilized by intrinsic ligaments (which are part of the joint capsule), extrinsic ligaments, and by the **medial** and **lateral menisci**. Four important extrinsic ligaments include the following:

1. **Anterior cruciate ligament (ACL).** The anterior cruciate ligament, or ACL, extends from the anterior tibial plateau to the lateral femoral condyle. Its function is to prevent hyperextension of the knee.

2. **Posterior cruciate ligament (PCL).** The posterior cruciate ligament, or PCL, extends from the posterior tibial plateau to the medial femoral condyle. It crosses under the ACL, and the two together form an "X." The PCL prevents posterior displacement of the tibia on the femur.

3. **Medial collateral and lateral collateral ligaments.** The medial and lateral collateral ligaments (MCL and LCL) extend from the medial tibia and the lateral fibula to the femur, respectively. They resist varus and valgus (medial and lateral) stresses.

There are also a number of **bursae**, fluid-filled sacs, around the knee joint (**Figure 9.5B**). These bursae reduce the friction around the knee joint as it moves.

## Procedure Identifying Structures of the Knee Joint

Identify the following structures of the knee joint on anatomical models or fresh specimens. Check off each structure as you identify it.

- ☐ Joint capsule
- ☐ Ligaments
  - Lateral collateral ligament
  - Medial collateral ligament
  - Anterior cruciate ligament (ACL)
  - Posterior cruciate ligament (PCL)
  - Patellar ligament (tendon)
- ☐ Menisci
  - Medial meniscus
  - Lateral meniscus
- ☐ Medial and lateral femoral condyles
- ☐ Tibial plateau

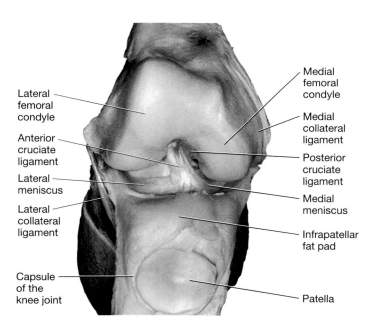

FIGURE **9.5** Right knee joint: **(A)** anterior view; *(continues)*

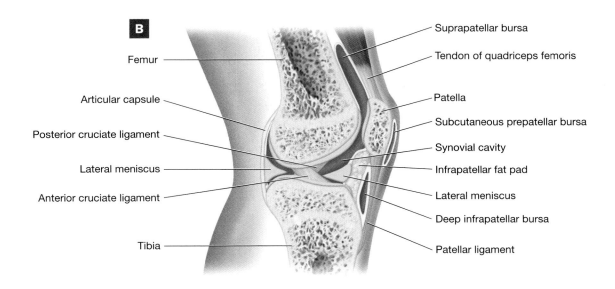

Labels (left side, top to bottom):
Femur
Articular capsule
Posterior cruciate ligament
Lateral meniscus
Anterior cruciate ligament
Tibia

Labels (right side, top to bottom):
Suprapatellar bursa
Tendon of quadriceps femoris
Patella
Subcutaneous prepatellar bursa
Synovial cavity
Infrapatellar fat pad
Lateral meniscus
Deep infrapatellar bursa
Patellar ligament

FIGURE **9.5** Right knee joint *(cont.)*: **(B)** midsagittal section

## Procedure Testing the Integrity of the ACL and PCL

One of the more common knee injuries is a torn ACL and/or PCL. To assess the integrity of the ACL and PCL, tests known as the **anterior drawer** and **posterior drawer**, respectively, are performed. A normal result of both tests is a minimum amount of motion when the tibia is moved anteriorly and posteriorly. When you have completed the tests, answer Check Your Understanding questions 5 and 6 (p. 222).

**1** Have your partner sit with the knees bent at 90° and relaxed.

**2** Grasp your partner's leg with both hands around the proximal tibia and fibula.

**3** Gently pull the leg anteriorly, being careful not to extend the knee joint.

▌ Amount of motion: _____

▌ Which ligament was being assessed (ACL or PCL)? _____

**4** Now gently push the leg posteriorly. Be careful not to flex the knee joint.

▌ Amount of motion: _____

▌ Which ligament was being assessed (ACL or PCL)? _____

# Exercise 9-3

## Motions of Synovial and Cartilaginous Joints

Each time you move your body in a seemingly routine fashion (such as walking or climbing stairs), you are producing motion at a tremendous number of joints. Many possible motions can occur at synovial and cartilaginous joints. These motions are illustrated in **Figure 9.6.**

In the following procedure you will determine which joints you are moving with some commonly performed actions and which motions are occurring at each joint.

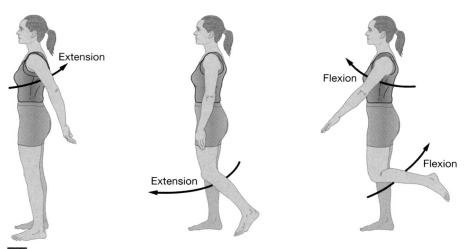

**A** Angular movements: Extension and flexion at the shoulder and knee

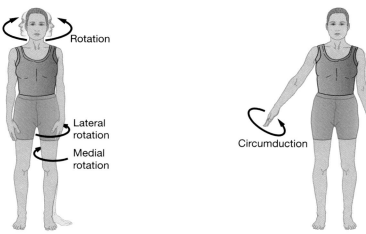

**B** Rotation of the head, neck, and lower limb

**C** Angular movements: adduction, abduction, and circumduction of the upper limb at the shoulder

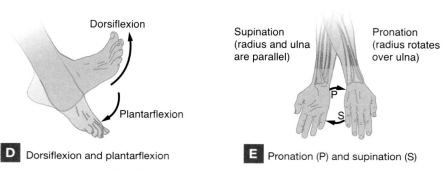

**D** Dorsiflexion and plantarflexion

**E** Pronation (P) and supination (S)

FIGURE **9.6** Motions of synovial and cartilaginous joints

## **Procedure** Identifying Joint Motions of Common Movements

Team up with a partner, and have your partner perform each of the following actions. Watch carefully as the actions are performed, and list the joints that are in motion.

Ask your instructor if he or she wants you to use the technical name or the common name for each joint (e.g., glenohumeral versus shoulder joint). Some of the joints, such as the hip joint and knee joint, will be obvious. Others, such as the radioulnar joint, the fingers and toes, and the sacroiliac joint, are less obvious and easily overlooked.

Once you have listed the joints being used, determine which motions are occurring at each joint. Keep in mind the type and the range of motion of each joint as you answer each question.

**1** Walking up stairs

Joints moving:                                    Motions occurring:

_____

_____

_____

_____

_____

_____

_____

_____

**2** Doing jumping jacks

Joints moving:                                      Motions occurring:

_____

_____

_____

_____

_____

_____

_____

**3** Answering the telephone

Joints moving:                          Motions occurring:

_____          _____
_____          _____
_____          _____
_____          _____
_____          _____
_____          _____
_____          _____
_____          _____
_____          _____

**4** Jumping rope

Joints moving:                          Motions occurring:

_____          _____
_____          _____
_____          _____
_____          _____
_____          _____
_____          _____
_____          _____
_____          _____

## Check Your Recall

**1** Joints that permit no motion are called
   a. synarthroses.
   b. amphiarthroses.
   c. cartilaginous joints.
   d. diarthroses.

**2** Freely moveable joints are called
   a. cartilaginous joints.
   b. synarthroses.
   c. amphiarthroses.
   d. fibrous joints.
   e. diarthroses

**3** Most cartilaginous joints allow _____ and are _____ joints.
   a. no motion; synarthrotic
   b. some motion; diarthrotic
   c. no motion; amphiarthrotic
   d. some motion; amphiarthrotic

**4** Synovial joints are filled with _____, which _____ the joint.
   a. serous fluid; lubricates
   b. synovial fluid; lubricates
   c. serous fluid; increases range of motion of
   d. synovial fluid; increases range of motion of

**5** *Mark the following statements as true (T) or false (F). If the statement is false, correct it so it is a true statement.*

   _____ a. The articulating ends of the bones in a synovial joint are covered by a synovial membrane.

   _____ b. Articular discs improve the fit between two bones in a synovial joint.

   _____ c. The joint capsule of a synovial joint is lined with a synovial membrane.

   _____ d. Ligaments provide a smooth, nearly frictionless surface for articulation.

   _____ e. Fluid-filled sacs called bursae often lie between tendons and the joint capsule of a synovial joint.

9

**6** Define the following terms:

Nonaxial joint _____

_____

Uniaxial joint _____

_____

Biaxial joint _____

_____

Multiaxial joint _____

_____

**7** Which of the following describes a plane joint correctly?
   a. The convex articular surface of one bone fits into a concave articular surface of another bone.
   b. One bone rotates around another bone.
   c. The flat articular surfaces of two bones glide past one another.
   d. The rounded, ball-like end of one bone fits into a concave depression of another bone.

**8** Which of the following correctly describes a pivot joint?
   a. The convex articular surface of one bone fits into a concave articular surface of another bone.
   b. One bone rotates around another bone.
   c. The flat articular surfaces of two bones glide past one another.
   d. The rounded, ball-like end of one bone fits into a concave depression of another bone.

**9** Label the following parts of the knee joint in **Figure 9.7**.

☐ Anterior cruciate ligament           ☐ Medial meniscus
☐ Lateral collateral ligament          ☐ Patellar ligament
☐ Lateral meniscus                     ☐ Posterior cruciate ligament
☐ Medial collateral ligament

**9**

FIGURE **9.7** Knee joint

**10** *Matching:* Match the following terms with the correct description.

_____ Flexion

_____ Circumduction

_____ Adduction

_____ Extension

_____ Dorsiflexion

_____ Abduction

_____ Rotation

_____ Pronation

A.  Movement of a body part toward the midline

B.  Movement around a central axis

C.  Turning the palm over to face down

D.  Decreasing the angle between two bones

E.  Movement of a body part away from the midline

F.  Increasing the angle between two bones

G.  Movement of the ankle that decreases the angle between the foot and leg

H.  Movement in a circle

Name _____

Section _____ Date _____

# Check Your Understanding

**Critical Thinking and Application Questions**

**1** Would fibrous and cartilaginous joints be able to perform their functions if they had joint cavities? Explain.

_____

_____

_____

_____

**2** Your father plans to visit an alternative medicine practitioner for "cranial therapy." You read the brochure for this proposed treatment and note that it involves the movement and manipulation of the cranial sutures, which the practitioner claims will cure nearly all diseases. What kind of joint is a cranial suture, both structurally and functionally? Given these facts, what do you think of the claims made about "cranial therapy"?

_____

_____

_____

_____

**3** The term *double-jointed* describes individuals who have an abnormally large range of motion in a given joint. This excess range of motion is not because of the presence of a second joint but, instead, weakness of the ligaments surrounding the joint. Why would weakness of ligaments lead to a greater range of motion at a joint?

_____

_____

_____

_____

**4** Compare the structure of the hip joint and the shoulder joint. Which joint do you think would be more likely to dislocate? Why?

_____

_____

_____

_____

**5** Your friend is on the basketball court when she suddenly hears a loud popping sound and hyperextends her knee. What likely has happened? What results would you see from the anterior drawer test?

_____

_____

_____

_____

**6** Meniscal tears are very common knee injuries. How could a torn meniscus affect the function of the knee joint?

_____

_____

_____

_____

# Muscle Tissue

**10**

## OBJECTIVES

*Once you have completed this unit, you should be able to:*

- Describe the microscopic anatomy of skeletal muscle fibers.

- Identify structures of skeletal muscle fibers.

- Identify structures associated with the neuromuscular junction.

- Describe the sequence of events of the skeletal muscle fiber action potential and the sliding filament mechanism of contraction.

- Describe the length-tension relationship of skeletal muscle fibers.

- Distinguish between smooth and cardiac muscle tissue.

# PRE-LAB EXERCISES

Complete the following exercises prior to coming to lab, using your textbook and lab manual for reference.

## Pre-Lab Exercise **10-1**

✎ Key Terms

You should be familiar with the following terms before coming to lab.

| Term | Definition |
|------|------------|

**General Skeletal Muscle Structures**

Epimysium _____

_____

Fascicle _____

_____

Perimysium _____

_____

Muscle fiber _____

_____

Endomysium _____

_____

**Structures of the Skeletal Muscle Fiber**

Sarcolemma _____

_____

T-tubule _____

_____

Sarcoplasmic reticulum _____

_____

Terminal cisternae _____

_____

Myofibril _____

_____

**10**

Myofilament _____

_____

Thick filament _____

_____

Thin filament _____

_____

## Components of the Sarcomere

A band _____

_____

I band _____

_____

Z disk _____

_____

H zone _____

_____

M line _____

_____

## Terms of Muscle Physiology

Action potential _____

_____

Crossbridge cycle _____

_____

Excitation-contraction coupling _____

_____

Sliding filament mechanism _____

_____

Length-tension relationship _____

_____

Label and color the basic structures of skeletal muscles depicted in **Figure 10.1** with the terms from Exercise 10-1 (p. 229). Use your text and Exercise 10-1 in this unit for reference.

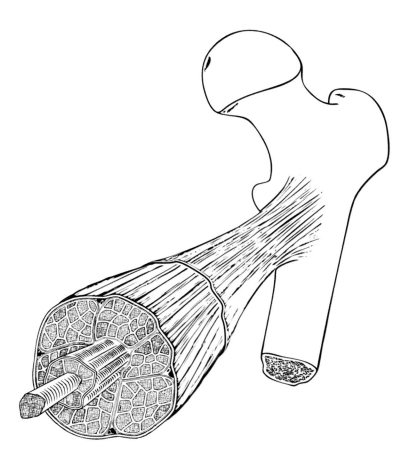

FIGURE **10.1** Basic skeletal muscle structure

## Pre-Lab Exercise **10-3**

### Muscle Fiber Microanatomy

Label and color the microscopic anatomy of the skeletal muscle fiber depicted in **Figure 10.2** with the terms from Exercise 10-1 (p. 229). Use your text and Exercise 10-1 in this unit for reference.

FIGURE **10.2** Skeletal muscle fiber

# Pre-Lab Exercise **10-4**

### Neuromuscular Junction

Label and color the microscopic anatomy of the neuromuscular junction depicted in **Figure 10.3** with the terms from Exercise 10-2 (p. 233). Use your text and Exercise 10-2 in this unit for reference.

FIGURE **10.3** The neuromuscular junction

# Pre-Lab Exercise **10-5**

## Types of Muscle Tissue

You learned about the three types of muscle tissue in Unit 5 (p. 106). Review the properties of each type of muscle tissue by filling in **Table 10.1**.

TABLE **10.1** The Three Types of Muscle Tissue

| Muscle Type | Striated or Nonstriated | One or Multiple Nuclei | Size and Shape of Cells | Voluntary or Involuntary | Special Features |
|---|---|---|---|---|---|
| Cardiac muscle | | | | | |
| Skeletal muscle | | | | | |
| Smooth muscle | | | | | |

# EXERCISES

You already began to explore muscle tissue in Unit 5 (p. 106), where you were introduced to the three types of muscle tissue: skeletal muscle, smooth muscle, and cardiac muscle. In this unit we delve deeper into muscle tissue's structure and function. Our primary focus will be on skeletal muscle tissue to prepare us for our examination of skeletal muscles in Unit 11. The structure of skeletal muscle cells is one of the most complicated topics you will study in this book, but it can become easier if you remember that their structure follows their function, all the way down to the tiny proteins that make up the majority of a muscle fiber.

The first two exercises in this unit explore the gross structure and the microscopic structure of skeletal muscles and the nerve cells that supply them. In the third exercise we will apply what we learned about skeletal muscle structure to skeletal muscle physiology. In the final exercise we will examine the structure of smooth and cardiac muscle tissue to help you differentiate these tissue types on microscope slides.

## Exercise **10-1**

### Skeletal Muscle Anatomy

#### MATERIALS

- ❏ Skeletal muscle models
- ❏ Skeletal muscle fiber model
- ❏ Skeletal muscle tissue slide
- ❏ Light microscope with oil-immersion objective
- ❏ Oil
- ❏ Colored pencils

A skeletal muscle is composed of skeletal muscle cells, also called **muscle fibers,** arranged into groups called **fascicles (Figure 10.4).** The muscle as a whole is covered by a connective tissue sheath called the **epimysium,** which blends with the thick, superficial fascia that binds muscles into groups. The epimysium also blends with the fibers of tendons and aponeuroses, which connect the muscle to bones or soft tissue. Each fascicle is surrounded by another connective tissue sheath called the **perimysium.** Individual skeletal muscle fibers are surrounded by their extracellular matrix, known as the **endomysium.**

Skeletal muscle fibers are long, cylindrical cells wrapped by their plasma membrane, known as the **sarcolemma (Figures 10.5** and **10.6).** About 80% of the skeletal muscle fiber's cytoplasm, also called its **sarcoplasm,** is filled with small cylindrical organelles called **myofibrils.** The remainder of the sarcoplasm contains abundant mitochondria, multiple nuclei, and a modified endoplasmic reticulum called the **sarcoplasmic reticulum** (SR) that wraps around the myofibrils.

**10**

FIGURE **10.4** (**A**) Basic skeletal muscle structure; (**B**) dissected muscle

As we will discuss shortly, the SR stores calcium ions required for muscle to contract. Note that at certain points along the myofibril, the SR swells to form **terminal cisternae**. Running down the middle of each terminal cisterna is an inward extension of the sarcolemma known as a **T-tubule**. A group of two terminal cisternae and a T-tubule is called a **triad**.

Myofibrils are composed of protein subunits called **myofilaments**. The two types of myofilaments involved in contraction are: (a) **thick filaments**, composed of the contractile protein **myosin**, and (b) **thin filaments**, composed of the contractile protein **actin** and the regulatory proteins **troponin** and **tropomyosin**. The arrangement of myofilaments within

FIGURE **10.5** Skeletal muscle fiber

FIGURE **10.6** Skeletal muscle tissue, photomicrograph **(A)** 40× objective; **(B)** 100× objective (oil immersion)

the myofibrils is what gives skeletal muscle its characteristic **striated** appearance. The dark regions of the striations, called **A bands**, are dark because this is where the thick and thin filaments overlap. The light regions, called **I bands**, appear light because they contain only thin filaments.

On closer inspection, there are more lines and bands than simply the A and I bands. For example, bisecting the A band is a lighter region called the **H zone** (which itself is bisected by a line called the **M line** that contains proteins that stabilize and support the thick filaments). Also, bisecting the I band is a dark line called the **Z disk**. The terms of this alphabet soup are used to describe the fundamental unit of contraction: the **sarcomere**. A sarcomere, defined as the space from one Z disk to the next Z disk, consists of a full A band and two half-I bands.

## Procedure  Model Inventory for Skeletal Muscle and Skeletal Muscle Fibers

Identify the following structures of skeletal muscle and skeletal muscle fibers on models and diagrams, using your textbook and this unit for reference. As you examine the anatomical models and diagrams, record on the model inventory in **Table 10.2** the name of the model and the structures you were able to identify.

### Skeletal Muscle Anatomy

1. Connective tissue coverings
   a. Epimysium
   b. Perimysium
2. Fascicle
3. Muscle fiber (cell)
4. Tendon
5. Aponeurosis

### Muscle Fiber Microanatomy

1. Sarcolemma
2. Sarcomere
   a. A band
   b. I band
   c. H zone
   d. Z disk
   e. M line
3. Transverse tubules
4. Sarcoplasmic reticulum
5. Terminal cisternae
6. Triad
7. Endomysium

TABLE **10.2** Model Inventory for Skeletal Muscle Anatomy and Microanatomy

| Model/Diagram | Structures Identified |
|---|---|
|  |  |
|  |  |
|  |  |

## **Procedure** Microscopy of Skeletal Muscle Tissue

View a prepared slide of skeletal muscle tissue.

**1** First examine the slide with the regular high-power objective of your light microscope.

**2** Draw and color what you see, and label your drawing with the terms indicated.

**3** Then switch to an oil-immersion lens (your instructor may have one set up as a demonstration), and identify the structures of the sarcomere.

### **High-Power (40×) Objective**

1.  Striations
    a.  A band
    b.  I band
2.  Sarcolemma
3.  Nuclei
4.  Endomysium

### **Oil-Immersion Lens**

1.  Sarcomere
2.  A band
3.  I band
4.  Z disk
5.  H zone
6.  M line

10

# Exercise **10-2**

## The Neuromuscular Junction

### MATERIALS

- ❑ Neuromuscular junction models
- ❑ Neuromuscular junction slide
- ❑ Light microscope
- ❑ Colored pencils

Skeletal muscle fibers are voluntary, which means that they contract under the conscious control of the nervous system. Each skeletal muscle fiber has a **motor neuron** that triggers it to contract. The place where the nerve meets the muscle fiber is called the **neuromuscular junction** (NMJ) (**Figure 10.7**). The neuromuscular junction consists of three parts:

1. **Axon terminal.** Each motor neuron splits into many swollen axon terminals, each of which contacts one muscle fiber. Each axon terminal has **synaptic vesicles** in its cytosol that contain the neurotransmitter **acetylcholine**.

2. **Synaptic cleft.** The axon terminal doesn't come into direct contact with the muscle fiber; instead, there is a space between the axon terminal and the muscle fiber called the synaptic cleft.

3. **Motor end plate.** The motor end plate is a specialized region of the sarcolemma that contains **acetylcholine receptors**.

In the next exercise we will examine the physiology of the NMJ and its role in muscle contraction.

FIGURE **10.7** The neuromuscular junction

## **Procedure** Model Inventory for the Neuromuscular Junction

Identify the following structures of the neuromuscular junction on models and diagrams, using your textbook and this unit for reference. As you examine the anatomical models and diagrams, record on the model inventory in **Table 10.2** the name of the model and the structures you were able to identify (note that this is the model inventory in Exercise 10-1).

- ▌ Motor neuron
- ▌ Axon terminal
- ▌ Synaptic vesicles
- ▌ Synaptic cleft
- ▌ Motor end plate
- ▌ Acetylcholine receptors

## Procedure Microscopy of the Neuromuscular Junction

Obtain a prepared slide of a neuromuscular junction. Use your colored pencils to draw what you see, and label your drawing with the following terms.

1. Motor neuron
2. Axon terminal
3. Skeletal muscle fiber
4. Striations

# Exercise 10-3

## Muscle Physiology

### MATERIALS

- ❏ Hand dynamometer
- ❏ Model/diagram of the contraction cycle
- ❏ Colored pencils

Skeletal muscle contraction begins when a skeletal muscle fiber is stimulated by a motor neuron, in a relationship known as **excitation-contraction coupling**. Notice in **Figure 10.8** what happens when a nerve impulse (**neuronal action potential**) reaches the axon terminal: Acetylcholine is released into the synaptic cleft by exocytosis of the synaptic vesicles. Acetylcholine then binds and opens cation channels in the motor end plate, which causes sodium ions to rush into the sarcoplasm. As the cell accumulates positive charges, the electrical difference across the sarcolemma becomes more positive, a change called **depolarization**. This initiates a series of electrical events across the sarcolemma of the muscle fiber, collectively called a **muscle action potential**.

The depolarization that begins at the motor end plate propagates through the sarcolemma and dives into the interior of the muscle fiber along the T-tubules. As the action potential spreads along the T-tubules, calcium ion channels open in the adjacent terminal cisternae, and calcium ions flood the sarcoplasm. Before calcium ions enter the sarcoplasm, actin and myosin are prevented from interacting by the regulatory protein tropomyosin. When calcium ions flood the sarcoplasm, they bind to the regulatory protein troponin, which causes the tropomyosin to shift its position away from the actin-myosin binding site. This allows a contraction to begin.

A skeletal muscle fiber contracts by a mechanism known as the **sliding filament mechanism**, in which the thick and thin filaments slide past one another during a contraction. The key event of this mechanism is the **crossbridge cycle**, during which actin and myosin bind and detach repeatedly from one another. The steps of a crossbridge cycle are as follows (**Figure 10.9**):

1. ATP is hydrolyzed, and this energy "cocks" the myosin molecule at its hinge into an upright position. The ADP and P$_i$ remain associated with the myosin head.

2. Actin binds to the myosin head, forming a **crossbridge**.

3. ADP and P$_i$ dissociate from the myosin head, and the myosin molecule moves forward into its relaxed position. As it moves, it pulls the actin strand with it in a motion known as a **power stroke**.

4. A new ATP molecule binds to myosin, which causes actin and myosin to dissociate. ATP is hydrolyzed again, the myosin molecule is recocked, and the cycle repeats.

This cycle will continue as long as acetylcholine remains in the synaptic cleft (provided that calcium ions and ATP remain available). Therefore, to terminate a contraction, the body must degrade the acetylcholine with an enzyme called **acetylcholinesterase**. When stimulation ceases, the calcium ions are pumped back into the SR, and troponin and tropomyosin move back into their original positions. At this point, actin and myosin can no longer bind, and the muscle fiber returns to its resting length.

**1.** A neuronal action potential travels through the motor neuron to the axon terminals.

**2.** The action potential triggers release of acetylcholine from the synaptic vesicles by exocytosis.

**3.** Acetylcholine binds acetylcholine receptors in the motor end plate and triggers a muscle action potential.

Action potential

FIGURE **10.8** Events at the neuromuscular junction

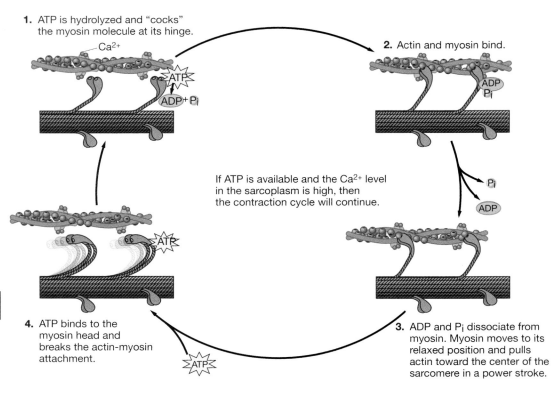

1. ATP is hydrolyzed and "cocks" the myosin molecule at its hinge.

Ca²⁺

ATP

ADP+Pᵢ

2. Actin and myosin bind.

ADP
Pᵢ

If ATP is available and the Ca²⁺ level in the sarcoplasm is high, then the contraction cycle will continue.

Pᵢ

ADP

ATP

4. ATP binds to the myosin head and breaks the actin-myosin attachment.

ATP

3. ADP and Pᵢ dissociate from myosin. Myosin moves to its relaxed position and pulls actin toward the center of the sarcomere in a power stroke.

FIGURE **10.9** Single crossbridge cycle showing the sliding filament mechanism of contraction

10

## **Procedure** Time to Trace

Now let's trace the entire series of events from the time a neuronal action potential reaches the axon terminal to the end of a muscle contraction when tropomyosin returns to the position where it blocks actin's binding site. As you trace, you will include the events that occur at the neuromuscular junction, the electrical changes that occur across the sarcolemma, the events of the crossbridge cycle, and the events that terminate the muscle contraction. Hints are included to ensure that you stay on the right track as you trace. When you have completed the activity, answer Check Your Understanding questions 1–4 (p. 247).

**Start:** neuronal action potential reaches the axon terminal →

_____ is released from synaptic vesicles by _____ →

_____ diffuses across the _____ →

acetylcholine binds to receptors on the motor end plate →

_____ channels open in the _____ →

_____ diffuses into the _____ and begins an _____ →

the _____ propagates along the _____ and _____ →

_____ ions are released from the _____ →

calcium ions bind to troponin → troponin pulls _____ away from _____ →

_____ is hydrolyzed, which _____; actin binds _____ →

_____ dissociate from _____ and it undergoes a _____ →

_____ is hydrolyzed, which breaks the _____ attachment and recocks the _____ →

crossbridge cycle repeats → nerve stimulation ceases and _____ in the _____ is degraded →

_____ ions are pumped back into the _____ →

**tropomyosin moves back into its position blocking actin's binding site. End**

## **Procedure** Diagramming Your Tracing

Take the pathway that you just traced, and turn it into a diagram that contains the neuromuscular junction, a skeletal muscle fiber, and the thick and thin filaments in a myofibril. Your diagram does not have to be a work of art. A simple schematic representation of each structure is all that is necessary. You may use the figures in this unit for reference, but it is best to compose and organize your diagram in a way that makes sense to you so you may return to it for study purposes.

## **Procedure** Measuring the Length-Tension Relationship

The amount of tension that can be generated during a contraction is influenced by how much the thick and thin filaments overlap when the muscle fiber is at rest, a relationship known as the **length-tension relationship** (**Figure 10.10**). The overlap of the thick and thin filaments is determined by the resting length of the sarcomeres. When the muscle is stretched, the sarcomeres are longer and have less overlap. When the muscle is shortened, the sarcomeres are shorter and have more overlap.

Notice in **Figure 10.10** that the muscle fiber can generate the maximum amount of tension when its sarcomeres are at about 100%–120% of their natural length. Conversely, the amount of tension generated during a contraction diminishes greatly both when the muscle is overly stretched and when it is excessively shortened.

In this experiment you will be measuring the ability of your digital flexor muscles to generate tension at different resting muscle lengths. You will take your measurements with an instrument called a **hand dynamometer** that measures the force you apply when it is squeezed. When you have finished the experiment, answer Check Your Understanding question 5 (p. 248).

FIGURE **10.10** The length-tension relationship

**1** Position your arm and wrist as shown in **Figure 10.11A**, with your wrist flexed at a 90° angle.

**2** Grip the dynamometer, and squeeze it with your maximum effort. Record in **Table 10.3** the amount of force you generated.

**3** Position your arm and wrist as shown in **Figure 10.11B**, with your wrist extended at a 0° angle.

**4** Grip the dynamometer, and squeeze it with your maximum effort. Record in **Table 10.3** the amount of force you generated.

**5** Position your arm and wrist as shown in **Figure 10.11C**, with your wrist hyperextended at a −90° angle.

**6** Grip the dynamometer, and squeeze it with your maximum effort. Record in **Table 10.3** the amount of force you generated.

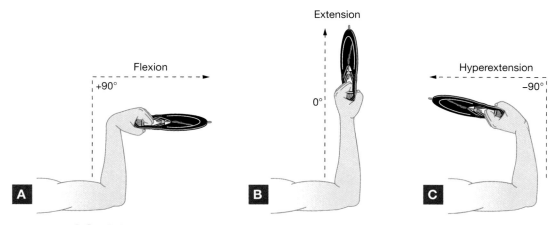

FIGURE **10.11** Wrist positions for the dynamometer experiment: (**A**) flexed at 90°; (**B**) extended at 0°; (**C**) hyperextended at −90°

TABLE **10.3** Results of Dynamometer Experiment

| Wrist Position | Amount of Force Generated |
|---|---|
| Flexed at 90° | |
| Extended at 0° | |
| Hyperextended at −90° | |

**7** Interpret your results:

    **a** In which wrist position were the muscles' sarcomeres overly stretched?

    **b** In which wrist position were the muscles' sarcomeres excessively shortened?

    **c** Which wrist position yielded the most forceful contraction? Why?

# Exercise 10-4

## Smooth and Cardiac Muscle Tissues

### MATERIALS
❏ Smooth muscle slide
❏ Cardiac muscle slide
❏ Light microscope
❏ Colored pencils

Although smooth and cardiac muscle tissues differ from skeletal muscle tissue in many ways, all three share the same basic function: to generate tension. This functional similarity brings with it structural similarities, as well. Cardiac muscle tissue is structurally the most similar to skeletal muscle tissue. As you can see in **Figure 10.12A**, cardiac muscle tissue is also striated and consists of alternating A bands and I bands arranged into sarcomeres.

The two tissue types have notable differences, though. For one, the cells of cardiac muscle tissue, called **cardiac myocytes**, are shorter than skeletal muscle fibers, branching, and typically have only a single nucleus. In addition, adjacent cardiac myocytes are joined by specialized structures called **intercalated disks**, which contain gap junctions and desmosomes. The intercalated disks connect all cardiac myocytes physically and electrically and allow the heart to contract as a unit. Finally, cardiac myocytes are involuntary and **autorhythmic**: They depolarize and contract without stimulation from the nervous system because of specialized cardiac myocytes called **pacemaker cells.**

Smooth muscle tissue, shown in **Figure 10.12B**, is not striated and, therefore, has no A bands, I bands, or sarcomeres. Smooth muscle cells tend to be long, thin, and spindle-shaped with a single, centrally located nucleus. All smooth muscle cells are involuntary, and most require extrinsic innervation to contract. Smooth muscle is found in many places in the body, including the lining of all hollow organs, the arrector pili muscles in the dermis, and the iris and ciliary muscle of the eye.

Intercalated disks

Nuclei

Cardiac myocytes

Nuclei of smooth muscle fibers

Smooth muscle tissue

FIGURE **10.12** Other types of muscle tissue: (**A**) cardiac muscle tissue; (**B**) smooth muscle tissue

## How to Distinguish Between Types of Muscle Tissue

With a bit of practice, distinguishing between the three types of muscle tissue on microscope slides becomes fairly easy. Following are some hints to help you differentiate these three tissues:

**Striations**  If striations are present, the muscle is either skeletal or cardiac. If there are no striations, it is smooth muscle.

**Shape of the cells**  Skeletal muscle fibers are long and thin, whereas cardiac myocytes are short, wide, and branching. Smooth muscle cells are thin and flat. My past students have remarked that smooth muscle cells somewhat resemble stratified squamous epithelium.

**Location of the nucleus**  Each skeletal muscle fiber has multiple nuclei located next to the sarcolemma. Cardiac myocytes usually have a single, large nucleus located among the myofibrils. The single, flattened nucleus of a smooth muscle cell is usually in the middle of the cell.

**Other features**

ⓘ  The intercalated disks of cardiac muscle tissue appear as black lines oriented parallel to the striations. No other tissue type has a similar feature, so intercalated disks are an immediate giveaway that you are looking at cardiac muscle tissue.

ⓘ  Smooth muscle tissue is typically found on a slide with other types of tissue because it lines hollow organs. Scroll through the slide to see if you can find any additional tissue types, such as epithelial and/or connective tissue. If these other tissue types are on top of or below the muscle cells, there is a good chance you are examining smooth muscle tissue.

10

## **Procedure** Microscopy of Cardiac and Smooth Muscle Tissue

**1** Obtain prepared slides of cardiac and smooth muscle tissue.

**2** Use your colored pencils to draw what you see, and label your drawing with the following terms.

After you have completed the activity, answer Check Your Understanding questions 6 and 7 (p. 248).

**1** Cardiac muscle
   a. Striations
     (1) A band
     (2) I band
   b. Sarcolemma
   c. Nucleus
   d. Intercalated disk
   e. Cardiac myocyte

**10**

**2** Smooth Muscle
   a. Nucleus
   b. Smooth muscle cell
   c. Blood vessels (or other tissues)

Name _____

Section _____ Date _____

**UNIT 10 REVIEW**

## Check Your Recall

**1** Label the following terms on **Figure 10.13.**

- ☐ Epimysium
- ☐ Fascicle
- ☐ Muscle fiber
- ☐ Perimysium
- ☐ Tendon

10

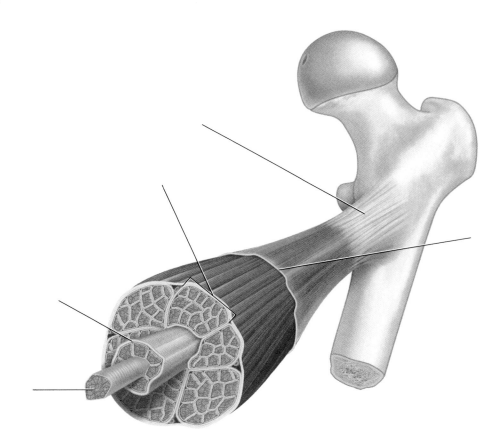

FIGURE **10.13** Basic skeletal muscle structure

**2** Label the following terms on **Figure 10.14.**

☐ A band                    ☐ Nucleus                    ☐ Sarcoplasmic reticulum
☐ I band                    ☐ Sarcolemma                 ☐ T-tubule
☐ Myofibril                 ☐ Sarcomere                  ☐ Z disk

FIGURE **10.14** Skeletal muscle fiber

**3** The striations in skeletal muscle fibers are attributable to
   a. light and dark pigments found in the sarcoplasm.
   b. the arrangement of thick and thin filaments in the myofibril.
   c. overlapping Z disks.
   d. overlapping adjacent skeletal muscle fibers.

**4** The sarcoplasmic reticulum
   a. stores sodium ions.
   b. forms from inward extensions of the sarcolemma.
   c. stores calcium ions.
   d. is made of contractile proteins.

**5** *Fill in the blanks:* Thick filaments are composed of the protein _____, and thin filaments are composed of the proteins _____, _____, and _____.

**6** Which of the following is *not* a part of the neuromuscular junction?

a. Axon terminal

b. Dendritic receptor

c. Synaptic cleft

d. Motor end plate

**7** Excitation-contraction coupling refers to

a. the muscle fiber's depolarization.

b. the relationship between length of the sarcomere and the tension generated.

c. the sliding of thick and thin filaments past one another.

d. the relationship between excitation at the sarcolemma and release of calcium ions from the sarcoplasmic reticulum.

**8** Number the following events of skeletal muscle contraction in the proper order, with number 1 for the first event and number 9 for the last event.

_____ Actin binds to the myosin head, and a power stroke occurs as ADP and $P_i$ dissociate from myosin.

_____ The terminal cisternae of the sarcoplasmic reticulum release calcium ions into the sarcoplasm because of the action potential.

_____ The sarcoplasmic reticulum removes calcium ions from the sarcoplasm, and tropomyosin returns to its original position.

_____ The energy from ATP hydrolysis "cocks" the myosin crossbridge into an upright position.

_____ Acetylcholinesterase degrades acetylcholine in the synaptic cleft.

_____ Calcium ions bind to troponin, which pulls tropomyosin away from actin.

_____ Acetylcholine binds to receptors on the motor end plate and opens cation channels.

_____ ATP binds and causes the actin and myosin to dissociate from one another, after which it is hydrolyzed and myosin is recocked. The cycle repeats.

_____ Sodium ions flood the sarcoplasm and initiate an action potential that spreads along the sarcolemma and into the muscle fiber along the T-tubules.

**9** Maximal tension in a skeletal muscle contraction is generated

a. when the sarcomeres are at about 100–120% of their natural length.

b. when the muscle fiber is stretched before the contraction.

c. when the muscle fiber is shortened before the contraction, and there is maximal overlap of the filaments.

d. when the sarcoplasmic reticulum releases fewer calcium ions.

**10** Mark the following statements as applying to *skeletal* muscle tissue, *cardiac* muscle tissue, *smooth* muscle tissue, or any combination of the three.

_____ The primary function is to generate tension.

_____ The tissue lines hollow organs.

_____ The cells are uninucleate.

_____ The cells are involuntary.

_____ The cells are autorhythmic.

_____ The cells must be stimulated by a neuron to contract.

_____ The cells are striated in appearance.

_____ The cells lack striations and sarcomeres.

_____ The cells are multinucleate.

# Check Your Understanding

### Critical Thinking and Application Questions

**1** Would a muscle cell be able to produce a functional contraction if it lacked T-tubules? Explain.

_____

_____

_____

_____

_____

**2** The drug Botox® is made from the toxin of the bacterium *Clostridium botulinum*. It is injected subdermally to minimize fine lines and creases in the face. The toxin prevents motor neurons from releasing acetylcholine. How would this produce the desired cosmetic results? What could happen if this toxin were absorbed systemically?

_____

_____

_____

_____

_____

**3** *Myasthenia gravis* is an autoimmune disease in which the immune system produces antibodies that attack the acetylcholine receptor on the motor end plate. What symptoms would you expect with this disease? Explain.

_____

_____

_____

_____

_____

**4** The condition *rigor mortis* develops several hours after death because of a lack of ATP. It is characterized by muscular rigidity. Explain why a lack of ATP would cause the sustained muscle contraction of rigor mortis.

_____

_____

_____

_____

**5** You are at the gym with a friend who complains that she cannot lift as much weight on a new machine as she could on the old machine. You examine the machine and note that the starting position for the exercise puts your friend's muscles in a stretched position. How does this explain her seeming loss of strength?

_____

_____

_____

_____

_____

**6** How does the structure of cardiac myocytes and intercalated disks follow the function of cardiac muscle tissue and the heart?

_____

_____

_____

_____

**7** When you are hungry, the muscle cells lining your intestines and stomach contract, producing characteristic "stomach growling." Would you expect these muscle cells to be striated or smooth? Explain your reasoning.

_____

_____

_____

_____

# Muscular System

**11**

## OBJECTIVES

*Once you have completed this unit, you should be able to:*

- Identify muscles of the upper and lower limbs, trunk, head, and neck.

- Describe the origin, insertion, and action of selected muscles.

- List the muscles required to perform common movements.

# PRE-LAB EXERCISES

Complete the following exercises prior to coming to lab, using your textbook and lab manual for reference.

## Pre-Lab Exercise 11-1

### ✎ Key Terms

You should be familiar with the following terms before coming to lab. For the skeletal muscles, describe the muscles' location and appearance. Please note that this list is not all-inclusive, as the terminology for the skeletal muscles is extensive.

| Term | Definition |
|------|------------|
| **General Terms** | |
| Origin | |
| Insertion | |
| Muscle action | |
| **Skeletal Muscles of the Head, Neck, and Thorax** | |
| Muscles of facial expression | |
| Muscles of mastication | |
| Sternocleidomastoid | |
| Trapezius | |
| Erector spinae | |
| Intercostal muscles | |
| Diaphragm | |
| Rectus abdominis | |
| **Skeletal Muscles That Move the Upper Limb** | |
| Deltoid | |

**11**

Latissimus dorsi _____

Pectoralis major _____

Rotator cuff muscles _____

Biceps brachii _____

Triceps brachii _____

Brachioradialis _____

Wrist/digit flexors _____

Wrist/digit extensors _____

**Skeletal Muscles That Move the Lower Limb**

Iliopsoas _____

Gluteus muscles _____

Sartorius _____

Quadriceps femoris group _____

Adductor group _____

Hamstrings group _____

Gastrocnemius _____

Soleus _____

Ankle/foot flexors _____

Ankle/foot extensors _____

Skeletal Muscle Anatomy

Label and color the muscles depicted in **Figure 11.1** with the terms from Exercise 11-1 (p. 257). Use your text and Exercise 11-1 in this unit for reference.

A

FIGURE **11.1** Human musculature: (**A**) anterior view *(continues)*

11

B

FIGURE **11.1** **Human musculature** *(cont.)*: **(B) posterior view** *(continues)*

FIGURE **11.1** Human musculature *(cont.)*: **(C)** lateral view of the face; **(D)** abdominal muscles *(continues)*

**E**

FIGURE **11.1** Human musculature *(cont.)*: **(E) posterior torso**

# Pre-Lab Exercise **11-3**

## Muscle Origins, Insertions, and Actions

Complete **Table 11.1** with the origins, insertions, and *main* actions for the listed muscles. You may
notice in your textbook that the origin and insertion of a muscle often are extensive. For example,
the latissimus dorsi muscle originates from T7–L5, the lower three or four ribs, the inferior angle of the scapula, and the
thoracolumbar fascia. This obviously is quite a bit to write and to remember, so I suggest shortening it to something that
makes sense to you (e.g., "lower back and scapula"). You also will notice that most muscles have more than one action,
which you may wish to simplify as well. If your instructor prefers that you learn the more technical version, learning a
simplified version first may be helpful.

TABLE **11.1** Muscle Origins, Insertions, and Actions

| Muscle | Origin | Insertion | Action(s) |
|---|---|---|---|
| **Muscles of the Head and Neck** | | | |
| Sternocleidomastoid | | | |
| Trapezius | | | |
| **Muscles of the Thorax, Abdomen, and Back** | | | |
| Rectus abdominis | | | |
| External oblique | | | |
| Internal oblique | | | |
| Erector spinae | | | |
| Deltoid | | | |
| Pectoralis major | | | |
| Latissimus dorsi | | | |
| **Muscles of the Upper Limb** | | | |
| Biceps brachii | | | |
| Triceps brachii | | | |
| Brachioradialis | | | |
| **Muscles of the Lower Limb** | | | |
| Iliopsoas | | | |
| Gluteus maximus | | | |
| Sartorius | | | |
| Rectus femoris | | | |
| Adductor group | | | |
| Biceps femoris | | | |
| Semitendinosus | | | |
| Semimembranosus | | | |
| Gastrocnemius | | | |
| Tibialis anterior | | | |

11

# EXERCISES

The human body has nearly 700 **skeletal muscles** that range dramatically in size and shape from the large trapezius muscle to the tiny corrugator supercilii muscle. Luckily for you, we will be learning only about 70 muscles rather than the full 700.

A muscle begins at its **origin**, generally the more stationary part, and attaches to its **insertion**, generally the part that the muscle moves. We are most familiar with muscles that insert into and move bones. Many muscles, however, insert into structures other than bones. Examples include the diaphragm, which inserts into its own central tendon, and the muscles of facial expression, which insert into skin or other muscles.

The exercises in this unit help you become familiar with the main muscle groups and their component muscles. The first exercise introduces you to muscle anatomy, the second exercise focuses on muscle origins and insertions, and the third and final exercise explores muscle actions.

## Exercise **11-1**

### Skeletal Muscles

---

**MATERIALS**

☐ Muscle models: upper limb, lower limb, trunk, head, neck

---

In this exercise, we divide skeletal muscles into muscle groups that have similar functions. For example, we classify the latissimus dorsi muscle with the muscles of the upper limb, rather than the muscles of the back or trunk, because it moves the upper limb. We will use the following groupings of skeletal muscles in this unit:

1. **Muscles that move the head, neck, and face.** We can subdivide the muscles that move the head, neck, and face into the muscles of **facial expression**, the muscles of **mastication**, and the muscles that move the head. The muscles of facial expression control the various facial expressions humans are capable of making, and they all insert into skin or other muscles. Examples include the **orbicularis oculi**, which closes and squints the eye, the **orbicularis oris**, which purses the lips, and the **zygomaticus major** and **minor**, which pull the corners of the mouth laterally during smiling. The other muscles of facial expression are illustrated in **Figure 11.2**. The muscles of mastication are involved in chewing. They include the **masseter**, a thick muscle over the lateral jaw, and the fan-shaped **temporalis muscle**, which rests over the lateral skull.

Several muscles move the head, including the **trapezius**, which holds the head upright and hyperextends the neck (note that it also elevates the shoulders), and the **sternocleido-mastoid**, a strap-like muscle in the neck that flexes the head and neck, rotates the neck laterally, and flexes the neck laterally. In addition, two muscles called the **splenius capitis** and **splenius cervicis** extend and hyperextend the head (**Figure 11.3**).

2. **Muscles that move the trunk.** The muscles that move the trunk are the muscles of the thorax, muscles of the abdominal wall, and postural muscles of the back (**Figures 11.4, 11.5 and 11.6**).

   a. The muscles of the thorax are generally involved in the muscle movements that produce ventilation. The small **pectoralis minor**

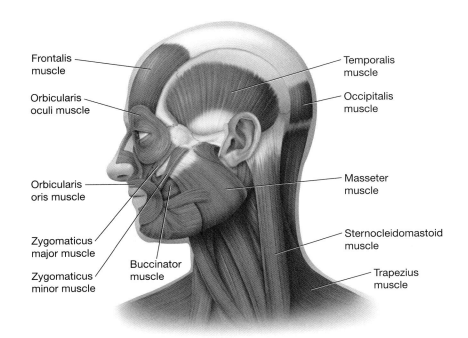

FIGURE **11.2** Facial musculature, lateral view

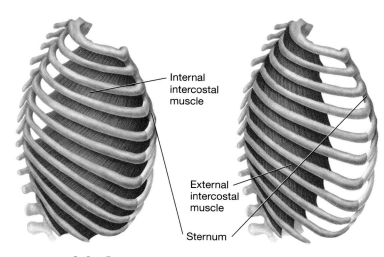

FIGURE **11.3** Splenius muscles of the neck

Labels for figure 11.3:
- Mastoid process
- Splenius capitis muscle
- Spinous processes of the vertebrae
- Splenius cervicis muscle

FIGURE **11.4** Internal and external intercostal muscles, lateral view

Labels for figure 11.4:
- Internal intercostal muscle
- External intercostal muscle
- Sternum

**A**

Labels:
- Pectoralis major muscle
- Serratus anterior muscle
- External oblique muscle
- Rectus abdominis muscle
- Transversus abdominis muscle
- Internal oblique muscle

FIGURE **11.5** Muscles of the trunk: (**A**) anterior view of the abdomen; (**B**) lateral view of the individual abdominal muscles

**B**

Superficial labels:
- External oblique muscle
- Iliac crest

Intermediate labels:
- Xiphoid process
- Rectus abdominis muscle
- Lumbar fascia muscle
- Internal oblique muscle
- Pubic tubercle

Deep labels:
- Lumbar fascia muscle
- Transversus abdominis muscle
- Inguinal ligament

**Superficial**     **Intermediate**     **Deep**

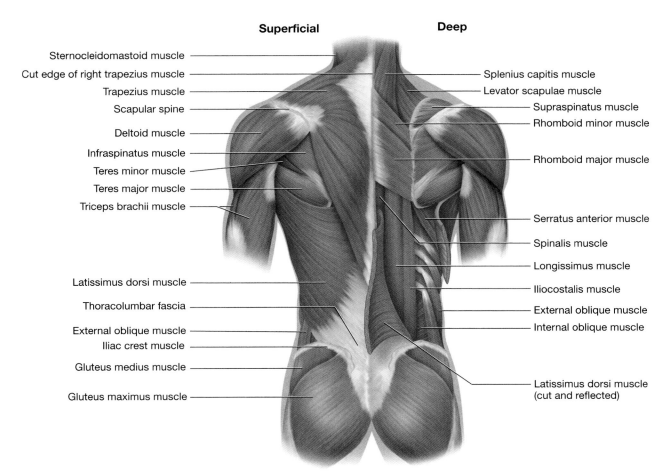

**Superficial**      **Deep**

Sternocleidomastoid muscle

Cut edge of right trapezius muscle

Trapezius muscle

Scapular spine

Deltoid muscle

Infraspinatus muscle

Teres minor muscle

Teres major muscle

Triceps brachii muscle

Latissimus dorsi muscle

Thoracolumbar fascia

External oblique muscle

Iliac crest muscle

Gluteus medius muscle

Gluteus maximus muscle

Splenius capitis muscle

Levator scapulae muscle

Supraspinatus muscle

Rhomboid minor muscle

Rhomboid major muscle

Serratus anterior muscle

Spinalis muscle

Longissimus muscle

Iliocostalis muscle

External oblique muscle

Internal oblique muscle

Latissimus dorsi muscle
(cut and reflected)

FIGURE **11.6** Muscles of the trunk, posterior view

draws the scapula anteriorly and the rib cage superiorly during forced inspiration and expiration, the **internal** and **external intercostal** muscles are located between the ribs and are involved in both quiet and forced inspiration and expiration, and the circular **diaphragm** is the main muscle for inspiration. Note that the diaphragm and pectoralis minor are not visible in **Figure 11.5**.

b. The abdominal muscles move the vertebral column and increase intra-abdominal pressure. The **rectus abdominis** is the central and superficial muscle that flexes the vertebral column, the **internal** and **external obliques** are located laterally and rotate the vertebral column, and the deep **transversus abdominis** compresses the abdominal contents to increase intra-abdominal pressure.

c. The three postural muscles of the back are all part of the long muscle known as the **erector spinae**. The three components of the erector spinae include the lateral **iliocostalis**, the middle **longissimus**, and the medial **spinalis**.

3. **Muscles that move the shoulder**. Only three muscles are "prime movers" of the shoulder: the **deltoid**, or "shoulder muscle," the prime abductor of the shoulder; the **pectoralis major**, or "chest muscle," the prime muscle of shoulder flexion; and the posterior **latissimus dorsi**, the prime muscle of shoulder extension. The **serratus anterior**, located deep to the pectoralis major, assists in shoulder abduction and raising the shoulder. The other shoulder muscles, the **rotator cuff muscles**, are the **infraspinatus**, the **subscapularis**, the **supraspinatus**, and the **teres minor**. The tendons of these muscles unite with the fibrous capsule of the shoulder joint, and their main function is to reinforce this capsule to prevent dislocation of the humerus from the scapula.

4. **Muscles that move the forearm and wrist**. The four main muscles that move the forearm are the anterior **biceps brachii**, the anterior and deep **brachialis**, the anterior and lateral **brachioradialis**, and the posterior **triceps brachii** (**Figures 11.7** and **11.8**). The first three muscles play a role in forearm flexion, and the triceps brachii extends the forearm. The remainder of the muscles of the upper limb act on the wrist, the hand, and the digits. Muscles that flex the wrist and the digits, located on the anterior forearm, are the **flexor carpi radialis**, **flexor carpi ulnaris**,

FIGURE **11.7** Anterior view of the muscles of the right upper limb

Labels (Figure 11.7):
Clavicle
Deltoid muscle
Sternum
Pectoralis major muscle
Biceps brachii muscle
Coracobrachialis muscle
Triceps brachii muscle:
  Lateral head
  Long head
  Medial head
Brachialis muscle
Pronator teres muscle
Tendon of biceps brachii muscle
Brachioradialis muscle
Extensor carpi radialis longus muscle
Flexor pollicis longus muscle
Pronator quadratus muscle
Flexor retinaculum muscle
Medial epicondyle of humerus
Flexor carpi radialis muscle
Palmaris longus muscle
Flexor carpi ulnaris muscle
Flexor digitorum superficialis muscle
Palmar aponeurosis

FIGURE **11.8** Posterior view of the muscles of the right upper limb

Labels (Figure 11.8):
Supraspinatus muscle
Spine of scapula
Deltoid muscle (cut)
Greater tubercle of humerus
Infraspinatus muscle
Teres minor muscle
Teres major muscle
Triceps brachii muscle:
  Lateral head
  Long head
Latissimus dorsi muscle
Insertion of triceps brachii muscle
Brachialis muscle
Humerus
Brachioradialis muscle
Extensor carpi radialis longus muscle
Olecranon process of ulna
Anconeus muscle
Extensor carpi radialis brevis muscle
Extensor digitorum muscle
Abductor pollicis longus muscle
Extensor pollicis brevis muscle
Extensor pollicis longus muscle
Tendons of extensor digitorum muscle
Flexor carpi ulnaris muscle
Extensor carpi ulnaris muscle
Extensor digiti minimi muscle
Extensor indicis muscle

flexor digitorum superficialis, and flexor digitorum profundus muscles (Figure 11.9). Muscles that extend the wrist and the digits, located on the posterior forearm, are the extensor carpi radialis longus, extensor digitorum, and extensor carpi ulnaris (Figure 11.10).

5. **Muscles that move the hip and knee.** Muscles that move the hip joint exclusively include the deep **iliopsoas**, the anterior **pectineus**, the anterior **adductor group**, and the posterior **gluteus maximus** and **gluteus medius** (**Figures 11.11** and **11.12**). The iliopsoas is the prime flexor of the hip joint and is assisted by the pectineus. As implied by their names, the muscles in the adductor group are the prime adductors of the hip joint. The gluteus maximus is the prime extensor of the hip joint, and the gluteus medius is a major abductor of the hip joint. Other posterior muscles are illustrated in **Figure 11.12**. The remainder of the thigh muscles generally move both the knee and hip joints. The anterior thigh muscles include the four muscles of the **quadriceps femoris group** (**rectus femoris, vastus lateralis, vastus intermedius**, and **vastus medialis** muscles), the **sartorius**, and the medial **gracilis**. Most of these muscles have the dual function of flexion at the hip joint and extension at the knee joint. The posterior thigh muscles include the three muscles of the **hamstrings group**: the lateral **biceps femoris** and the medial **semitendinosus** and **semimembranosus** muscles. These three muscles produce both extension at the hip joint and flexion at the knee joint.

FIGURE **11.9** Muscles of the anterior right forearm: (**A**) superficial muscles; (**B**) intermediate muscles; (**C**) deep muscles

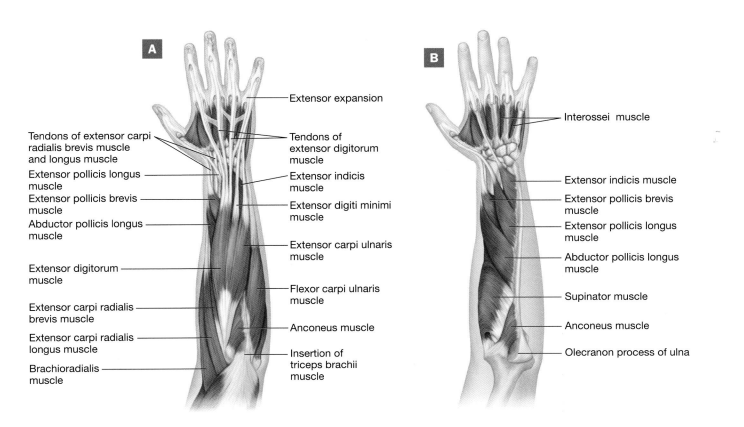

FIGURE **11.10** Muscles of the posterior right forearm: (**A**) superficial muscles; (**B**) deep muscles

**A**

12th rib

Quadratus lumborum muscle

Psoas major muscle

Iliacus muscle

Iliopsoas muscle

Tensor fasciae latae muscle

Quadriceps femoris muscles:
Rectus femoris
Vastus lateralis
Vastus medialis

Tendon of quadriceps femoris muscle

Patella

Patellar ligament

Fibularis longus muscle

Tibialis anterior muscle

Extensor digitorum longus muscle

Extensor hallucis longus muscle

Extensor digitorum brevis muscle

Fibularis tertius muscle

12th thoracic vertebra

Psoas minor muscle

Iliac crest

Anterior superior iliac spine

Pectineus muscle
Adductor magnus muscle
Gracilis muscle
Sartorius muscle
Adductor longus muscle

Gastrocnemius muscle

Tibia

Soleus muscle

Superior and inferior extensor retinacula

Extensor hallucis brevis muscle

**B**

Gluteus medius muscle

Gluteus maximus muscle

Adductor magnus muscle

Gracilis muscle

Hamstrings:
Biceps femoris muscle:
Long head
Short head

Semimembranosus muscle

Semitendinosus muscle

Gastrocnemius muscle medial head (cut)

Tibialis posterior muscle

Flexor digitorum longus muscle

Tendon of tibialis posterior muscle

Medial malleolus

Calcaneous

Iliotibial tract

Plantaris muscle (cut)

Gastrocnemius muscle lateral head (cut)

Popliteus muscle

Fibula

Soleus muscle (cut)

Flexor hallucis longus muscle

Fibularis brevis muscle

Calcaneal tendon (cut)

**FIGURE 11.11** Muscles of the right lower limb: **(A)** anterior lower limb; **(B)** posterior lower limb

6. **Muscles that move the ankle and foot.** The most obvious muscle of the posterior leg is the large **gastrocnemius** muscle, also known as the "calf muscle." This two-headed muscle originates on the distal femur and inserts into the posterior calcaneus via the **calcaneal tendon**, more commonly called the **Achilles tendon**. Deep to the gastrocnemius is the **soleus** muscle, which unites with the gastrocnemius and contributes to the calcaneal tendon. Together these two muscles are the prime muscles that produce plantarflexion at the ankle joint. The gastrocnemius also produces some flexion at the knee joint. Deep to these two muscles are muscles that plantarflex the foot and the digits, as well as the **tibialis posterior**, the main inverter of the foot. On the lateral leg we find two muscles, the **fibularis longus** and **fibularis brevis**, the main everters of the foot. Anteriorly are the extensors such as the **extensor digitorum longus**, which extends the toes, and the **tibialis anterior**, which dorsiflexes the foot and the ankle joint. The muscles of the leg are best seen in **Figures 11.11** and **11.13**.

**Figures 11.14** and **11.15** provide whole body views of many of the muscles we have just discussed.

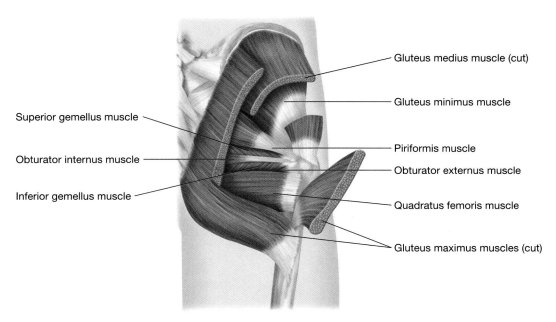

Gluteus medius muscle (cut)

Gluteus minimus muscle

Superior gemellus muscle

Piriformis muscle

Obturator internus muscle

Obturator externus muscle

Inferior gemellus muscle

Quadratus femoris muscle

Gluteus maximus muscles (cut)

FIGURE **11.12** Deeper muscles of the gluteal region

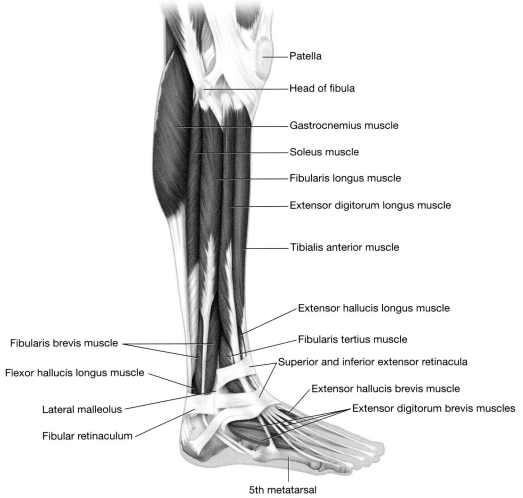

Patella

Head of fibula

Gastrocnemius muscle

Soleus muscle

Fibularis longus muscle

Extensor digitorum longus muscle

Tibialis anterior muscle

Extensor hallucis longus muscle

Fibularis brevis muscle

Fibularis tertius muscle

Superior and inferior extensor retinacula

Flexor hallucis longus muscle

Extensor hallucis brevis muscle

Extensor digitorum brevis muscles

Lateral malleolus

Fibular retinaculum

5th metatarsal

FIGURE **11.13** Lateral view of muscles of the leg

Frontalis muscle

Orbicularis oculi muscle

Orbicularis oris muscle

Sternocleidomastoid muscle

Serratus anterior muscle

External abdominal oblique muscle

Umbilicus muscle

Rectus sheath

Tensor fasciae latae muscle

Abductor pollicis brevis muscle

Abductor digiti minimi muscle

Sartorius muscle

Gracilis muscle

Tendon of rectus femoris muscle

Patella

Patellar ligament

Gastrocnemius muscle

Soleus muscle

Zygomaticus muscles

Masseter muscle

Trapezius muscle

Deltoid muscle

Pectoralis major muscle

Brachialis muscle

Biceps brachii muscle

Rectus abdominis muscle

Brachioradialis muscle

Flexor carpi radialis muscle

Palmaris longus muscle

Flexor retinaculum muscle

Pectineus muscle

Adductor longus muscle

Rectus femoris muscle

Vastus lateralis muscle

Vastus medialis muscle

Fibularis longus muscle

Tibialis anterior muscle

Extensor digitorum longus muscle

Fibularis brevis muscle

FIGURE **11.14** Muscles of the body, anterior view

Occipitalis muscle

Sternocleidomastoid muscle

Infraspinatus muscle

Teres minor muscle

Teres major muscle

Latissimus dorsi muscle

External abdominal
oblique muscle

Gluteus medius muscle

Gluteus maximus muscle

Gracilis muscle

Adductor magnus muscle

Semimembranosus muscles

Trapezius muscle

Deltoid muscle

Long head of
triceps brachii muscle

Lateral head of
triceps brachii muscle

Medial head of
triceps brachii muscle

Brachioradialis muscle

Extensor carpi muscle

Flexor carpi muscle

Extensor digitorum muscle

Extensor retinaculum muscle

Dorsal interosseus muscle

Vastus lateralis muscle

Long head of
biceps femoris muscle

Semitendinosus muscle

Short head of
biceps femoris muscle

Popliteal fossa

Medial head of
gastrocnemius muscle

Lateral head of
gastrocnemius muscle

Soleus muscles

Tendo calcaneus
(Achilles tendon)

FIGURE **11.15** Muscles of the body, posterior view

11

## Procedure Model Inventory for the Skeletal Muscles

Identify the following muscles on models and diagrams, using your textbook and this unit for reference. As you examine the anatomical models and diagrams, record on the model inventory in **Table 11.2** the name of the model and the structures you were able to identify. When you have completed the activity, answer Check Your Understanding question 1 (p. 279).

### Muscles That Move the Head, Neck, and Face

1. Muscles of facial expression
   a. Epicranius (frontalis and occipitalis)
   b. Orbicularis oculi
   c. Zygomaticus
   d. Buccinator
   e. Orbicularis oris

2. Muscles of mastication
   a. Temporalis
   b. Masseter

3. Muscles of head and neck
   a. Platysma
   b. Sternocleidomastoid
   c. Trapezius

### Muscles That Move the Trunk

1. Muscles of the thorax
   a. Pectoralis minor
   b. External intercostals
   c. Internal intercostals
   d. Diaphragm

2. Muscles of the abdominal wall
   a. Rectus abdominis
   b. External obliques
   c. Internal obliques
   d. Transversus abdominis

3. Postural muscles of the back
   a. Erector spinae
      (1) Iliocostalis
      (2) Longissimus
      (3) Spinalis

### Muscles That Move the Shoulder

1. Muscles of the shoulder
   a. Deltoid
   b. Pectoralis major
   c. Latissimus dorsi
   d. Serratus anterior

2. Rotator cuff muscles
   a. Infraspinatus
   b. Subscapularis
   c. Supraspinatus
   d. Teres minor

### Muscles That Move the Forearm and Wrist

1. Muscles of the arm
   a. Biceps brachii
   b. Triceps brachii
   c. Brachialis
   d. Brachioradialis

2. Pronator teres
3. Flexor carpi radialis
4. Flexor carpi ulnaris
5. Flexor digitorum superficialis
6. Flexor digitorum profundus

7. Extensor carpi radialis longus
8. Extensor digitorum
9. Extensor carpi ulnaris

### Muscles That Move the Hip and Knee

1. Muscles of the pelvic girdle
   a. Iliopsoas (iliacus and psoas major)
   b. Gluteus maximus
   c. Gluteus medius

2. Muscles of the thigh
   a. Sartorius
   b. Tensor fascia lata
   c. Pectineus
   d. Quadriceps femoris group
      (1) Rectus femoris
      (2) Vastus medialis
      (3) Vastus intermedius
      (4) Vastus lateralis

   e. Adductor group
      (1) Adductor magnus
      (2) Adductor longus
      (3) Adductor brevis
   f. Hamstring muscles
      (1) Biceps femoris
      (2) Semimembranosus
      (3) Semitendinosus

## Muscles That Move the Ankle and Foot

1. Gastrocnemius
2. Soleus
3. Flexor digitorum longus
4. Fibularis (peroneus) longus
5. Fibularis (peroneus) brevis
6. Tibialis posterior
7. Tibialis anterior
8. Extensor digitorum longus

Your instructor may wish to omit certain muscles included above or add muscles not included in these lists. List any additional structures below:

**11**

TABLE **11.2** Model Inventory for Skeletal Muscles

| Model/Diagram | Structures Identified |
|---|---|
| | |
| | |
| | |
| | |
| | |

## Exercise **11-2**

### Muscle Origins and Insertions

**MATERIALS**
- ❑ Small skeleton
- ❑ Modeling clay

Before you can fully understand a muscle's actions, you must understand a muscle's origin and insertion. As we discussed in the introduction to this unit, the **origin** of a muscle is the part from which it originates, which generally is the more stationary part, and the muscle typically **inserts** into the part that it moves. For example, you can see in **Figure 11.16** that the biceps brachii muscle originates on the coracoid process of the scapula and crosses the elbow joint, where it inserts into the radial tuberosity. Notice also that the triceps brachii originates from the humerus and the inferior scapula and crosses the elbow joint to insert into the olecranon process.

Once you determine a muscle's origin and insertion, figuring out its actions becomes easy. Let's examine the biceps brachii and triceps brachii muscles again. The biceps brachii inserts into the radial tuberosity, so we know it will move the forearm. Given how it crosses the anterior elbow joint, we can conclude it will cause forearm flexion at the elbow joint. The triceps brachii inserts into the olecranon process, so we know it also will move the forearm. Given how it crosses the posterior elbow joint, we can conclude that it will cause forearm extension at the elbow joint. Now wasn't that easy?

FIGURE **11.16** Origin and insertion of the biceps brachii and triceps brachii muscles, posterior view

### **Procedure** Build Muscles

In this exercise you will use small skeletons and modeling clay to build specific muscle groups. This may sound easy, but there is a catch: You must determine the actions of the muscle by looking only at the origin and insertion of each muscle you build. Use Pre-Lab Exercise 11-3 (p. 255) as a guide to each muscle's origin and insertion.

**1** Obtain a small skeleton and five colors of modeling clay.

**2** Build the indicated muscles, using a different color of clay for each muscle. As you build, pay careful attention to the origin and insertion of each muscle.

**3** Determine the primary actions for each muscle you have built by looking *only* at the origin and insertion. Record this information in **Tables 11.3–11.5**.

TABLE **11.3** Muscles and Actions for Group 1

| Muscle | Actions |
|---|---|
| Biceps femoris | |
| Semitendinosus | |
| Semimembranosus | |
| Gracilis | |
| Adductor group | |

TABLE **11.4** Muscles and Actions for Group 2

| Muscle | Actions |
|---|---|
| Biceps brachii | |
| Triceps brachii | |
| Brachioradialis | |
| Deltoid | |
| Latissimus dorsi | |

TABLE **11.5** Muscles and Actions for Group 3

| Muscle | Actions |
|---|---|
| Pectoralis major | |
| Pectoralis minor | |
| Trapezius | |
| Masseter | |
| Temporalis | |

This exercise should look familiar, because you did a similar exercise in Unit 9 (Articulations, p. 214). As before, you will be performing various activities to demonstrate which joints are moving with each activity. In this exercise, however, rather than focusing on the motions that occur at each joint, you will be determining *which muscles* are causing the motions at each joint. Following is an example:

## Walking in Place

### Part 1

First list all motions occurring at each joint (as before, the joints are listed with their common names, and the lists are certainly *not* all-inclusive):

- Vertebral column and neck: extension
- Shoulder: flexion/extension
- Elbow: flexion
- Hip: flexion/extension, abduction/adduction
- Knee: flexion/extension
- Ankle: dorsiflexion/plantarflexion, inversion/eversion

### Part 2

Now refer to **Table 11.1** in Pre-Lab Exercise 11-3 (p. 256) to determine which muscles are causing each joint to move. Please note that these lists are far from complete. With nearly 700 muscles in the human body, a complete list could take all day!

- Vertebral column and neck
  - Extension: erector spinae and trapezius
- Shoulder
  - Flexion: pectoralis major
  - Extension: latissimus dorsi
- Elbow
  - Flexion: biceps brachii, brachialis, and brachioradialis
- Hip
  - Flexion: iliopsoas, sartorius
  - Extension: biceps femoris, semitendinosus, semimembranosus, and gluteus maximus
  - Abduction: tensor fasciae latae, sartorius, and gluteus medius
  - Adduction: gracilis and adductor
- Knee
  - Flexion: biceps femoris, semitendinosus, semimembranosus, and gastrocnemius
  - Extension: rectus femoris, vastus medialis, vastus intermedius, and vastus lateralis
- Ankle
  - Dorsiflexion: tibialis anterior
  - Plantarflexion: gastrocnemius and soleus
  - Inversion: tibialis anterior and tibialis posterior
  - Eversion: fibularis brevis and fibularis longus

Okay, so now you're thinking, "Wow, that's a lot of work!" Following are a few hints that will make this task less daunting:

- You will notice that the following activities are the same as the activities from Exercise 9-3 in Unit 9 (p. 214). To save yourself some work, reference that exercise to complete Part 1 of this exercise.

▌ As before, perform the indicated activity. If you don't, you'll miss some things!

▌ Unless your instructor asks you to do otherwise, try to keep it simple. List only the main muscles for each action. For example, when listing the muscles to flex the hip, we easily could include 10 different muscles. To make it simpler, list only those muscles we discussed that flex the hip as their *primary action* (as in the example on page 270).

Ready to give it a try?

## **Procedure** Determine the Muscles Involved in Common Movements

Determine the joints being moved for Part 1, and then determine the muscles moving the joints for Part 2. Refer to Unit 9 (p. 201) for help. After you have completed the activity, answer Check Your Understanding questions 2–4 (pp. 279–280).

**1** Walking up the stairs

Part 1:

Part 2:

**2** Doing jumping jacks

Part 1:

Part 2:

**3** Answering the telephone

Part 1: _____

Part 2: _____

_____

_____

_____

_____

_____

_____

_____

_____

**4** Jumping rope

Part 1: _____

Part 2: _____

_____

_____

_____

_____

_____

_____

## Check Your Recall

**1** Label the following muscles on **Figure 11.17.**

☐ Buccinator
☐ Frontalis
☐ Masseter
☐ Occipitalis
☐ Orbicularis oculi

☐ Orbicularis oris
☐ Sternocleidomastoid
☐ Temporalis
☐ Zygomaticus major and minor

11

FIGURE **11.17** Facial musculature, lateral view

**2** Label the following muscles on Figure 11.18.

- ☐ External oblique
- ☐ Internal oblique
- ☐ Rectus abdominis
- ☐ Serratus anterior
- ☐ Transversus abdominis

FIGURE **11.18** Muscles of the trunk, anterior view

**3** Label the following muscles on **Figure 11.19.**

☐ Deltoid
☐ Erector spinae
☐ Gluteus maximus
☐ Gluteus medius
☐ Infraspinatus

☐ Latissimus dorsi
☐ Supraspinatus
☐ Trapezius
☐ Triceps brachii

**FIGURE 11.19** Muscles of the trunk, posterior view

**4** Label the following muscles on **Figure 11.20.**

- ☐ Biceps brachii
- ☐ Brachioradialis
- ☐ Gracilis
- ☐ Pectoralis major
- ☐ Rectus femoris
- ☐ Sartorius
- ☐ Tibialis anterior
- ☐ Vastus lateralis
- ☐ Vastus medialis

FIGURE **11.20** Muscles of the body, anterior view

Name _____

Section _____ Date _____

**5** Label the following muscles on **Figure 11.21.**

☐ Biceps femoris
☐ Gastrocnemius
☐ Gluteus maximus
☐ Semimembranosus
☐ Semitendinosus
☐ Soleus
☐ Triceps brachii

FIGURE **11.21** Muscles of the body, posterior view

**6** *Fill in the blanks:* A muscle's _____ is generally the more stationary part, and a muscle _____ into the part that it moves.

**7** Which of the following muscles is *not* a forearm flexor at the elbow?
a. Brachioradialis
b. Triceps brachii
c. Biceps brachii
d. Brachialis

**8** Which of the following muscles does *not* extend the hip and flex the knee?
a. Sartorius
b. Biceps femoris
c. Semimembranosus
d. Semitendinosus

**9** Which of the following muscles is the prime abductor of the arm at the shoulder joint?
a. Pectoralis major
b. Latissimus dorsi
c. Pectoralis minor
d. Deltoid

**10** Which of the following muscles is primarily a postural muscle?
a. Rectus abdominis
b. Transversus abdominis
c. Erector spinae
d. Latissimus dorsi

# Check Your Understanding

**Critical Thinking and Application Questions**

**1** Muscle strain, or "pulling a muscle," may result from overuse injuries or from trauma. Typically muscle strain results in pain around the muscle with movement and with pressure. Predict which muscle or muscles may be strained from pain in each of the following locations:

**a** Lateral thigh _____

**b** Posterior arm _____

**c** Lateral neck _____

**d** Abdomen just medial to the midline _____

**e** Anterolateral leg _____

**f** Thoracic region just medial to the vertebral column _____

**2** Neuromuscular diseases may lead to a condition called *drop foot,* in which a patient is unable to dorsiflex his or her foot. Which muscle(s) do you think is/are involved in this condition? Which muscles may have to compensate for lack of dorsiflexion during walking? Explain.

_____

_____

_____

_____

_____

**3** One of the most commonly performed orthopedic surgeries is rotator cuff repair. Often this involves removing and reattaching the deltoid muscle at its insertion. What functional impairment would be present after this procedure? What exercises could a physical therapist use to recondition the deltoid muscle?

_____

_____

_____

_____

_____

11

**4** Radical mastectomy is a type of surgery for breast cancer that involves removing all of the breast tissue, the lymph nodes, and the pectoralis major muscle. What functional impairment would be present following this surgery? Which other muscles might a physical therapist train to enable the patient to regain some of this lost function?

**5** A stroke, caused by a clot in a blood vessel of the brain, may lead to a loss of function of certain muscles. Which motions would an individual be unable to perform if the following muscles lost function?

**a** Orbicularis oculi

**b** Sternocleidomastoid

**c** Gluteus maximus

**6** The pectoralis minor normally originates on the third to fifth ribs and inserts into the coracoid process of the scapula. Predict its action from this information.

**7** When a person is in respiratory distress, the origin and insertion of the pectoralis minor actually switch, and the part of the muscle attached to the coracoid process acts as the origin. How will this change the muscle's action?

# Nervous Tissue

## OBJECTIVES

*Once you have completed this unit, you should be able to:*

■ Describe the microanatomy of nervous tissue and the parts of a synapse.

■ Identify structures of the neuron and neuroglial cells.

■ Describe the sequence of events of a synapse and a neuronal action potential.

Name _____ Section _____ Date _____

# PRE-LAB EXERCISES

Complete the following exercises prior to coming to lab, using your textbook and lab manual for reference.

## Pre-Lab Exercise **12-1**

 Key Terms

You should be familiar with the following terms before coming to lab.

| Term | Definition |
|------|------------|
| **Parts of the Neuron** | |
| Neuron | |
| Cell body | |
| Axon | |
| Dendrite | |
| Myelin sheath | |
| Node of Ranvier | |
| **Structures of the Synapse** | |
| Synapse | |
| Axon terminal | |
| Synaptic vesicle | |
| Synaptic cleft | |
| Presynaptic neuron | |

Postsynaptic neuron _____

_____

## Neurophysiology Terms

Action potential _____

_____

Depolarization _____

_____

Repolarization _____

_____

Local potential _____

_____

Hyperpolarization _____

_____

Label and color the microscopic anatomy of nervous tissue depicted in **Figure 12.1** with the terms from Exercise 12-1 (p. 285). Use your text and Exercise 12-1 in this unit for reference.

FIGURE **12.1** A multipolar neuron

# EXERCISES

We begin our study of the nervous system in this unit, and we will explore it further over the next three units. The nervous system is one of the major homeostatic systems in the body. It regulates cellular activities by sending nerve impulses, also called **action potentials**, via cells called **neurons**. Neurons are supported by a variety of smaller cells collectively called **neuroglial cells**, and together these two cell types make up **nervous tissue**.

In the following exercises you will examine the cells and structure of nervous tissue. Then you will apply what you learned about the structure of these cells to their physiology with a tracing exercise.

## Exercise 12-1

### Neurons and Neuroglia

> **MATERIALS**
> - ❑ Neuron models
> - ❑ Synapse models
> - ❑ Modeling clay in four colors
> - ❑ Nervous tissue slide
> - ❑ Myelin sheath slide
> - ❑ Light microscope
> - ❑ Colored pencils

## 12

There are two types of cells within nervous tissue: **neurons** and **neuroglial cells** (**Figure 12.2**). Neurons are large cells that transmit and generate messages in the form of nerve impulses, or **neuronal action potentials**. Although they vary widely in size and structure, most have the following features in common (**Figure 12.3**):

1. **Cell body.** The cell body is the biosynthetic center of the neuron, containing the nucleus and many of the organelles. Its cytoskeleton consists of densely packed microtubules and **neurofilaments** that compartmentalize the rough endoplasmic reticulum into dark-staining structures called **Nissl bodies**. Cell bodies are often found in clusters. In the central nervous system (CNS), clusters of cell bodies are called **nuclei**, and in the peripheral nervous system (PNS), such clusters are called **ganglia**.

2. **Axon.** A single axon exits the cell body to transmit messages in the form of action potentials to other neurons, muscles, and glands. An axon generates action potentials across its plasma membrane, known as the **axolemma**, at the **axon hillock**. The action potential propagates the length of the axon and down its multiple terminal branches, called **telodendria**. At the end of each telodendrion is an **axon terminal**. Recall from Unit 10 that the axon terminal contains synaptic vesicles with neurotransmitters that send messages to the axon's target.

**FIGURE 12.2 Nervous tissue, photomicrograph (motor neuron smear)**

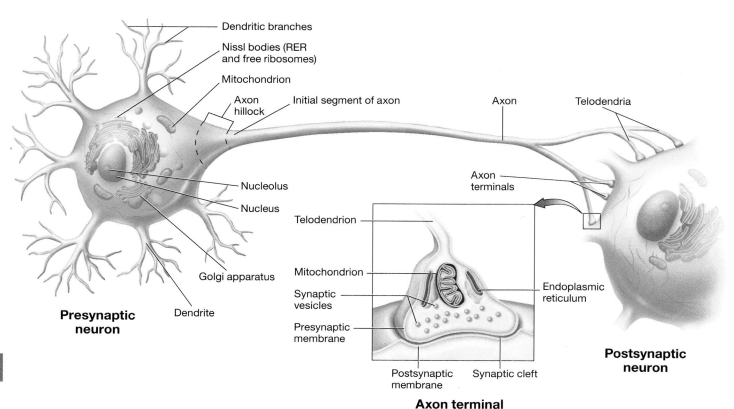

FIGURE **12.3** A multipolar neuron

3. **Dendrites.** Most neurons have one or more branching processes called dendrites that receive messages from other neurons. They transmit these messages to the neuron's cell body, but they are not capable of generating action potentials.

Neurons are classified structurally on the basis of the number of processes extending from the cell body (**Figure 12.4**). The most common type of neuron is the large **multipolar neuron,** which has three or more processes—specifically, one axon and two or more dendrites. Most multipolar neurons resemble highly branched trees and have hundreds to thousands of dendrites.

Both **bipolar neurons** and **pseudounipolar neurons** are rarer than multipolar neurons and are associated with the senses. Bipolar neurons have one axon and one dendrite and are found in special sense organs such as the olfactory epithelium and the retina of the eye. Pseudounipolar neurons have only one process and are found in the skin, where they detect and transmit sensations such as pain and temperature. A short distance from the cell body, the process branches like a "T" into a **central process,** which travels from the cell body to the CNS, and a **peripheral process,** which travels from a sensory receptor to the cell body. Both the central process and the peripheral process can generate action potentials and are therefore considered axons.

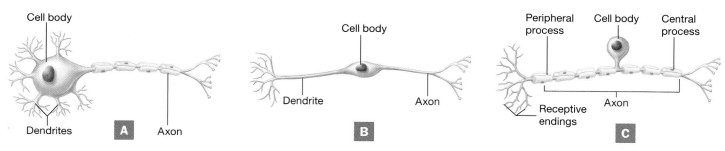

FIGURE **12.4** Types of neurons: (**A**) multipolar neuron; (**B**) bipolar neuron; (**C**) pseudounipolar neuron

Neurons communicate with one another and with their target cells at a junction called a **synapse**. In Unit 10 you learned about a type of synapse called the *neuromuscular junction* (NMJ). The synapse between two neurons is similar to the NMJ and consists of three parts:

1. **Presynaptic neuron**. The neuron that sends the message is called the presynaptic neuron. It contains neurotransmitters packaged in synaptic vesicles in its axon terminals.

2. **Synaptic cleft**. As we saw in the NMJ, the two neurons in a neuronal synapse are separated by a small space. This space is filled with extracellular fluid through which the neurotransmitters diffuse when they are released from the synaptic vesicles.

3. **Postsynaptic neuron**. The neuron that receives the message is the postsynaptic neuron. Its plasma membrane contains receptors for neurotransmitters. When the neurotransmitters bind to these receptors, the membrane potential of the postsynaptic neuron changes.

Neuroglial cells are much smaller than neurons, and they outnumber neurons about 50 to 1—no small feat considering that the nervous system contains about a trillion neurons! Neuroglial cells include the following (**Figure 12.5**; note that this figure shows only neuroglial cells of the CNS):

1. **Schwann cells**. Schwann cells form a structure called the **myelin sheath** around the axons of certain neurons in the PNS (**Figures 12.6 and 12.7**). As you can see in **Figure 12.6A**, Schwann cells can myelinate only one axon, and they do so by wrapping clockwise around the axon. The outer edge of the Schwann cell, called the **neurilemma**, contains most of its cytoplasm and the nucleus. The myelin sheath protects and insulates the axons and speeds up conduction of action potentials. Because the sheath is made up of individual neuroglial cells, there are small gaps between the cells where the plasma membrane of the axon is exposed. These gaps are called **nodes of Ranvier**, and the myelin-covered segments between the nodes are called **internodes**.

2. **Satellite cells**. Satellite cells surround the cell bodies of neurons in the PNS. These cells are believed to enclose and support the cell bodies, although their precise function is unknown.

3. **Astrocytes**. Astrocytes are the most numerous neuroglial cell type in the CNS. These star-shaped cells have many functions, including anchoring neurons and blood vessels in place with processes called **perivascular feet** and

FIGURE **12.5** Neuroglial cells of the CNS

FIGURE **12.6** Myelin sheath: (**A**) Schwann cells around a PNS axon; (**B**) an oligodendrocyte around multiple CNS axons

FIGURE **12.7** Myelin sheath, photomicrograph

regulating the extracellular environment of the brain. In addition, they facilitate the formation of the **blood-brain barrier,** created by tight junctions in the brain capillaries that prevent many substances in the blood from entering the brain tissue.

4. **Oligodendrocytes**. Oligodendrocytes, shown in **Figure 12.6B,** have long extensions that wrap around the axons of certain neurons in the CNS to form the myelin sheath. Note that one oligodendrocyte can myelinate several axons.

5. **Microglial cells.** The small microglial cells are very active phagocytes that clean up debris surrounding the neurons. Microglia also degrade and ingest damaged or dead neurons.

6. **Ependymal cells.** The ciliated ependymal cells line the hollow spaces of the brain and spinal cord. They assist in forming the fluid that bathes the brain and spinal cord, **cerebrospinal fluid,** and circulate it with their cilia.

## Procedure Model Inventory for Nervous Tissue

Identify the following structures of nervous tissue on models and diagrams using your textbook and this unit for reference. As you examine the anatomical models and diagrams, record on the model inventory in Table 12.1 the name of the model and the structures you were able to identify. When you have completed the activity, answer Check Your Understanding questions 1 and 2 (p. 299).

1. Neuron
   a. Axon
      (1) Axolemma
      (2) Axon hillock
      (3) Telodendria
   b. Dendrite(s)
   c. Cell body (soma)
   d. Nissl bodies
   e. Neurofilaments

2. Neuroglial cells
   a. Oligodendrocytes
   b. Schwann cells
   c. Satellite cells
   d. Microglial cells
   e. Astrocytes
   f. Ependymal cells

3. Other structures
   a. Myelin sheath
   b. Node of Ranvier
   c. Nucleus
   d. Ganglion

4. Parts of the synapse
   a. Presynaptic neuron
   b. Axon terminal
   c. Synaptic vesicle
   d. Synaptic cleft
   e. Postsynaptic neuron
   f. Neurotransmitter receptors

Table **12.1** Model Inventory for Nervous Tissue

| Model/Diagram | Structures Identified |
|---|---|
|  |  |
|  |  |
|  |  |
|  |  |

## Procedure Building a Myelin Sheath

In this exercise you will demonstrate the difference between the methods by which Schwann cells and oligodendrocytes myelinate the axons of neurons in the peripheral nervous system and the central nervous system, respectively. Both types of cells create the myelin sheath by wrapping themselves around the axon repeatedly (as many as 100 times). Recall, however, that Schwann cells can myelinate only one axon, whereas oligodendrocytes send out "arms" to myelinate several axons in their vicinity. When you have completed the activity, answer Check Your Understanding question 3 (p. 299).

**1** Obtain four colors of modeling clay (blue, green, yellow, and red, if available).

**2** Use the following color code to build three central nervous system (CNS) axons, one peripheral nervous system (PNS) axon, one oligodendrocyte, and two Schwann cells:

   a. CNS axons: blue      b. PNS axon: green      c. Oligodendrocyte: yellow      d. Schwann cells: red

**3** Build your oligodendrocyte so it reaches out to myelinate the three CNS axons.

**4** Build your Schwann cells so they myelinate the single PNS axon, making sure to leave a gap for the node of Ranvier.

Let's now identify some of the structures from the previous procedures on microscope slides. You will examine two tissue sections:

1. **Motor neuron smear.** On this slide you should see many multipolar neurons surrounded by many much smaller neuroglial cells (they may look like purple dots; refer to **Figure 12.2**). As you scan the slide with the low-power objective, find a well-stained neuron, and move the objective lens to high power. Look for Nissl bodies, and try to identify the single axon and the axon hillock. (This may be difficult, so don't get discouraged if all of the processes look similar.)

2. **Myelin sheath.** A longitudinal section of axons stained with a stain specific for the components of myelin (often osmium) is useful for seeing the sheath itself and also the nodes of Ranvier (refer back to **Figure 12.7**). You will likely need to use the high-power objective to see the nodes.

Obtain prepared slides of a motor neuron smear and the myelin sheath. Use your colored pencils to draw what you see, and label your drawing with the following terms.

**1** Nervous Tissue
   a. Cell body
   b. Axon
   c. Dendrites
   d. Neuroglial cells

**2** Myelin Sheath
   a. Axon with myelin sheath
   b. Node of Ranvier

# Exercise 12-2

## Nervous Tissue Physiology

We introduced some of the basic principles of nervous system physiology and electrophysiology in Unit 10, with muscle tissue physiology, and also earlier in this unit. In Unit 10 you learned that muscle fibers have excitable plasma membranes that can be stimulated by a neuron to elicit a muscle fiber action potential. Neurons also have excitable plasma membranes, and the sequence of events of a neuronal action potential closely resembles the sequence of events of a muscle fiber action potential. In addition, like a muscle fiber, a neuron does not generate action potentials spontaneously—postsynaptic neurons must be stimulated first by presynaptic neurons.

Despite these similarities, the physiology of the two cell types has key differences. One key difference between the two is that stimulation of a neuron doesn't always lead to an action potential. Recall that two neurons communicate at a synapse, and the presynaptic neuron releases neurotransmitters that bind to receptors on the postsynaptic membrane. These neurotransmitters cause one of two effects: (1) it may be excitatory and cause a small depolarization in the post-synaptic membrane that makes an action potential more likely, or (2) it may be inhibitory and cause a small **hyperpolarization** in the postsynaptic membrane that makes an action potential less likely. Both effects are termed **local potentials**.

If a postsynaptic neuron is stimulated to depolarize enough times by one or more presynaptic neurons, these small depolarizations will **summate** and spread through the neuron's plasma membrane. When the depolarizations spread to the axon hillock and reach sufficient magnitude, a value called **threshold**, an action potential (or *nerve impulse*) is generated. The steps of a neuronal action potential are as follows:

1. Voltage-gated sodium ion channels in the axolemma open, and sodium ions flood the cell, causing depolarization.

2. Voltage-gated sodium channels close, and voltage-gated potassium ion channels open.

3. Potassium ions exit the cell, causing repolarization.

4. Potassium ions continue to exit the cell even after the cell returns to its resting membrane potential, causing hyperpolarization (the cell becomes more negative than at rest).

Once the action potential is generated at the axon hillock, it is **propagated** down the length of the axon to its axon terminals in much the same way electricity is conducted along a wire. When the action potential reaches the axon terminal, it triggers voltage-gated calcium ion channels to open. The influx of calcium ions into the axon terminal causes exocytosis of synaptic vesicles, and neurotransmitter molecules are released into the synaptic cleft. The effects of these neurotransmitters will depend upon the type of neurotransmitter released and the type of target cell (another neuron, a muscle cell, or a gland cell). If the target cell is another postsynaptic neuron, the neurotransmitters will cause a local potential in the postsynaptic neuron's membrane.

**12**

Let's now trace the entire series of events through a three-neuron loop, including the steps of synaptic transmission between neurons 1 and 2, the steps of the action potential that occur in neuron 2, the propagation of the action potential down the axon of neuron 2, and the steps of synaptic transmission that occur between neuron 2 and 3. Hints are included to ensure that you stay on the right track as you trace. You may wish to consult your textbook for more details on this process. When you have completed the activity, answer Check Your Understanding questions 4–6 (pp. 299–300).

**Start:** **neuronal action potential reaches the axon terminal of neuron 1** →

_____ are released from _____

by _____ →  _____

diffuse across the _____ → neurotransmitters bind to the postsynaptic

membrane of neuron 2 and generate an excitatory local potential → excitatory local potentials _____

_____ → local potentials spread through the _____

to the _____ → axon hillock is depolarized to threshold →

_____ →

open, causing _____ → _____

close and _____ open → _____

exit the cytosol, causing _____ →

continue to exit until the membrane is _____ → the action potential is propagated

along the axon → the action potential reaches the _____ of _____ →

_____ are released from the _____ by _____ →

_____ diffuse across the _____ →

**neurotransmitters cause an excitatory or inhibitory local potential in the postsynaptic membrane of neuron 3. End**

Take the pathway that you just traced, and turn it into a diagram that contains the three neurons. Your diagram does not have to be a work of art—only a simple schematic representation of each neuron and each process is necessary. You may use the figures in this unit and your text for reference, but it is best to compose and organize your diagram in a way that makes sense to you so you may return to it for study purposes.

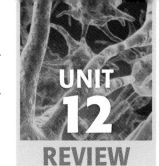

## Check Your Recall

**1** Label the following structures on **Figure 12.8.**

☐ Axon                    ☐ Cell body              ☐ Synaptic vesicles
☐ Axon hillock          ☐ Dendrites              ☐ Telodendria
☐ Axon terminals       ☐ Synaptic cleft

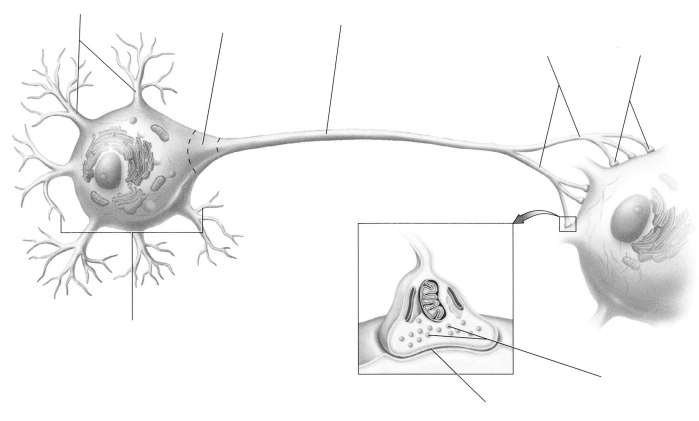

FIGURE **12.8** A neuron

**2** The neuron pictured in **Figure 12.8** is a
   a. pseudounipolar neuron.
   b. bipolar neuron.
   c. multipolar neuron.
   d. unipolar neuron.

**3** *Matching:* Match the neuroglial cell with its correct function.

_____ Oligodendrocytes

_____ Astrocytes

_____ Microglial cells

_____ Schwann cells

_____ Satellite cells

_____ Ependymal cells

A. Create the myelin sheath in the PNS

B. Ciliated cells in the CNS that form and circulate cerebrospinal fluid

C. Surround the cell bodies of neurons in the PNS

D. Anchor neurons and blood vessels, maintain extracellular environment around neurons, assist in the formation of the blood-brain barrier

E. Phagocytic cells of the CNS

F. Form the myelin sheath in the CNS

**4** What is the function of the myelin sheath?

_____

_____

**5** Label the following structures on **Figure 12.9**.

☐ Internode
☐ Myelin
☐ Neurilemma

☐ Node of Ranvier
☐ Oligodendrocyte
☐ Schwann cell

FIGURE **12.9** The myelin sheath

**6** Which part of the neuron is capable of generating an action potential?

   a.  Dendrite

   b.  Axon

   c.  Myelin sheath

   d.  Cell body

**7** Put the following events of the neuronal action potential in the correct order, placing number 1 next to the first event, number 2 next to the second event, and so on.

   _____  Voltage-gated sodium ion channels close, and voltage-gated potassium ion channels open.

   _____  Potassium ions continue to exit the cell, causing hyperpolarization.

   _____  Voltage-gated sodium ion channels open, and the axon depolarizes.

   _____  Potassium ions exit the cell, causing repolarization.

**8** Neurotransmitters that bind the postsynaptic membrane generally generate a/an

   a.  local potential.

   b.  action potential.

   c.  resting membrane potential.

   d.  pacemaker potential.

**9** Where are synaptic vesicles located?

   a.  Axon terminals

   b.  Dendrites

   c.  Cell body

   d.  Both a and b are correct.

   e.  All of the above.

**10** What triggers exocytosis of synaptic vesicles?

   a.  Arrival of a local potential at the cell body

   b.  Hyperpolarization of the postsynaptic membrane

   c.  Neurotransmitters binding to the postsynaptic membrane

   d.  The influx of calcium ions into the axon terminal

12

## Check Your Understanding

### Critical Thinking and Application Questions

**1** Neurons are amitotic, which means that after a certain stage, they do not divide further. Most tumor cells are characterized by a rapid rate of mitosis. Considering this, of which cell types (neurons or neuroglia) must brain tumors be composed? Why?

_____

_____

_____

**2** A traumatic brain injury may result in a large number of damaged or dying neurons. In such a case, which neuroglial cell would you expect to be present in large numbers? Explain.

_____

_____

_____

**3** Multiple sclerosis is a *demyelinating* disease, in which the patient's immune system attacks and destroys the cells that form the myelin sheath in the central nervous system. What types of symptoms would you expect from such a disease? Why? Would Schwann cells or oligodendrocytes be affected?

_____

_____

_____

**4** Local anesthetic drugs such as lidocaine block sodium ion channels in the neuron plasma membrane from opening. What effect would this have on action potential generation and conduction?

_____

_____

_____

**5** Certain inhaled anesthetic agents are thought to open $Cl^-$ channels in the membranes of postsynaptic neurons in the brain, an effect that causes inhibitory local potentials. Why might this action put a person "to sleep" during anesthesia?

_____

_____

_____

_____

_____

**6** Predict what would happen if the calcium ion channels in the axon terminal were blocked. Explain.

_____

_____

_____

_____

_____

12

# Central Nervous System

# 13

## OBJECTIVES

*Once you have completed this unit, you should be able to:*

- Describe the gross anatomy of the brain.
- Identify structures of the brain.
- Identify structures of the spinal cord.

# PRE-LAB EXERCISES

Complete the following exercises prior to coming to lab, using your textbook and lab manual for reference.

## Pre-Lab Exercise 13-1

✎ Key Terms

You should be familiar with the following terms before coming to lab.

| Term | Definition |
|------|------------|

**Structures of the Brain**

Cerebral hemispheres _____

_____

Cerebral cortex _____

_____

Basal nuclei _____

**13**  _____

Corpus callosum _____

_____

Diencephalon _____

_____

Thalamus _____

_____

Hypothalamus _____

_____

Midbrain _____

_____

Pons _____

_____

Medulla oblongata _____

_____

Cerebellum _____

_____

Dura mater _____

_____

Arachnoid mater _____

_____

Pia mater _____

_____

Ventricles _____

_____

Dural sinuses _____

_____

## Structures of the Spinal Cord

Gray matter horns (anterior, posterior, lateral) _____

_____

Funiculi (anterior, posterior, lateral) _____

_____

Conus medullaris _____

_____

Filum terminale _____

_____

Cauda equina _____

_____

13

Label and color the diagrams of the brain depicted in **Figures 13.1, 13.2, 13.3,** and **13.4** with the terms from Exercise 13-1 (p. 307). Use your text and Exercise 13-1 in this unit for reference.

FIGURE **13.1** Brain, lateral view

FIGURE **13.2** Brain, inferior view

FIGURE **13.3** Brain, midsagittal section

13

FIGURE **13.4** Brain and meninges, frontal and parasagittal section

## Pre-Lab Exercise **13-3**

### Spinal Cord Anatomy

Label and color the diagrams of the spinal cord depicted in **Figures 13.5** and **13.6** with the terms from Exercise 13-2 (p. 313). Use your text and Exercise 13-2 in this unit for reference.

FIGURE **13.5** Brain and spinal cord, posterior view

FIGURE **13.6** Spinal cord, transverse section

# EXERCISES

The **central nervous system** (**CNS**) consists of two organs: the **brain** and the **spinal cord**. The anatomy of the brain and the spinal cord may seem complex, but really they simply consist of a hollow, folded tube with an enlarged end (the brain). Together these two organs control many of our homeostatic functions, our sensation, our movement, and our higher brain functions. The anatomy of both organs will be explored in the three exercises in this unit.

## Exercise 13-1

### Anatomy of the Brain

**MATERIALS**
- ❏ Brain models: whole and sectioned
- ❏ Ventricle models
- ❏ Brainstem models
- ❏ Dural sinus model

The brain is divided into four regions: the **cerebral hemispheres** (collectively called the **cerebrum**), the **diencephalon**, the **brainstem**, and the **cerebellum**. As just discussed, the brain is a hollow organ with spaces called **ventricles** filled with a fluid similar to plasma called **cerebrospinal fluid** (**CSF**). As you can see in **Figure 13.7**, the largest of the ventricles, called the **lateral ventricles**, are located in the right and left cerebral hemispheres. Note in **Figure 13.7A** that the lateral ventricles resemble rams' horns when viewed from the anterior side. The smaller **third ventricle** is housed within the diencephalon. It is continuous with the **fourth ventricle**, found in the brainstem, via a small canal called the **cerebral aqueduct**. The fourth ventricle is continuous with a canal that runs down the central spinal cord called the **central canal**.

Within each of the four ventricles we find collections of blood vessels known as **choroid plexuses**. As blood flows through the choroid plexuses, fluid filters out into the ventricles, and at that point is called CSF. The largest choroid plexuses are within the lateral ventricles.

Recall from Unit 12 that the ventricles are lined by neuroglial cells called ependymal cells, whose cilia beat to circulate CSF. One of the main functions of CSF is to reduce brain weight (the brain is buoyant in the CSF). Without CSF, your brain literally would crush itself under its own weight!

The cerebrum is the most superior portion of the brain (**Figure 13.8**). Its surface consists of elevated ridges called **gyri** and shallow grooves called **sulci**. Deep grooves, called **fissures**, separate major regions of the cerebral hemispheres. The **longitudinal fissure** separates the right and left hemispheres. The cerebrum consists of five lobes: the **frontal**, **parietal**, **temporal**, **occipital**, and deep **insula** lobes (remember this last one by the mnemonic "the *insula* is *insula*ted"). The cerebral

**13**

**A** Anterior view    **B** Left lateral view

FIGURE **13.7** Ventricles: (**A**) anterior view; (**B**) lateral view hemispheres are responsible for the brain's cognitive functions, including learning and language, conscious interpretation of sensory information, conscious planning of movement, and personality.

The cell bodies and unmyelinated axons and dendrites of the cerebral neurons lie in the cerebrum's outer 2 millimeters in a region called the **cerebral cortex**. These portions of the neurons are unmyelinated, which gives the cerebral cortex a gray color, and for this reason it is called **gray matter**. The cell bodies and processes of the cerebral cortex communicate with other parts of the nervous system by bundles of myelinated axons called **white matter**. The largest tract of cerebral white matter is called the **corpus callosum,** and it connects the right and left cerebral hemispheres.

Gray matter isn't confined to the cerebral cortex. Clusters of cell bodies called **nuclei** are found throughout the cerebrum. An important group of nuclei, the **basal nuclei**, monitors voluntary motor functions. The neurons of these nuclei are connected to other parts of the nervous system by various tracts of white matter in the cerebrum.

Deep to the cerebral hemispheres in the central core of the brain we find the **diencephalon**, composed of three main parts (**Figure 13.9**):

1. **Thalamus.** The thalamus is a large, egg-shaped mass of gray and white matter that constitutes 80% of the diencephalon. It is a major integration and relay center that edits and sorts information going into the cerebrum. It essentially functions as the "gateway" into the cerebrum.

2. **Hypothalamus.** The hypothalamus is located on the anterior and inferior aspect of the diencephalon. It is a deceptively small structure that contains the nuclei whose neurons carry out many of the body's homeostatic functions including helping to regulate the endocrine system; monitoring the sleep-wake cycle; controlling thirst, hunger, and body temperature; and helping to monitor the autonomic nervous system (good things do come in small packages, after all!). An endocrine organ called the **pituitary gland** is connected to the hypothalamus by a stalk called the **infundibulum.**

3. **Epithalamus.** The epithalamus is located on the posterior and superior aspect of the diencephalon. It contains an endocrine organ called the **pineal gland** that secretes the hormone **melatonin**, which helps to regulate the sleep-wake cycle.

The third major portion of the brain, the **brainstem**, influences the automatic functions of the body, such the rhythm for breathing, heart rate, blood pressure, and certain reflexes. The most superior portion of the brainstem is the **midbrain**, and inferior to it we find the rounded **pons**, which bulges anteriorly. The last segment of the brainstem, the **medulla oblongata** (or simply *medulla*), is continuous inferiorly with the spinal cord.

The fourth major component of the brain is the large posterior **cerebellum**. It consists of two highly convoluted lobes connected by a piece called the **vermis**. Like the cerebral hemispheres, the cerebellum has an outer **cerebellar cortex** composed of gray matter and inner white matter. The cerebellar white matter is called the **arbor vitae** because of its resemblance to the branches of a tree. The cerebellum coordinates and plans ongoing motor activities and is critical in reducing and preventing motor error with movement.

FIGURE **13.8** Brain: (**A**) lateral view; (**B**) inferior view

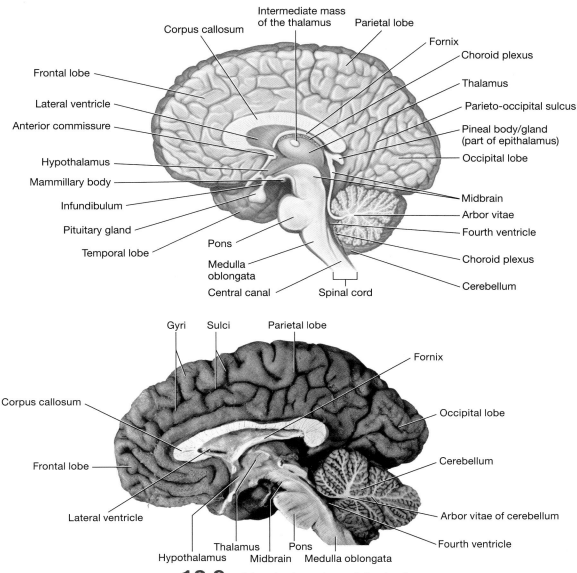

FIGURE **13.9** Midsagittal section of the brain

Note in **Figure 13.10** that a set of three membranes, collectively called the **meninges** (singular: *meninx*), surrounds the brain. The meninges include the following:

1. **Dura mater.** The outermost meninx is the thick, leathery, double-layered dura mater. The superficial periosteal dural layer is fused to the skull, and the deeper meningeal layer is continuous with the dura mater of the spinal cord. The two layers of the dura are fused, but in three regions the deep layer separates from the superficial layer and dives into the brain to form three structures:
   a. the **falx cerebri,** which forms a partition between the right and left cerebral hemispheres,
   b. the **falx cerebelli,** which separates the two cerebellar hemispheres, and
   c. the **tentorium cerebelli,** which separates the cerebrum from the cerebellum.

   At these locations there are spaces between the two dural layers collectively called the **dural sinuses** (see **Figure 17.8**, p. 424) All deoxygenated blood from the brain drains into the dural sinuses, which in turn drain into veins exiting the head and neck.

2. **Arachnoid mater.** The middle meninx, the arachnoid mater, is separated from the dura by a space called the **subdural space.** This space contains little bundles of the arachnoid mater called the **arachnoid granulations,** vascular structures that project into the dural sinuses and allow CSF to reenter the blood. You can see the pattern of CSF circulation from its formation by the choroid plexuses to its return to the blood in **Figure 13.11.**

**FIGURE 13.10** Brain and meninges, frontal and parasagittal section

**CSF circulation**

❶ CSF is produced by the choroid plexus of each ventricle.

❷ CSF flows through the ventricles and into the subarachnoid space via the median and lateral apertures. Some CSF flows through the central canal of the spinal cord.

❸ CSF flows through the subarachnoid space.

❹ CSF is absorbed into the dural venous sinuses via the arachnoid villi.

Wastes and unnecessary solutes absorbed

CSF forms as a filtrate containing glucose, oxygen, vitamins, and ions ($Na^+$, $Cl^+$, $Mg^{2+}$, etc.)

**CSF formation by choroid plexuses**

**FIGURE 13.11** Circulation of CSF through the brain and spinal cord

3. **Pia mater.** The thinnest, innermost meninx is the pia mater. The pia mater clings to the surface of the cerebral hemispheres and is richly supplied with blood vessels. A space between the pia mater and the arachnoid mater, called the **subarachnoid space**, is filled with CSF.

## Procedure Model Inventory for the Brain

Identify the following structures of the brain on models and diagrams, using your textbook and this unit for reference. As you examine the anatomical models and diagrams, record on the model inventory in **Table 13.1** the name of the model and the structures you were able to identify. The brain's structure is fairly complex, and it's best to examine models in as many different planes of section as possible. When you have completed the activity, answer Check Your Understanding questions 1–3 (p. 325).

1. Cerebrum
   a. Cerebral hemispheres
   b. Corpus callosum (cerebral white matter)
   c. Cerebral cortex (gray matter)
   d. Basal nuclei (these will only be visible on certain sections of the brain)
   e. Lobes of the cerebrum
      (1) Frontal lobe
      (2) Parietal lobe
      (3) Occipital lobe
      (4) Temporal lobe
      (5) Insula lobe
   f. Fissures
      (1) Longitudinal fissure
      (2) Transverse fissure
   g. Sulci
      (1) Central sulcus
         (a) Precentral gyrus
         (b) Postcentral gyrus
      (2) Lateral sulcus
      (3) Parieto-occipital sulcus
   h. Fornix
2. Diencephalon
   a. Thalamus
   b. Hypothalamus
      (1) Infundibulum
      (2) Pituitary gland
   c. Epithalamus
   d. Pineal gland
3. Brainstem
   a. Midbrain
   b. Pons
   c. Medulla oblongata
4. Cerebellum
   a. Vermis
   b. Arbor vitae
5. Brain coverings
   a. Dura mater
      (1) Subdural space
      (2) Falx cerebri
      (3) Falx cerebelli
      (4) Tentorium cerebelli
   b. Arachnoid mater
      (1) Subarachnoid space
   c. Pia mater
6. Ventricles
   a. Lateral ventricles
   b. Third ventricle
   c. Fourth ventricle
   d. Cerebral aqueduct
7. Vascular structures
   a. Dural sinuses
      (1) Superior sagittal sinus
      (2) Transverse sinus
      (3) Cavernous sinus
   b. Choroid plexus
   c. Arachnoid granulations

13

TABLE **13.1** Model Inventory for the Brain

| Model | Structures Identified |
|---|---|
|  |  |
|  |  |
|  |  |
|  |  |
|  |  |

13

# Exercise 13-2

## The Spinal Cord

### MATERIALS

❑ Spinal cord models: whole and sectioned

The medulla oblongata passes through the foramen magnum of the occipital bone and becomes the **spinal cord** (**Figure 13.12**). Note in **Figure 13.12A** that the spinal cord does not extend the entire length of the vertebral column; rather, it ends between the first and second lumbar vertebra. At this point it tapers to form the end of the spinal cord, called the **conus medullaris**, which gives off a tuft of nerve roots called the **cauda equina** (meaning "horse's tail"). The cauda equina fills the remainder of the vertebral column to the sacrum and exits out of the appropriate foramina to become spinal nerves.

The cranial meninges are continuous with spinal meninges and are similar in name and structure (**Figure 13.12B** and **13.13**). The CSF that surrounds the brain and fills the spaces between the cranial meninges also fills the spaces between the spinal meninges. But there are a few notable differences between the cranial and spinal meninges. For one, the spinal dura mater consists of only *one* layer rather than two like the cranial dura. This

FIGURE **13.12** Spinal cord: (**A**) posterior view in the vertebral column; (**B**) transverse section of the spinal cord in the vertebral column; (**C**) inferior spinal cord dissection

13

single dural layer does not attach to the vertebral column, which creates a space between the spinal dura and the interior vertebral foramen called the **epidural space**. Because the cranial dura is fused to the interior of the skull, there is no epidural space around the brain.

Another difference between the cranial and spinal meninges is the presence of small extensions of pia mater in the spinal cord called **denticulate ligaments**. These tiny ligaments secure the spinal cord to the vertebral column. In addition, the pia mater continues long after the spinal cord ends and forms a long, fibrous extension called the **filum terminale** that eventually attaches to the coccyx.

Internally, the spinal cord consists of a butterfly-shaped core of gray matter that surrounds the CSF-filled **central canal** (**Figure 13.13**). The gray matter is divided into regions, or **horns**. Anteriorly are the **anterior** (or *ventral*) **horns**, which contain the cell bodies of motor neurons. The axons of the neurons of the anterior horn exit the spinal cord and form the ventral root, which eventually becomes part of a spinal nerve. Also part of the spinal nerve, the posterior (or dorsal) root contains bundles of sensory axons. The posterior root's cell bodies are located just lateral to the spinal cord in a swollen knob called the posterior (or dorsal) root ganglion. These neurons receive input from sensory axons in the peripheral nervous system. In this region are the posterior (or dorsal) horns, and in the thoracic and lumbar regions of the spinal cord are the lateral horns, which contain the cell bodies of autonomic neurons.

As you can see in **Figure 13.13**, it is possible to discern the anterior from the posterior spinal cord by looking at the shapes of the anterior and posterior horns. The anterior horns are broad and flat on the ends, whereas the posterior horns are more tapered, and they extend farther out toward the edge.

Surrounding the spinal gray matter is the spinal white matter, which can be divided into three **funiculi** (or *columns*): the **anterior, posterior,** and **lateral funiculi**. Each funiculus contains myelinated axons grouped into bundles called **tracts**. Tracts contain axons that have the same beginning and end points and the same general function. **Ascending tracts** carry sensory information from sensory neurons to the brain, and **descending tracts** carry motor information from the brain to motor neurons.

**13**

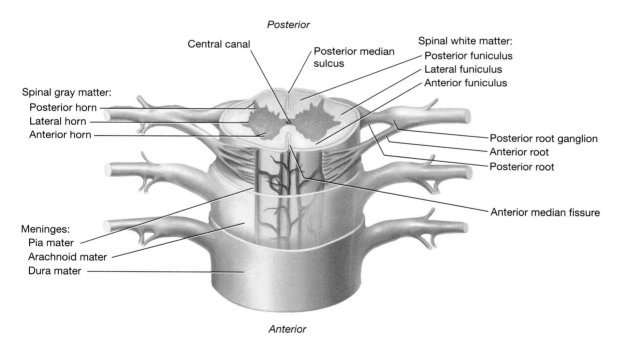

FIGURE **13.13** Transverse section of the spinal cord

## Procedure  Model Inventory for the Spinal Cord

Identify the following structures of the spinal cord on models and diagrams, using your textbook and this unit for reference. As you examine the anatomical models and diagrams, record on the model inventory in Table 13.2 the name of the model and the structures you were able to identify. When you have completed the activity, answer Check Your Understanding questions 4 and 5 (p. 326).

1. Meninges
   a. Dura mater
   b. Arachnoid mater
   c. Pia mater
      (1) Denticulate ligaments
2. Spinal gray matter
   a. Anterior horn
   b. Lateral horn
   c. Posterior (dorsal) horn
3. Spinal white matter
   a. Anterior funiculus
   b. Lateral funiculus
   c. Posterior funiculus

4. Spinal nerve roots:
   a. Posterior root
      (1) Posterior root ganglion
   b. Anterior root
5. Conus medullaris
6. Filum terminale
7. Cauda equina
8. Cervical enlargement
9. Lumbar enlargement
10. Central canal
11. Anterior median fissure
12. Posterior median sulcus

TABLE **13.2** Model Inventory for the Spinal Cord

| Model | Structures Identified |
|-------|----------------------|
|       |                      |
|       |                      |
|       |                      |
|       |                      |
|       |                      |

# Exercise 13-3

## Brain and Spinal Cord Dissection

### MATERIALS

❏ Sheep brain and spinal cord

❏ Hammer and chisel (if the brain is still in the skull)

❏ Dissection equipment and trays

Often structures of the brain and spinal cord are difficult to see on anatomical models. This exercise will allow you to examine these structures more closely by dissecting a preserved sheep brain. You will note that certain structures, such as the frontal lobes of the cerebral hemispheres, are proportionally smaller in the sheep than in the human brain. Note that the process of preservation makes many structures of the brain and spinal cord much tougher than they would be in a fresh specimen.

## Procedure Dissection

**1** If the brain is still encased in the skull, you have your work cut out for you. The best way to approach extracting it from the skull is to take a hammer and chisel and gently (at least as gently as one can with a hammer and chisel!) remove it piece by piece.

**Safety Note**

⚠ Goggles and gloves are required!

**2** As you gently hack away at the skull, you will note a thick membrane holding the skull in place. This is the dura mater, and it can make removal of the skull somewhat difficult. Ideally, you would like to preserve the dura, but you may end up cutting through it as you remove the brain.

**3** Once you have removed most of the skull, gently lift out the brain. (If you're careful, you may be able to get the brain out with the pituitary gland still attached.) You may have to loosen the remaining attachments of the dura with your finger.

**4** Once the brain is out, note the thick part of the dura covering the longitudinal fissure. If you cut through this with scissors, you will enter the superior sagittal sinus.

**5** Next remove the dura to reveal the thin membrane on top of the brain. This is the arachnoid mater.

**6** Remove an area of the arachnoid mater to see the shiny inner membrane—the pia mater—directly touching the surface of the brain. Note that the pia mater follows the convolutions of the gyri and sulci.

**7** Examine the surface anatomy of both the superior and the inferior surfaces of the sheep brain (**Figures 13.14** and **13.15**).

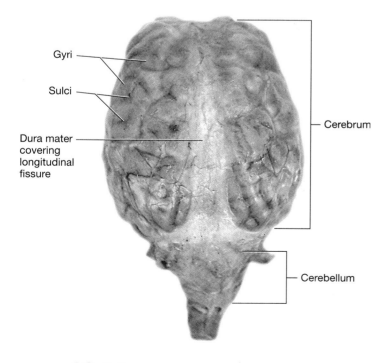

FIGURE **13.14** Superior view of the sheep brain

FIGURE **13.15** Inferior view of the sheep brain

Below, draw what you see on both surfaces, and label your drawing with the following structures:

- ☐ Arachnoid mater
- ☐ Cerebellum
- ☐ Cerebrum
- ☐ Dura mater

- ☐ Gyri
- ☐ Longitudinal fissure
- ☐ Medulla oblongata
- ☐ Olfactory bulb

- ☐ Optic chiasma
- ☐ Pituitary gland
- ☐ Pons
- ☐ Sulci

**8** Note the size of the olfactory bulbs. Are they larger or smaller than those you observed in the human brain? Why do you think this is so?

**9** Spread the two cerebral hemispheres, and identify the corpus callosum.

**10** Make a cut down the brain's midsagittal plane to separate the two cerebral hemispheres.

**11** Examine the brain's internal anatomy (**Figure 13.16**), and stick your finger in the lateral ventricle. You will see (or feel) that it is much larger than it appears.

FIGURE **13.16** Lateral view of the sheep brain; midsagittal section

Draw what you see in the space below, and label your drawing with the following structures:

- ☐ Arbor vitae
- ☐ Choroid plexus
- ☐ Corpus callosum
- ☐ Fornix

- ☐ Hypothalamus
- ☐ Lateral ventricle
- ☐ Medulla oblongata
- ☐ Midbrain

- ☐ Pineal gland
- ☐ Pons
- ☐ Thalamus

**12** Section one of the halves of the brain in the frontal plane, approximately along the central sulcus. Note the outer cerebral cortex (the gray matter) and the inner white matter. From this view you also can see the lateral and third ventricles. Again draw what you see in the space below and label your drawing with the following structures:

☐ Cerebral cortex      ☐ Falx cerebri      ☐ Thalamus
☐ Cerebral white matter      ☐ Lateral ventricle      ☐ Third ventricle

13

**13** If the brain on which you are working contains a segment of the spinal cord, move to it now, and identify the single-layered dura surrounding it. Peel it back to see the arachnoid mater.

**14** Section the cord in the transverse plane. Note the butterfly-shaped inner gray matter and the outer white matter. Draw what you see in the space below, and label your drawing with the following structures:

- ☐ Anterior funiculus
- ☐ Anterior horn
- ☐ Central sulcus

- ☐ Lateral funiculus
- ☐ Posterior funiculus
- ☐ Posterior horn

- ☐ Spinal arachnoid mater
- ☐ Spinal dura mater

13

## Check Your Recall

**1** Label the following parts of the brain in **Figures 13.17** and **13.18.**

- ☐ Central sulcus
- ☐ Cerebellum
- ☐ Corpus callosum
- ☐ Frontal lobe
- ☐ Hypothalamus
- ☐ Lateral sulcus
- ☐ Medulla oblongata
- ☐ Midbrain
- ☐ Occipital lobe
- ☐ Parietal lobe
- ☐ Pineal gland
- ☐ Pons
- ☐ Postcentral gyrus
- ☐ Precentral gyrus
- ☐ Temporal lobe
- ☐ Thalamus

13

FIGURE **13.17** Brain, lateral view

FIGURE **13.18** Brain, midsagittal section

**2** Identify the following structures in **Figure 13.19.**

☐ Central canal
☐ Cerebral aqueduct
☐ Fourth ventricle
☐ Lateral ventricle
☐ Third ventricle

FIGURE **13.19** Brain, left lateral view

**3** Label the following parts of the spinal cord in **Figure 13.20.**

☐ Anterior funiculus
☐ Anterior horn
☐ Anterior median fissure
☐ Central canal

☐ Lateral funiculus
☐ Lateral horn
☐ Posterior funiculus
☐ Posterior horn

☐ Posterior median sulcus
☐ Spinal arachnoid mater
☐ Spinal dura mater

FIGURE **13.20** Spinal cord, transverse section

**4** Which of the following is *not* one of the four main regions of the brain?

a. Cerebral hemispheres

b. Brainstem

c. Cerebral aqueduct

d. Cerebellum

e. Diencephalon

**5** Where is the cerebral cortex located, and what is located here?

_____

_____

**6** White matter consists of

a. unmyelinated cell bodies, dendrites, and axons.

b. myelinated axons.

c. unmyelinated axons.

d. myelinated cell bodies, dendrites, and axons.

**7** *Matching:* Match the following terms with the correct definition.

| | |
|---|---|
| _____ Thalamus | A. Most inferior portion of the brainstem |
| _____ Pia mater | B. Posterior part of the brain that controls and monitors ongoing movement |
| _____ Medulla oblongata | C. Innermost and thinnest meninx |
| _____ Epithalamus | D. Most superior part of the brainstem |
| _____ Cerebellum | E. Largest, egg-shaped component of the diencephalon; edits and sorts information coming into the cerebrum |
| _____ Hypothalamus | F. Middle part of the brainstem that bulges anteriorly |
| _____ Dura mater | G. Middle meninx |
| _____ Pons | H. Inferior part of the diencephalon that controls many aspects of homeostasis |
| _____ Midbrain | I. Contains the pineal gland and secretes the hormone melatonin |
| _____ Arachnoid mater | J. Outermost and thickest meninx |

**8** *Fill in the blanks:* The spinal cord extends from the _____ of the occipital bone to the

_____ vertebra. It terminates as the _____ and gives off a bundle

of nerve roots called the _____.

13

**9** Which of the following spaces is found around the spinal cord but *not* around the brain?

a. Epidural space

b. Subdural space

c. Subarachnoid space

d. Suprarachnoid space

**10** The _____ of the spinal cord contains the cell bodies of motor neurons, whereas the _____ of the spinal cord contains the cell bodies of neurons that receive information from sensory neurons.

a. lateral horn; anterior horn

b. lateral horn; posterior horn

c. anterior horn; lateral horn

d. anterior horn; posterior horn

13

Name _____

Section _____ Date _____

# Check Your Understanding

### Critical Thinking and Application Questions

**1** Traumatic brain injuries may result in damage to the blood vessels in the spaces between the meninges. These damaged blood vessels leak blood into either the subdural or subarachnoid spaces—a potentially deadly condition called a *subdural* or *subarachnoid hemorrhage*. Why is this condition so dangerous?

_____

_____

_____

_____

_____

_____

**2** Predict the effects of injuries to the following areas:

a. Cerebral cortex _____

_____

b. Brainstem _____

_____

c. Cerebellum _____

_____

d. Hypothalamus _____

_____

**3** Which of the injuries from question 2 do you think would be the most damaging to survival? Why?

_____

_____

_____

_____

_____

13

**4** A common way to deliver anesthesia for surgery and childbirth is to inject the anesthetic agent into the epidural space (*epidural anesthesia*). A possible complication of this procedure is a tear in the dura mater that causes CSF to leak out of the central nervous system. Why would a loss of cerebrospinal fluid be problematic? What symptoms do you predict with this condition? (*Hint*: Think about the function of cerebrospinal fluid.)

_____

_____

_____

_____

_____

_____

**5** During a procedure called a *lumbar puncture*, CSF surrounding the meninges is withdrawn with a needle. This procedure generally is performed between L3 and L5. Why do you think the fluid is withdrawn here rather than from higher up in the vertebral column (e.g., the cervical vertebrae)? Often, the sampled CSF is tested for bacteria or viruses if a brain infection is suspected. Why would CSF sampled from the spinal cord give you information about the condition of the brain?

_____

_____

_____

_____

_____

_____

**6** An individual sustains injuries to only the descending tracts of the spinal cord. Will this person experience deficits in movement, sensation, both, or neither? Explain.

_____

_____

_____

_____

_____

_____

# Peripheral and Autonomic Nervous System

# 14

## OBJECTIVES

*Once you have completed this unit, you should be able to:*

- Identify structures of a peripheral nerve.

- Identify, describe, and demonstrate functions of cranial nerves.

- Identify spinal nerves and plexuses.

- Describe a simple spinal reflex arc, and test the effects of mental concentration on the patellar tendon (knee-jerk) reflex.

- Describe the effects of the two branches of the autonomic nervous system on the body systems.

Name _____ Section _____ Date _____

# PRE-LAB EXERCISES

Complete the following exercises prior to coming to lab, using your textbook and lab manual for reference.

## Pre-Lab Exercise **14-1**

 Key Terms

You should be familiar with the following terms before coming to lab. Please note that the individual cranial and spinal nerves are covered in Pre-Lab Exercises 14-3 (p. 331) and 14-5 (p. 333), respectively.

| Term | Definition |
| --- | --- |
| **General Terms** | |
| Peripheral nerve | |
| Cranial nerve | |
| Spinal nerve | |
| Nerve plexus | |
| Autonomic nervous system | |
| Sympathetic nervous system | |
| Parasympathetic nervous system | |
| **Parts of a Peripheral Nerve** | |
| Fascicle | |
| Axon | |
| Epineurium | |
| Perineurium | |

Endoneurium _____

_____

**Spinal Nerve Plexuses**

Cervical plexus _____

_____

Brachial plexus _____

_____

Lumbar plexus _____

_____

Sacral plexus _____

_____

14

## Pre-Lab Exercise **14-2**
Peripheral Nerve Anatomy

Label and color the peripheral nerve diagram depicted in **Figure 14.1** with the terms from Exercise 14-1 (p. 335). Use your text and Exercise 14-1 in this unit for reference.

FIGURE **14.1** Anatomy of a peripheral nerve

# Pre-Lab Exercise 14-3

## Cranial Nerve Locations and Functions

Complete **Table 14.1** with the functions of each pair of cranial nerves, and indicate whether the nerve is motor, sensory, or mixed.

TABLE **14.1** The Cranial Nerves

| Cranial Nerve | Functions | Motor, Sensory, or Mixed |
|---|---|---|
| CN I: Olfactory nerve | | |
| CN II: Optic nerve | | |
| CN III: Oculomotor nerve | | |
| CN IV: Trochlear nerve | | |
| CN V: Trigeminal nerve | | |
| CN VI: Abducens nerve | | |
| CN VII: Facial nerve | | |
| CN VIII: Vestibulocochlear nerve | | |
| CN IX: Glossopharyngeal nerve | | |
| CN X: Vagus nerve | | |
| CN XI: Accessory nerve | | |
| CN XII: Hypoglossal nerve | | |

14

# Pre-Lab Exercise 14-4

## Nerve Plexus and Spinal Nerve Anatomy

Label the diagram of the nerve plexuses and their spinal nerves depicted in **Figure 14.2** with the terms from Exercise 14-3 (p. 343). Use your text and Exercise 14-3 in this unit for reference. Note that this diagram is presented in color to facilitate identification of the nerves.

FIGURE **14.2** The nerve plexuses and the ventral rami of the spinal nerves

# Pre-Lab Exercise 14-5

## Spinal Nerve Locations and Functions

The spinal nerves (with the exception of the intercostal nerves) can be traced to the specific nerve plexus from which they originate. Fill in **Table 14.2** with the nerve plexus from which each nerve originates and the functions of each nerve.

TABLE **14.2** Selected Spinal Nerves

| Spinal Nerve | Nerve Plexus | Functions |
|---|---|---|
| Phrenic nerve | | |
| Musculocutaneous nerve | | |
| Ulnar nerve | | |
| Median nerve | | |
| Femoral nerve | | |
| Sciatic nerve | | |
| Tibial nerve | | |
| Common peroneal (fibular) nerve | | |

14

# Pre-Lab Exercise 14-6

## Branches of the Autonomic Nervous System

The autonomic nervous system (ANS) has two subdivisions: (a) the sympathetic nervous system (SNS), and (b) the parasympathetic nervous system (PSNS). Fill in **Table 14.3** with the general characteristics of each branch of the ANS and the effects of each system on its target organs.

TABLE **14.3** Characteristics of the Sympathetic and Parasympathetic Nervous Systems

| Characteristic | SNS | PSNS |
|---|---|---|
| Main function | | |
| Location of nerve roots | | |
| Neurotransmitter(s) | | |
| **Effects on Target Organs and Physiological Processes** | | |
| Heart | | |
| Bronchioles (airway passages of the lungs) | | |
| Blood vessels to abdominal organs and skin | | |
| Blood vessels to skeletal muscle | | |
| Digestive and urinary functions | | |
| Metabolic rate | | |
| Pupils | | |

14

# EXERCISES

In Unit 13 we discussed the central nervous system (CNS), and now we turn to the other division of the nervous system—the **peripheral nervous system** (**PNS**). The PNS consists of organs called **peripheral nerves**, or simply *nerves*. There are two types of nerves classified by location:

1. Nerves that originate from the brain and the brainstem, called **cranial nerves**, and

2. nerves that originate from the spinal cord, called **spinal nerves**.

The PNS is divided broadly into two divisions based upon function:

1. The **motor** or **efferent division**, which innervates muscle cells and glands, and

2. the **sensory** or **afferent division**, which innervates the skin, joints, special sensory organs, and other sensory receptors.

The motor and sensory divisions may be further subdivided based upon the structures served: (a) the **somatic sensory** and **somatic motor divisions**, which serve skin and skeletal muscles, respectively, and (b) the **visceral sensory** and **visceral motor divisions,** which serve smooth muscle, cardiac muscle, and glands. The visceral motor division controls most of our automatic functions such as blood pressure, digestion, and urine production, and also is known as the **autonomic nervous system** (**ANS**).

In the following exercises, you will examine the structure and functions of the organs and divisions of the PNS. As you study the PNS, bear in mind that its functions are closely interwoven with the CNS, and the branches of the nervous system cannot function independently of one another.

## Exercise **14-1**

### Peripheral Nerve Anatomy

> **MATERIALS**
> ☐ Peripheral nerve models

A **nerve** is an organ consisting of myelinated and unmyelinated axons (also known as *nerve fibers*), blood vessels, lymphatic vessels, and connective tissue sheaths (**Figure 14.3**). Each individual axon is covered by its axolemma (plasma membrane), superficial to which we find a connective tissue sheath called the **endoneurium**. Axons are grouped into bundles called **fascicles,** covered with another connective tissue sheath called the **perineurium**. Groups of fascicles, blood vessels, and lymphatic vessels are surrounded by the most superficial connective tissue sheath: the **epineurium**.

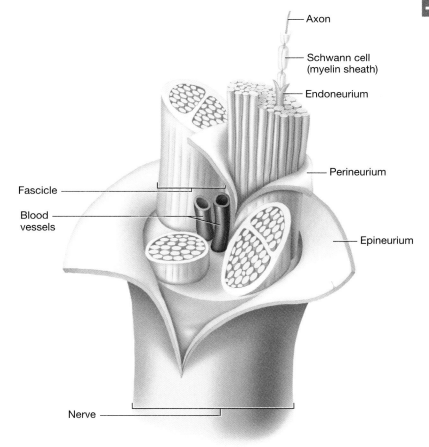

FIGURE **14.3** Anatomy of a peripheral nerve

## Procedure Model Inventory for a Nerve

Identify the following parts of a nerve on models and diagrams, using your textbook and this unit for reference. As you examine the anatomical models and diagrams, record on the model inventory in Table 14.4 the name of the model and the structures you were able to identify. When you have completed the activity, answer Check Your Understanding questions 1 and 2 (p. 355).

1. Nerve
2. Fascicle
3. Axon
4. Blood vessels

5. Nerve sheaths
   a. Epineurium
   b. Perineurium
   c. Endoneurium
   d. Myelin sheath

TABLE **14.4** Model Inventory for Peripheral Nerve Anatomy

| Model/Diagram | Structures Identified |
|---|---|
| | |
| | |
| | |

## Exercise 14-2

### The Cranial Nerves

**MATERIALS**

- ❑ Brain models
- ❑ Preserved brain specimens
- ❑ Penlight
- ❑ Snellen vision chart
- ❑ Tuning fork
- ❑ Unknown samples to smell
- ❑ PTC, thiourea, and sodium benzoate tasting papers

The 12 pairs of **cranial nerves** originate from the brain (**Figure 14.4**). Each nerve is given two names: (1) a sequential Roman numeral in order of its origin from the brain, and (2) a name that describes the nerve's location or function. For example, cranial nerve III is the third cranial nerve to arise from the brain. It is also called the oculomotor nerve because one of its functions is to provide motor fibers to some of the muscles that move the eyeball.

All cranial nerves innervate structures of the head and neck. Three cranial nerves are **sensory nerves** whose fibers have purely sensory functions, four are **mixed nerves** that contain both sensory and motor fibers, and five are **motor nerves** that contain primarily motor fibers. Following is an overview of the main functions of each cranial nerve (note that CN = "cranial nerve"):

1. **CN I: Olfactory Nerve.** The olfactory nerve is a purely sensory nerve that innervates the olfactory mucosa in the superior nasal cavity, where it provides for the sense of smell.

2. **CN II: Optic Nerve.** The optic nerve is also a purely sensory nerve that provides for the sense of vision. Its fibers emerge from the retina of the eye and meet at the **optic chiasma**, where the nerves partially exchange fibers before diverging to form the **optic tracts**.

3. **CN III: Oculomotor Nerve.** The oculomotor nerve is a motor cranial nerve that innervates four of the six extraocular muscles that move the eyeball, the muscle that opens the eyelid, the muscle that constricts the pupil, and the muscle that changes the shape of the lens for near vision, an adjustment called **accommodation**.

4. **CN IV: Trochlear Nerve.** The trochlear nerve is a small motor nerve that innervates one of the six extraocular muscles that move the eyeball (the *superior oblique muscle*).

5. **CN V: Trigeminal Nerve.** The trigeminal nerve is a large mixed nerve named for the three branches that together provide sensory innervation to the face and motor innervation to the muscles of mastication (chewing).

FIGURE **14.4** Inferior view of the brain with the cranial nerves

6. **CN VI: Abducens Nerve.** The abducens is a small motor nerve that innervates the final extraocular muscle (the *lateral rectus muscle*).

7. **CN VII: Facial Nerve.** The mixed fibers of the facial nerve innervate many structures and provide for the following: motor to the muscles of facial expression; taste to the anterior two-thirds of the tongue; motor to the glands that produce tears (the lacrimal glands), mucus, and saliva; and sensory to part of the face and mouth.

8. **CN VIII: Vestibulocochlear Nerve.** The final sensory nerve, the vestibulocochlear nerve, innervates the structures of the inner ear and provides for the senses of hearing and equilibrium.

9. **CN IX: Glossopharyngeal Nerve.** The small mixed glossopharyngeal nerve provides motor fibers to the muscles of the pharynx (throat) involved in swallowing and sensory fibers to the posterior one-third of the tongue for taste sensation.

10. **CN X: Vagus Nerve.** The mixed vagus nerve is the only cranial nerve that "wanders" outside of the head and neck ("vagus" means "wanderer"). In the head and the neck it provides some sensory fibers to the skin of the head and the pharynx, motor fibers to muscles involved in speech and swallowing, and motor fibers to certain salivary glands. Outside the head and the neck it innervates most of the thoracic and abdominal viscera as the main nerve of the parasympathetic nervous system.

11. **CN XI: Accessory Nerve.** The accessory nerve is the only cranial nerve that has both a cranial component originating from the brainstem and a spinal component originating from the spinal cord. Its motor fibers innervate the muscles that move the head and the neck, such as the trapezius and sternocleidomastoid muscles.

12. **CN XII: Hypoglossal Nerve.** The hypoglossal nerve is a small motor nerve that innervates the muscles that move the tongue. Note that the hypoglossal nerve *moves* the tongue but does not provide any taste sensation to the tongue.

## HINTS & TIPS

### Remembering the Order of the Cranial Nerves

Many cranial nerve mnemonics have been created over the years to help students remember their correct order. Following is one of my favorite mnemonics, but if this one doesn't stick for you, try making up your own or doing an Internet search for "cranial nerve mnemonics":

> **O**h (Olfactory)
> **O**nce (Optic)
> **O**ne (Oculomotor)
> **T**akes (Trochlear)
> **T**he (Trigeminal)
> **A**natomy (Abducens)
> **F**inal (Facial)
> **V**ery (Vestibulocochlear)
> **G**ood (Glossopharyngeal)
> **V**acations (Vagus)
> **A**re (Accessory)
> **H**appening (Hypoglossal)

You also can help yourself remember the olfactory and optic nerves by reminding yourself that you have one nose (CN I, the *olfactory* nerve) and two eyes (CN II, the *optic* nerve).

## Procedure Model Inventory for Cranial Nerves I–XII

Identify cranial nerves I–XII and associated structures on anatomical models and/or preserved specimens of the brain. List in **Table 14.6** in Exercise 14-3 (p. 345) the nerves and structures you identify.

1. Olfactory nerves
   a. Olfactory bulbs
2. Optic nerves
   a. Optic chiasma
   b. Optic tract
3. Oculomotor nerves
4. Trochlear nerves
5. Trigeminal nerves
6. Abducens nerves
7. Facial nerves
8. Vestibulocochlear nerves
9. Glossopharyngeal nerves
10. Vagus nerves
11. Accessory nerves
12. Hypoglossal nerves

## **Procedure** Testing the Cranial Nerves

A component of every complete physical examination performed by health-care professionals is the cranial nerve exam. In this exercise you will put on your "doctor hat," and perform the same tests of the cranial nerves done during a physical examination. Pair up with another student, and take turns performing the following tests. For each test, first document your observations (in many cases, this will be "able to perform" or "unable to perform"). Then state which cranial nerve(s) you have checked with each test. Keep in mind that some tests check more than one cranial nerve, some cranial nerves are tested more than once, and each nerve is tested in this exercise at least once. When you have completed the activity, answer Check Your Understanding questions 1 and 2 (p. 355).

**1** Have your partner perform the following actions individually (not all at once—your partner may have difficulty smiling and frowning at the same time!): smile, frown, raise his or her eyebrows, and puff his or her cheeks.

*Observations:* _____

_____

_____

_____

*CN(s) tested:* _____

**2** Have your partner open and close his or her jaw and clench his or her teeth.

*Observations:* _____

_____

_____

_____

*CN(s) tested:* _____

**3** Have your partner elevate and depress his or her shoulders and turn his or her head to the right and the left.

*Observations:* _____

_____

_____

_____

*CN(s) tested:* _____

**4** Draw a large, imaginary "Z" in the air with your finger. Have your partner follow your finger with his or her eyes without moving his or her head. Repeat the procedure, this time drawing the letter "H" in the air with your finger.

*Observations:* _____

_____

_____

_____

*CN(s) tested:* _____

14

**5** Test pupillary response:

  **a** Dim the lights in the room about halfway.

  **b** Place your hand vertically against the bridge of your partner's nose as illustrated in **Figure 14.5**. This forms a light shield to separate the right and left visual fields.

  **c** Shine the penlight indirectly into the left eye from an angle as illustrated in **Figure 14.5**. Watch what happens to the pupil in the left eye.

  **d** Move the penlight away, and watch what happens to the left pupil.

  **e** Shine the light into the left eye again, and watch what happens to the pupil in the right eye. Move the penlight away, and watch what happens to the right pupil.

  **f** Repeat this process with the right eye.

  **g** Record your results in **Table 14.5**.

  *CN(s) tested:* _____

  _____

FIGURE **14.5** Method for testing the pupillary response

TABLE **14.5** Pupillary Response Results

| Action | Response of Left Pupil | Response of Right Pupil |
|---|---|---|
| Light shined into left eye | | |
| Light removed from left eye | | |
| Light shined into right eye | | |
| Light removed from right eye | | |

**6** Have your partner focus on an object on the far side of the room (e.g., the chalkboard or a chart) for 1 minute. Then have your partner switch his or her focus to an object in your hand (e.g., a pencil). Watch your partner's pupils carefully as the point of focus is changed from far to near.

*Observations:* _____

_____

_____

*CN(s) tested:* _____

14

**7** Place your hand lightly on your partner's throat, and have him or her swallow and speak. Feel for symmetrical movement of the larynx (throat).

*Observations:* _____

_____

_____

_____

*CN(s) tested:* _____

**8** Have your partner protrude his or her tongue. Check for abnormal deviation or movement (e.g., does the tongue move straight forward or does it move to one side?).

*Observations:* _____

_____

_____

_____

*CN(s) tested:* _____

**9** Test your partner's vision by having him or her stand 20 feet from a Snellen chart and read the chart, starting at the largest line and progressing to the smallest line he or she is able to see clearly. Record the ratio (e.g., 20/30) next to the smallest line your partner can read.

*Observations:* _____

_____

_____

_____

*CN(s) tested:* _____

**10** Hold a tuning fork by its handle, and strike the tines with a rubber mallet (or just tap it lightly on the lab table). Touch the stem to the top of your partner's head along the midsagittal line in the manner shown in **Figure 14.6**. This is called the Weber test. Ask your partner if he or she hears the vibration better in one ear, or if he or she hears the sound equally well in both ears.

*Observations:* _____

_____

_____

_____

_____

_____

*CN(s) tested:* _____

FIGURE **14.6** Weber test for hearing

**11** Have your partner stand with his or her eyes closed and arms at his or her sides for several seconds. Evaluate his or her ability to remain balanced.

*Observations:* _____

_____

_____

_____

*CN(s) tested:* _____

**12** Hold an unknown sample near your partner's nose, and fan the odor toward him or her by waving your hand over the container. Have him or her identify the substance by its scent.

*Observations:* _____

_____

_____

_____

*CN(s) tested:* _____

**13** Evaluate your partner's ability to taste using tasting papers. Have your partner place a piece of PTC paper on his or her tongue and determine if he or she can taste it (the ability to taste PTC is genetically determined; about half of the population can taste it). If your partner cannot taste the PTC, try the thiourea paper instead (a word of warning—thiourea tastes bad). If your partner cannot taste either of these papers, try the sodium benzoate paper.

*Observations:* _____

_____

_____

_____

*CN(s) tested:* _____

# Exercise 14-3

## Spinal Nerves and Reflexes

### MATERIALS
- ❏ Model of the spinal nerves
- ❏ Reflex hammer

Each of the 31 pairs of **spinal nerves** forms from the fusion of the dorsal and ventral roots of the spinal cord. Recall that the ventral roots carry motor fibers emerging from the spinal cord, and the dorsal roots carry sensory fibers to the spinal cord. Because each spinal nerve carries both motor and sensory fibers, all spinal nerves are mixed nerves.

Shortly after the dorsal and ventral roots fuse to form the spinal nerve, it splits into three branches: a **dorsal ramus**, a **ventral ramus**, and a small **meningeal branch**. The dorsal rami serve the skin, joints, and musculature of the posterior trunk. The meningeal branches reenter the vertebral canal to innervate spinal structures. The larger ventral rami travel anteriorly to supply the muscles of the upper and lower limbs, the anterior thorax and abdomen, and part of the back. The distribution of the ventral rami is illustrated in **Figure 14.7**.

The ventral rami of the thoracic spinal nerves travel anteriorly as 11 separate pairs of **intercostal nerves** that innervate the intercostal muscles, the abdominal muscles, and the skin of the chest and abdomen. The ventral rami of the cervical, lumbar, and sacral nerves combine to form four large **plexuses**, or networks, of nerves: the **cervical, brachial, lumbar**, and **sacral plexuses**. The major nerves of the cervical, brachial, thoracic, lumbar, and sacral plexuses are as follows:

1. **Cervical plexus**. The cervical plexus consists of the ventral rami of C1–C4 with a small contribution from C5. Its branches serve the skin of the head and the neck and certain neck muscles. Its major branch is the **phrenic nerve** (C3–C5), which serves the diaphragm.

2. **Brachial plexus**. The complicated-looking brachial plexuses consist of the ventral rami of C5–T1 (**Figure 14.8B**).

FIGURE **14.7** The nerve plexuses and the ventral rami of the spinal nerves

14

The first structures formed in the brachial plexus are its large trunks. Typically, C5 and C6 unite to form the superior trunk, C7 forms the middle trunk, and C8 and T1 unite to form the inferior trunk. Each trunk splits into an anterior division and a posterior division that become the plexus's cords. The anterior division of the inferior trunk forms the medial cord, which descends in the medial arm. The anterior divisions of the superior and middle trunks unite to form the lateral cord, which descends in the lateral arm. The posterior divisions of each trunk unite to form the posterior cord, which lies in the posterior arm. Several nerves originate from the brachial plexuses' cords and trunks, including the following:

a. **Axillary nerve.** The axillary nerve serves structures near the axilla, including the deltoid and teres minor muscles and the skin around this region.

b. **Musculocutaneous nerve.** The musculocutaneous nerve is located in the lateral arm and serves the anterior arm muscles (such as the biceps brachii) and the skin of the lateral forearm.

c. **Radial nerve.** The radial nerve is located in the posterior arm. It serves the posterior arm muscles and the forearm extensors as well as the skin in the lateral hand.

d. **Ulnar nerve.** The ulnar nerve, which you likely know as the "funny bone nerve," begins posteriorly but then crosses over to the anterior side of the arm as it curves around the medial epicondyle of the humerus. At this point the nerve is superficial and is easily injured when you smack your elbow on something. The ulnar nerve supplies certain forearm flexors, most of the intrinsic muscles of the hand, and the skin over the medial hand.

14

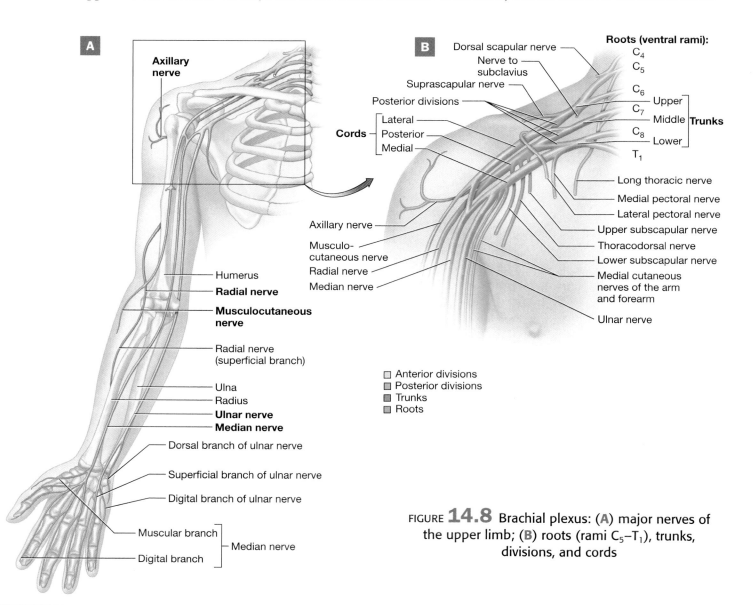

FIGURE **14.8** Brachial plexus: (**A**) major nerves of the upper limb; (**B**) roots (rami C$_5$–T$_1$), trunks, divisions, and cords

e. **Median nerve.** The median nerve is named as such because it travels approximately down the middle of the arm and forearm. It supplies most of the forearm flexors, certain intrinsic hand muscles, and the skin over the anterior and lateral hand. As the median nerve enters the wrist, it travels under a band of connective tissue called the *flexor retinaculum*. Occasionally the median nerve becomes trapped and inflamed under the flexor retinaculum, which results in **carpal tunnel syndrome**.

3. **Lumbar plexus.** Each lumbar plexus consists of the ventral rami of L1–L4 with a small contribution from T12 (**Figure 14.9A**). The largest nerve of this plexus is the **femoral nerve**, which provides motor innervation to most of the anterior thigh muscles and sensory innervation to the skin of the anterior and medial thigh, the leg, and the foot. Another smaller branch is the **lateral femoral cutaneous nerve**, which provides mainly sensory innervation to the anterolateral thigh.

4. **Sacral plexus.** The sacral plexus forms from ventral rami of L4–S4 (**Figure 14.9B**). Its largest nerve, and indeed the largest nerve in the body, is the **sciatic nerve**. The sciatic nerve travels in the posterior thigh, where it splits into two branches: the **tibial nerve** and the **common peroneal (fibular) nerve**. The tibial nerve provides motor innervation to the posterior muscles of the thigh, posterior leg, and foot, and sensory innervation to the posterior leg and foot. The common peroneal nerve provides motor and sensory innervation to the antero-lateral leg and the foot. A smaller branch of the sacral plexus is the **pudendal nerve**, which innervates the muscles of the pelvic floor and anogenital sphincters.

FIGURE **14.9** Nerve plexuses: (**A**) lumbar plexus; (**B**) sacral plexus

14

## Procedure Model Inventory for Nerves and Nerve Plexuses

Identify the following nerves and nerve plexuses on models and diagrams, using your textbook and this unit for reference. As you examine the anatomical models and diagrams, record on the model inventory in **Table 14.6** the name of the model and the structures you were able to identify. Note that you also used this model inventory for the cranial nerves in Exercise 14-2. When you have completed the activity, answer Check Your Understanding question 3 (p. 355).

1. Cervical plexus
   a. Phrenic nerve
2. Brachial plexus
   a. Axillary nerve
   b. Radial nerve
   c. Musculocutaneous nerve
   d. Ulnar nerve
   e. Median nerve
3. Thoracic (intercostal) nerves

4. Lumbar plexus
   a. Femoral nerve
   b. Lateral femoral cutaneous nerve
5. Sacral plexus
   a. Sciatic nerve
      (1) Tibial nerve
      (2) Common peroneal (fibular) nerve
   b. Pudendal nerve

TABLE **14.6** Model Inventory for Cranial Nerve and Spinal Nerve Anatomy

| Model/Diagram | Structures Identified |
|---|---|
|  |  |
|  |  |
|  |  |
|  |  |
|  |  |

## Procedure Testing Spinal Reflexes

A **reflex** is an involuntary, predictable motor response to a stimulus. The pathway through which information travels, shown below and in **Figure 14.10**, is called a **reflex arc**:

sensory receptor detects the stimulus → sensory neurons bring the stimulus to the CNS → the CNS processes and integrates the information → the CNS sends its output via motor neurons to an effector → the effector performs the triggered action

The human body has many different reflex arcs, one of the simplest of which is the **stretch reflex**. Stretch reflexes are important in maintaining posture and equilibrium and are initiated when a muscle is stretched. The stretch is detected by

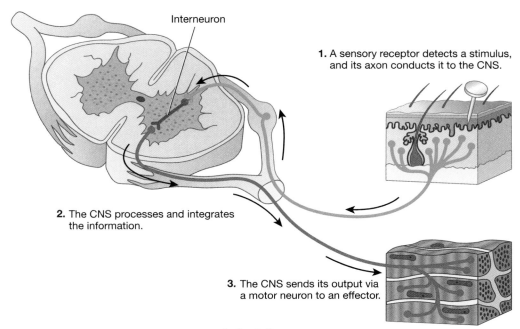

Interneuron

**1.** A sensory receptor detects a stimulus, and its axon conducts it to the CNS.

**2.** The CNS processes and integrates the information.

**3.** The CNS sends its output via a motor neuron to an effector.

FIGURE **14.10** A reflex arc

**muscle spindles**, specialized stretch receptors in the muscles, and this information is sent via sensory neurons to the CNS. The CNS then sends impulses down the motor neurons to the muscle that trigger a muscle contraction to counter the stretch.

You can demonstrate the stretch reflex easily: Sit down with your knees bent and relaxed, and palpate (feel) the musculature of your posterior thigh. How do the muscles feel (taut or soft)? Now stand up, and bend over to touch your toes. Palpate the muscles of your posterior thigh again. How do they feel? The reason they feel different in this position is that you stretched the hamstring muscles when you bent over to touch your toes. This triggered a stretch reflex that resulted in shortening (or tightening) of those muscles.

Technically, this reflex can be carried out without the help of the cerebral cortex and can be mediated solely by the spinal cord. We will see shortly, however, that the cortex is involved in even the simplest example of a stretch reflex—the patellar tendon (knee-jerk) reflex, shown in **Figure 14.11**. When you have completed the following activity, answer Check Your Understanding question 4 (p. 356).

**1** Have your lab partner sit in a chair with his or her legs dangling freely.

**2** Palpate your partner's patellar tendon between the tibial tuberosity and the patella.

**3** Tap this area with the flat end of a reflex hammer (sometimes a few taps are necessary to hit the right spot). What is the result?

**4** Now give your partner a difficult math problem to work (long division with decimals usually does the trick). As your partner works the problem, tap the tendon again. Is this response different from the original response? If yes, how, and why?

FIGURE **14.11** Patellar reflex

# Exercise 14-4

## The Autonomic Nervous System

### MATERIALS
- ❑ Automated or manual sphygmomanometer
- ❑ Stethoscope

The **autonomic nervous system** (ANS) is the largely involuntary branch of the peripheral nervous system charged with maintaining homeostasis in the face of changing conditions. The ANS has two branches—the **sympathetic nervous system** (SNS) and the **parasympathetic nervous system** (PSNS).

The cell bodies of the neurons of the SNS are located in the thoracic and lumbar regions of the spinal cord. Functionally, the SNS is often described in terms of the "fight-or-flight" response, in which the body prepares to respond to an emergency situation (e.g., running from a hungry alligator). This gives rise to a useful mnemonic: Your body is *sympathetic* to your emergency situation, so your *sympathetic* nervous system comes to the rescue. Although reaction to an emergency is a primary function of the SNS, it isn't the whole story. The SNS is activated by *any* excitement, emotion, or exercise, even something as simple as standing up. The axons of the SNS release the neurotransmitters **epinephrine**, **norepinephrine**, and **acetylcholine**. These neurotransmitters trigger the following effects in their target tissues:

- increased heart rate, which increases the blood pressure;
- constriction of the blood vessels serving the abdominal viscera and the skin, which also increases the blood pressure;
- dilation of the blood vessels serving the skeletal muscles;
- dilation of the airway passages in the lungs;
- decreased digestive and urinary functions;
- increased metabolic rate;
- increased release of fatty acids and glucose;
- dilation of the pupils; and
- the release of epinephrine from the adrenal medulla.

The cell bodies of the neurons of the PSNS are located in cranial nerve nuclei and in the sacral portion of the spinal cord. Functionally, the PSNS is discussed in terms of its "rest and recovery" role, in which it promotes functions associated with digestion, defecation, and diuresis (urine production). The axons of the PSNS release the neurotransmitter acetylcholine, which has the following effects on its target tissues:

- decreased heart rate, which decreases blood pressure;
- dilation of the blood vessels serving the skin and abdominal viscera, which decreases blood pressure (note that this is *not* a direct effect of the PSNS on the blood vessels. Instead, it is an indirect effect resulting from an inhibition of sympathetic nerves);
- mild constriction of the airway passages in the lungs;
- increased digestion and secretion of digestive products;
- increased urine production;
- storage of fats and glucose;
- constriction of the pupils; and
- adjustment of the lens for near vision.

Note that the effects of the SNS and PSNS are exactly opposite. Generally, the PSNS is subordinate to the SNS (digesting that doughnut isn't all that important when you're being chased by a hungry alligator); however neither branch of the ANS is ever completely quiet. The divisions work together constantly to provide the proper balance for any given situation.

## Procedure Testing Blood Pressure and Heart Rate Response to Exercise

One of the simplest ways to observe the effects of the SNS is with exercise. In this activity you will be measuring the effects of exercise on the heart rate and the blood pressure to see the SNS in action. When you have completed this activity, answer Check Your Understanding questions 5 and 6 (p. 356).

**1** Have your partner sit quietly for 2 minutes.

**2** After 2 minutes, measure your partner's resting heart rate by placing two fingers on his or her radial pulse (on the lateral side of the wrist). Count the number of beats for 1 minute. Record the rate in **Table 14.7**. Be sure to record the units (beats per minute).

**3** Obtain a stethoscope, and listen to your partner's heart. How does the heartbeat sound (quiet, loud, etc.)?

**4** Measure your partner's resting blood pressure using a sphygmomanometer (blood pressure cuff) and a stethoscope (instructions for taking blood pressures are found in Unit 18 (p. 443). You may also use an automated sphygmo-manometer if one is available. Record the blood pressure in **Table 14.7**. Be sure to record the units (mmHg).

**5** Have your partner perform vigorous exercise for several minutes (running up and down stairs works well). Note that even minor exercise will produce changes in blood pressure and pulse, but the changes are more apparent with vigorous exercise (especially for those who are novices at taking blood pressure).

**6** Immediately after the vigorous exercise, repeat the heart rate and blood pressure measurements, then listen to your partner's heart with a stethoscope. How does the heartbeat sound compared with the sounds you heard at rest?

**7** Have your partner rest while seated for 5 minutes. Repeat the heart rate and blood pressure measurements, and again listen to the heartbeat. Does the heart sound different after rest?

TABLE **14.7** Results of Exercise and SNS Experiment

| Situation | Blood Pressure | Heart Rate |
|---|---|---|
| Seated (at rest) | | |
| Immediately after exercise | | |
| After resting for 5 minutes | | |

14

**8** Interpret your results:

    **a** In which situation(s) did the SNS dominate? How do you know?

    _____

    _____

    _____

    _____

    **b** In which situation(s) did the PSNS dominate? How do you know?

    _____

    _____

    _____

    _____

14

## UNIT 14 REVIEW

## Check Your Recall

**1** Label the following structures on **Figure 14.12.**

- ☐ Axon
- ☐ Endoneurium
- ☐ Epineurium
- ☐ Fascicle
- ☐ Myelin sheath
- ☐ Nerve
- ☐ Perineurium

14

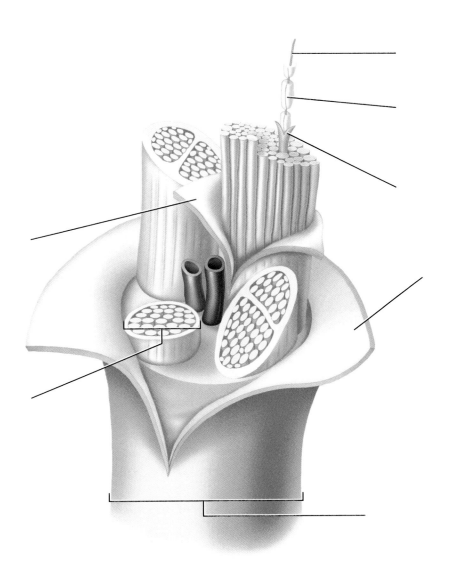

FIGURE **14.12** A peripheral nerve

**2** Another name for an axon is

   a. myelin sheath.

   b. endoneurium.

   c. nerve fiber.

   d. peripheral nerve.

**3** *Fill in the blanks:* Write the name of each cranial nerve next to its Roman numeral below.

CN I      _____

CN II     _____

CN III    _____

CN IV    _____

CN V     _____

CN VI    _____

CN VII   _____

CN VIII  _____

CN IX    _____

CN X     _____

CN XI    _____

CN XII   _____

**4** *Matching:* Match the cranial nerve with its main functions.

_____ CN I

_____ CN II

_____ CN III

_____ CN IV

_____ CN V

_____ CN VI

_____ CN VII

_____ CN VIII

_____ CN IX

_____ CN X

_____ CN XI

_____ CN XII

  A. Sensory to the face, motor to the muscles of mastication

  B. Motor to the trapezius and sternocleidomastoid muscles

  C. Hearing and equilibrium

  D. Olfaction (smell)

  E. Motor to the muscles of swallowing; taste to the posterior one-third of the tongue

  F. Motor to one-sixth extraocular muscles (superior oblique muscle)

  G. Motor to the tongue

  H. Vision

  I. Motor to the muscles of facial expression, taste to the anterior two-thirds of the tongue

  J. Motor to four-sixths extraocular muscles, dilates the pupil, opens the eye, changes the shape of the lens

  K. Motor to the muscles of swallowing and speaking, motor to the thoracic and abdominal viscera

  L. Motor to one-sixth extraocular muscles (lateral rectus muscle)

14

**5** Which of the following cranial nerves is not a purely sensory nerve?

a. Optic nerve

b. Vestibulocochlear nerve

c. Hypoglossal nerve

d. Olfactory nerve

**6** Label the following nerves and plexuses on **Figure 14.13**.

☐ Brachial plexus
☐ Cervical plexus
☐ Femoral nerve
☐ Lumbar plexus
☐ Median nerve
☐ Musculocutaneous nerve

☐ Radial nerve
☐ Sacral plexus
☐ Sciatic nerve
☐ Ulnar nerve

FIGURE **14.13** Nerve plexuses and the ventral rami of the spinal nerves

**7** *Matching:* Match each spinal nerve with the main structures it supplies.

_____ Phrenic nerve

_____ Median nerve

_____ Tibial nerve

_____ Radial nerve

_____ Femoral nerve

_____ Ulnar nerve

_____ Common fibular nerve

_____ Musculocutaneous nerve

A. Anterior and lateral leg muscles and skin
B. Posterior thigh and leg muscles, foot
C. Diaphragm
D. Some forearm flexors, most intrinsic hand muscles, skin on medial hand
E. Posterior arm muscles, forearm extensors, skin on lateral hand
F. Anterior arm muscles, skin on lateral forearm
G. Most forearm flexors, skin on anterior and lateral hand
H. Anterior thigh muscles, skin on anterior and medial thigh and leg

**8** Number the events of a reflex arc from 1 (first event) through 5 (last event).

_____ CNS sends output via motor neurons to an effector.

_____ Sensory neurons bring the stimulus to the CNS.

_____ The muscle contracts.

_____ CNS processes and integrates the information.

_____ Sensory receptor detects the stimulus.

**9** The receptor that detects the stretch in a stretch reflex is called a(an)
a. mitotic spindle.
b. muscle spindle.
c. capsular receptor.
d. efferent neuron.

**10** Mark each of the following as an effect of the sympathetic nervous system (SNS) or the parasympathetic nervous system (PSNS):

_____ Increased blood pressure

_____ Dilation of the pupils

_____ Changing the shape of the lens for near vision

_____ Dilation of the bronchioles

_____ Increased digestion

_____ Storage of fats and glucose

_____ Decreased urinary and digestive functions

_____ Increased heart rate

_____ Constriction of blood vessels serving skin and abdominal viscera

# Check Your Understanding

## Critical Thinking and Application Questions

**1** Damage to which cranial nerve(s) might produce the following results?

a. Inability to move the tongue _____

_____

b. Inability to taste _____

_____

c. Inability to move the eyes in any direction _____

_____

d. Inability to shrug the shoulders _____

_____

e. Inability to swallow _____

_____

**2** Dave was playing a hockey game when he was hit in the face by a puck. He suffered a broken nose during the injury and now finds that his sense of smell has diminished. What is likely causing this?

_____

_____

_____

_____

**3** Damage to which spinal nerve(s) might produce the following physical findings?

a. Inability to flex the forearm (with the biceps brachii muscle) _____

_____

b. Inability to breathe _____

_____

c. Inability to flex the hip and extend the knee with the anterior thigh muscles_____

_____

d. Inability to extend the hip and flex the knee with the posterior thigh muscles_____

_____

e. Inability to move the hand or feel the skin over the medial hand _____

_____

14

**4** Sometimes the reflex response is diminished or absent—a phenomenon termed *hyporeflexia*. Do you think hyporeflexia would be caused by disorders of the central nervous system or of the peripheral nervous system? Explain your reasoning.

**5** Drugs used to treat hypertension (high blood pressure) often work by blocking different components of the SNS. Explain how inhibiting the SNS may lower blood pressure.

**14**

**6** The disease *asthma* is characterized by increased airway resistance secondary to constriction of the airways, airway inflammation, and excessive secretion of mucus. One of the main classes of drugs used to treat asthma mimics the effects of the SNS on the airways. Explain how mimicking the SNS might help alleviate an asthma attack.

# General and Special Senses

# 15

## OBJECTIVES

*Once you have completed this unit, you should be able to:*

■ Identify structures of the eye.

■ Describe the extraocular muscles that move the eyeball.

■ Compare the functions of the rods and cones.

■ Identify structures of the ear.

■ Perform tests of hearing and equilibrium.

■ Identify structures of the olfactory and taste senses.

■ Determine the relative concentration of cutaneous sensory receptors in different regions of the body.

Name _____ Section _____ Date _____

# PRE-LAB EXERCISES

Complete the following exercises prior to coming to lab, using your textbook and lab manual for reference.

## Pre-Lab Exercise 15-1

 Key Terms

You should be familiar with the following terms before coming to lab.

| Term | Definition |
|------|-----------|

**Structures of the Eye**

Conjunctiva _____

_____

Lacrimal gland _____

_____

Sclera _____

_____

Cornea _____

_____

Iris _____

_____

Pupil _____

_____

Lens _____

_____

Ciliary body _____

_____

Choroid _____

_____

Retina _____

_____

Rods _____

_____

Cones _____

_____

**15**

**Structures of the Ear**

Auricle _____

_____

External auditory canal _____

_____

Tympanic membrane _____

_____

Auditory ossicles _____

_____

Pharyngotympanic tube _____

_____

Vestibule _____

_____

Semicircular canals _____

_____

Cochlea _____

_____

**Structures of Taste and Smell**

Chemosenses _____

_____

Olfactory epithelium _____

_____

Tongue papillae _____

_____

**15**

Label and color the structures of the eye depicted in **Figures 15.1** and **15.2** with the terms from Exercise 15-1 (p. 363). Use your text and Exercise 15-1 in this unit for reference.

**15**

FIGURE **15.1** Eyeball, sagittal section

FIGURE **15.2** External and accessory structures of the eye

# Pre-Lab Exercise **15-3**
Extraocular Muscles

The six muscles that move the eyeball are called **extraocular muscles**. First, label and color the extraocular muscles illustrated in **Figure 15.3**. Then, fill in **Table 15.1** with the location, action, and cranial nerve innervation for each of the extraocular muscles.

FIGURE **15.3** The extraocular muscles

TABLE **15.1** The Extraocular Muscles

| Muscle | Location | Action | Cranial Nerve Innervation |
|---|---|---|---|
| Superior rectus muscle | | | |
| Inferior rectus muscle | | | |
| Medial rectus muscle | | | |
| Lateral rectus muscle | | | |
| Superior oblique muscle | | | |
| Inferior oblique muscle | | | |

15

# Pre-Lab Exercise 15-4
## Anatomy of the Ear

Label and color the structures of the ear in **Figure 15.4** with the terms from Exercise 15-2 (p. 370).

FIGURE **15.4** The ear

# EXERCISES

**Sensation** is broadly defined as the detection of changes in the internal and external environments. Sensation may be conscious or subconscious, depending on the destination of the sensory information. For example, certain blood vessels have receptors that detect blood pressure. This information is taken to the brainstem, which makes changes as necessary to ensure that blood pressure remains relatively constant. This information never makes it to the cerebral cortex, so you are not consciously aware of it. However, information eventually taken to the cerebral cortex for integration and interpretation (e.g., the taste of your food or the level of light in a room) is something of which you are consciously aware. This is called **perception**, and it is the focus of this unit.

The following exercises ask you to examine the anatomy and physiology of the **special senses**: vision, hearing and equilibrium, taste, and smell. You also will examine the **general senses** in this unit, which include the senses of touch, pain, and temperature.

## Exercise **15-1**

### Anatomy of the Eye and Vision

> **MATERIALS**
> ❏ Eye models
> ❏ Preserved eyeballs
> ❏ Dissection equipment
> ❏ Dissection trays
> ❏ Snellen vision chart
> ❏ Dark green or blue paper
> ❏ Ruler

The eye is a complex organ consisting of three components:

1. External structures such as the **eyelids** (**Figure 15.5**),
2. Accessory structures such as the **lacrimal gland** and the extraocular muscles (**Figures 15.5** and **15.6**), and
3. The **eyeball** (**Figure 15.7**).

Many of the external and accessory structures of the eye protect the delicate eyeball.

Anteriorly, the eye is covered by the eyelids, or **palpebrae**. The eyelids meet medially and laterally at the **medial** and **lateral canthi**, respectively. The internal surface of the eyelids and much of the anterior eyeball are covered with a thin mucous membrane called the **conjunctiva**. Within the eyelids are the **meibomian glands**, whose oily secretions help to keep the eye lubricated.

One of the most prominent accessory structures of the eye is the **lacrimal apparatus**, which produces and drains tears. The lacrimal apparatus consists of the **lacrimal gland**, located in the superolateral

**15**

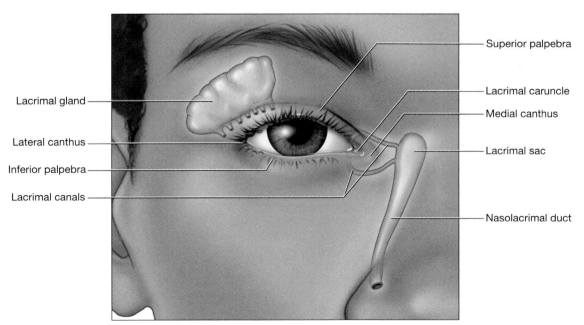

Lacrimal gland
Lateral canthus
Inferior palpebra
Lacrimal canals

Superior palpebra
Lacrimal caruncle
Medial canthus
Lacrimal sac
Nasolacrimal duct

FIGURE **15.5** External and accessory structures of the eye

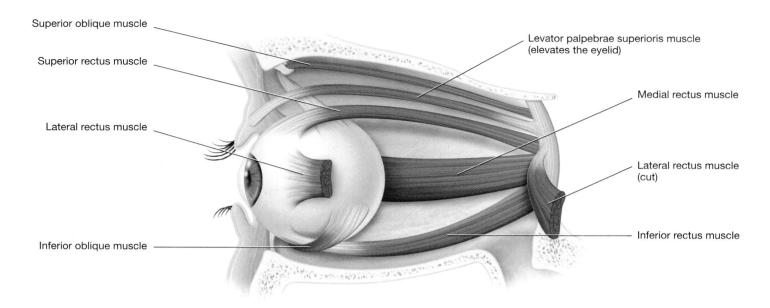

Superior oblique muscle

Superior rectus muscle

Lateral rectus muscle

Inferior oblique muscle

Levator palpebrae superioris muscle (elevates the eyelid)

Medial rectus muscle

Lateral rectus muscle (cut)

Inferior rectus muscle

FIGURE **15.6** The extraocular muscles

orbit, and the ducts that drain the tears it produces. The other major accessory structures are the **extraocular muscles,** which move the eyeball. As you learned in Unit 14, there are six extraocular muscles (**Figure 15.6**):

1. **Lateral rectus:** moves the eyeball laterally

2. **Medial rectus:** moves the eyeball medially

3. **Superior rectus:** moves the eyeball superiorly

4. **Inferior rectus:** moves the eyeball inferiorly

5. **Superior oblique:** moves the eyeball inferiorly and laterally

6. **Inferior oblique:** moves the eyeball superiorly and laterally

The eyeball itself is a hollow organ with three distinct tissue layers, or tunics (**Figure 15.7**):

1. **Fibrous tunic.** This outermost layer of the eyeball consists mostly of dense irregular connective tissue. It is avascular (lacks a blood supply) and consists of two parts:

   a. **Sclera.** The sclera is the white part of the eyeball, which makes up the posterior five-sixths of the fibrous tunic. It is white because of numerous collagen fibers that contribute to its thickness and toughness (in the same way a joint capsule or a ligament is tough and white).

   b. **Cornea.** The clear cornea makes up the anterior one-sixth of the fibrous tunic and is one of the refractory media of the eyeball (it bends light coming into the eye).

2. **Vascular tunic.** Also called the **uvea,** the vascular tunic carries most of the blood supply to the tissues of the eye. It is composed of three main parts:

   a. **Choroid.** The highly vascular choroid makes up the posterior part of the vascular tunic. The choroid is brown in color to prevent light scattering in the eye.

   b. **Ciliary body.** The ciliary body is located at the anterior aspect of the eye. It is made chiefly of the **ciliary muscle,** which controls the shape of the lens. The muscle attaches to the lens via small **suspensory ligaments.**

   c. **Iris.** The pigmented iris is the most anterior portion of the uvea. It consists of muscle fibers arranged around an opening called the **pupil.** As the fibers contract, the pupil either constricts or dilates.

15

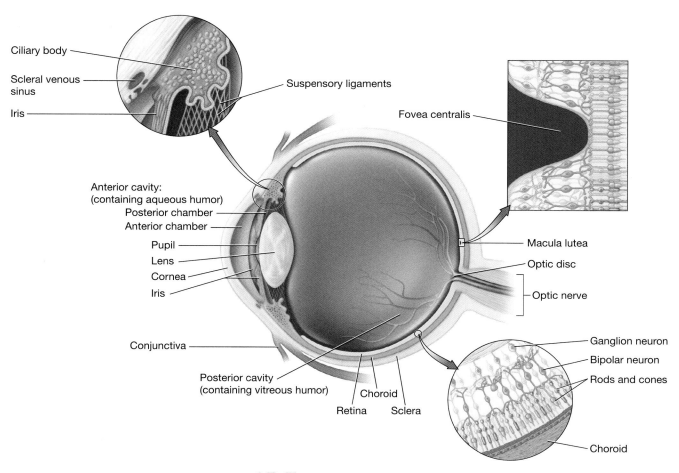

Ciliary body

Scleral venous sinus

Iris

Suspensory ligaments

Fovea centralis

Anterior cavity:
(containing aqueous humor)

Posterior chamber

Anterior chamber

Pupil

Lens

Cornea

Iris

Macula lutea

Optic disc

Optic nerve

Conjunctiva

Posterior cavity
(containing vitreous humor)

Choroid

Retina    Sclera

Ganglion neuron

Bipolar neuron

Rods and cones

Choroid

FIGURE **15.7** The eyeball, sagittal section

3. **Sensory tunic.** This layer consists of the **retina** and the **optic nerve.** The retina is a thin, delicate structure that contains **photoreceptors** called **rods** and **cones.**

   a. **Rods.** Rods are scattered throughout the retina and are responsible for vision in dim light and for peripheral vision.

   b. **Cones.** Cones are concentrated at the posterior portion of the retina and are found in highest numbers in an area called the **macula lutea.** At the center of the macula lutea is the **fovea centralis,** which contains only cones. Cones are responsible for color and high-acuity (sharp) vision in bright light.

   Note that there are no rods or cones at the posteriormost aspect of the eyeball where the optic nerve leaves the eyeball. This location is called the **optic disc,** or blind spot.

Another component of the eyeball is the **lens,** which allows for precise focusing of light on the retina. The lens divides the eyeball into the **anterior** and **posterior cavities** (sometimes called the anterior and posterior segments). The anterior cavity is filled with a watery fluid called **aqueous humor,** and the posterior cavity contains a thicker fluid called **vitreous humor.** Both help to refract light onto the retina.

15

## Procedure Model Inventory for the Eye

Identify the following structures of the eye and the eyeball on models and diagrams using your textbook and this unit for reference. As you examine the anatomical models and diagrams, record on the model inventory in **Table 15.2** the name of the model and the structures you were able to identify.

### Eyeball

1. Fibrous tunic
   a. Sclera
   b. Cornea
2. Vascular tunic (uvea)
   a. Choroid
   b. Ciliary body
   c. Suspensory ligaments
   d. Iris
   e. Pupil

3. Lens
4. Sensory tunic
   a. Retina
   b. Optic disc
   c. Macula lutea
   d. Fovea centralis
5. Optic nerve (cranial nerve II)
6. Optic chiasma

7. Anterior cavity
   a. Anterior chamber
   b. Posterior chamber
   c. Aqueous humor
   d. Scleral venous sinus
8. Posterior cavity
   a. Vitreous humor
   b. Hyaloid canal

### Accessory Structures

1. Palpebrae
2. Palpebral fissures
3. Medial and lateral canthi

4. Lacrimal apparatus
   a. Lacrimal gland
   b. Lacrimal caruncle
   c. Lacrimal canal
   d. Nasolacrimal duct

5. Conjunctiva
6. Extraocular muscles
   a. Superior oblique
   b. Inferior oblique

   c. Superior rectus
   d. Inferior rectus
   e. Medial rectus
   f. Lateral rectus

**15**

TABLE **15.2** Model Inventory for the Eye

| Model/Diagram | Structures Identified |
|---------------|----------------------|
|               |                      |
|               |                      |
|               |                      |
|               |                      |

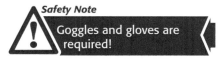

## Procedure Eyeball Dissection

In this exercise you will examine the structures of the eyeball on a fresh or preserved eyeball. I promise that eyeball dissection isn't as gross as it sounds!

**Safety Note**
⚠ Goggles and gloves are required!

**1** Examine the external anatomy of the eyeball (**Figure 15.8**), and record below the structures you can identify.

**2** Use scissors to remove the adipose tissue surrounding the eyeball. Identify the optic nerve.

**3** Hold the eyeball at its anterior and posterior poles, and use a sharp scalpel or scissors to make an incision in the frontal plane. Watch out, as aqueous humor and vitreous humor are likely to spill everywhere.

**4** Complete the incision, and separate the anterior and posterior portions of the eyeball (**Figure 15.9**). Take care to preserve the fragile retina—the thin, delicate yellow-tinted inner layer.

**5** List the structures you can identify in the anterior half of the eyeball (**Figure 15.10**):

**6** List the structures you can identify in the posterior half of the eyeball:

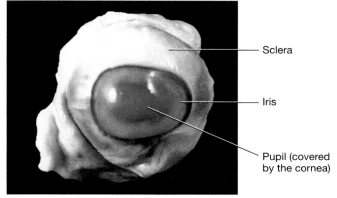

FIGURE **15.8** Anterior view of an eyeball

FIGURE **15.9** Frontal section of an eyeball showing the tunics

FIGURE **15.10** Posterior view of the anterior portion of the eyeball and lens

15

## Procedure Comparing the Distribution of Rods and Cones

We discussed earlier the unequal distribution of the photoreceptors in the retina. In this exercise you will see (no pun intended) firsthand the differences in vision produced by the rods and the vision produced by the cones. After you have completed the activity, answer Check Your Understanding questions 1 and 2 (p. 383).

**1** On a small sheet of paper, write the phrase "Anatomy is fun" in your regular-size print.

**2** Hold this piece of paper about 25 cm (10 inches) directly in front of your lab partner's eyes, and have your lab partner read the phrase.

Can your partner read the phrase clearly? _____

Which photoreceptors are producing the image? _____

**3** Now write a second phrase on the paper, and don't tell your partner what the phrase says. Hold the paper about 10 inches from your partner's peripheral vision field. Have him or her continue to stare forward and attempt to read what you have written.

Can your partner read the phrase clearly? _____

Which photoreceptors are producing the image? _____

**4** For the next test, dim the lights in the room. Have your partner stand 20 feet in front of a Snellen eye chart and read the chart. You should stand next to the chart to verify that the letters your partner reads are correct. This indicates your partner's vision relative to someone with perfect vision. For example, a person with 20/40 vision can see at 20 feet what someone with perfect vision could see at 40 feet. Record the number of the smallest line he or she can read without errors (e.g., 20/40).

Visual acuity: _____

**5** With the lights still dimmed and your partner standing in the same place, hold a piece of dark green or dark blue paper over the Snellen chart. Ask your partner to identify the color of the paper you are holding:

Paper color: _____

**6** Repeat the above processes with the lights illuminated:

Visual acuity: _____

Paper color: _____

**7** In which scenario were visual acuity and color vision better? Explain your findings.

_____

_____

_____

_____

_____

15

## Procedure Testing the Extraocular Muscles

In this procedure, you will determine which extraocular muscles are responsible for moving the eyeballs in each direction.

**1** Trace a line in the air about 1 foot in front of your partner's eyes, moving from your partner's left to right. Have your partner follow your finger without moving his or her head. Which extraocular muscles produce the movements you see for each eye?

Right eye _____

Left eye _____

**2** Now trace a diagonal line, starting at the upper right corner and moving to the lower left corner. Have your partner follow your finger again. Which extraocular muscles produce the movements you see for each eye?

Right eye _____

Left eye _____

**3** Again have your partner follow your finger, but trace a horizontal line from left to right. Which extraocular muscles produce the movements you see for each eye?

Right eye _____

Left eye _____

**4** Finally, trace another diagonal, this time from your partner's lower right to the upper left, and have your partner follow along (you should have traced out an hourglass shape overall). Which extraocular muscles produce the movements you see for each eye?

Right eye _____

Left eye _____

15

# Exercise 15-2

## Anatomy of the Ear, Hearing, and Equilibrium

### MATERIALS
❑ Ear models
❑ Tuning fork (500–1,000 Hz)
❑ Chalk

The ear contains structures both for hearing and equilibrium. It is divided into three regions: the outer, middle, and inner ear (**Figure 15.11**).

1. **Outer ear.** The outer ear begins with the **auricle**, or *pinna*, a shell-shaped structure composed of elastic cartilage that surrounds the opening to the **external auditory canal**. The external auditory canal extends about 2.5 cm into the temporal bone, where it ends in the **tympanic membrane**, which separates the outer ear from the middle ear.

2. **Middle ear.** The middle ear is a small air-filled cavity within the temporal bone that houses tiny bones called the **auditory ossicles**—the **malleus** (hammer), **incus** (anvil), and **stapes** (stirrup). The ossicles transmit vibrations from the tympanic membrane to the inner ear through a structure called the **oval window**. An additional structure in the middle ear is the **pharyngotympanic tube** (also called the Eustachian or auditory tube), which connects the middle ear to the pharynx (throat) and equalizes pressure in the middle ear.

3. **Inner ear.** The inner ear contains the sense organs for hearing and equilibrium. It consists of cavities collectively called the **bony labyrinth** filled with a fluid called **perilymph**. Within the perilymph is a series of membranes called the **membranous labyrinth** that contain a thicker fluid called **endolymph**. The bony labyrinth has three regions:

    a. **Vestibule.** The vestibule is an egg-shaped bony cavity that houses two structures responsible for equilibrium—the **saccule** and the **utricle**. Both structures transmit impulses down the vestibular portion of the vestibulocochlear nerve.

    b. **Semicircular canals.** Situated at right angles to one another, the semicircular canals house the **semicircular ducts** and the **ampulla,** which work together with the organs of the vestibule to maintain equilibrium. Their orientation allows them to sense rotational movements of the head and body. Like the saccule and utricle, the semicircular ducts and the ampulla transmit impulses down the vestibular portion of the vestibulocochlear nerve.

    c. **Cochlea.** The cochlea is a spiral bony canal that contains the **organ of Corti,** whose specialized **hair cells** transmit sound impulses to the cochlear portion of the vestibulocochlear nerve.

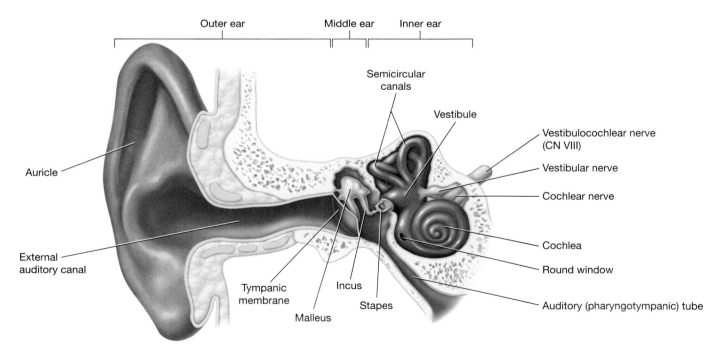

FIGURE **15.11** The anatomy of the ear

## Procedure  Model Inventory for the Ear

Identify the following structures of the ear on models and diagrams, using your textbook and this unit for reference. As you examine the anatomical models and diagrams, record on the model inventory in **Table 15.3** the name of the model and the structures you were able to identify. After you have completed the activity, answer Check Your Understanding questions 3 and 4 (p. 383).

1. Outer ear
   a. Auricle (pinna)
   b. External auditory canal
2. Middle ear
   a. Tympanic membrane
   b. Ossicles
      (1) Malleus
      (2) Incus
      (3) Stapes
   c. Oval window
   d. Round window
   e. Pharyngotympanic (auditory or Eustachian) tube

3. Inner ear
   a. Vestibule
      (1) Saccule
      (2) Utricle
   b. Semicircular canals
      (1) Semicircular duct
      (2) Ampulla
   c. Cochlea
      (1) Organ of Corti
      (2) Cochlear duct
   d. Vestibulocochlear nerve (cranial nerve VIII)
      (1) Vestibular nerve
      (2) Cochlear nerve

TABLE **15.3**  Model Inventory for the Ear

| Model/Diagram | Structures Identified |
|---|---|
|  |  |
|  |  |
|  |  |
|  |  |

15

## Hearing Acuity

There are two possible types of hearing loss:

1.  **Conductive hearing loss** results from interference of sound conduction through the outer and/or middle ear.

2.  **Sensorineural hearing loss** results from damage to the inner ear or the vestibulocochlear nerve.

Two clinical tests can help a health-care professional determine if hearing loss is conductive or sensorineural—the Weber test and the Rinne test. Both tests use tuning forks that vibrate at specific frequencies when struck. The tuning forks are placed directly on the bones of the skull to evaluate bone conduction—the ability to hear the vibrations transmitted through the bone. The forks are then held near the ear, not touching bone, to evaluate air conduction—the ability to hear the vibrations transmitted through the air. After you have completed both tests, answer Check Your Understanding question 5 (p. 384).

## Procedure Weber Test

**1** Obtain a tuning fork with a frequency of 500–1,000 Hz (cycles per second).

**2** Hold the tuning fork by the base, and strike it lightly with a mallet, or tap it on the edge of the table. The fork should begin ringing softly. If it is ringing too loudly, grasp the tines to stop it from ringing, and try again.

**3** Place the base of the vibrating tuning fork on the midline of your partner's head, as shown in **Figure 15.12**.

**4** Ask your partner if the sound is heard better in one ear or if the sound is heard equally in both ears. If the sound is heard better in one ear, this is called lateralization.

Was the sound lateralized? If yes, to which ear? _____

**5** To illustrate what it would sound like if the sound were lateralized, have your partner place his or her finger in one ear. Repeat the test.

**a** In which ear was the sound heard better? _____

**b** If a patient has conduction deafness, in which ear do you think the sound will be heard most clearly (the deaf ear or the good ear)? _____ Why? (If you are confused, think about your results when one ear was plugged.) _____

_____

_____

_____

**c** If a patient has sensorineural deafness, in which ear do you think the sound will be best heard?

_____

FIGURE **15.12** Weber test

15

## **Procedure** Rinne Test

**1** Strike the tuning fork lightly to start it ringing, as shown in **Figure 15.13**.

**2** Place the base of the tuning fork on your partner's mastoid process.

**3** Time the interval during which your partner can hear the sound. Your partner will have to tell you when he or she can no longer hear the ringing.

Time interval in seconds: _____

FIGURE **15.13** Rinne test

**4** Once your partner cannot hear the ringing, quickly move the still-vibrating tuning fork 1–2 cm lateral to the external auditory canal (the fork should not be touching your partner at this point).

**5** Time the interval from the point when you moved the tuning fork in front of the external auditory canal to when your partner can no longer hear the sound.

Time interval in seconds: _____

Which situation tested bone conduction? _____

Which situation tested air conduction? _____

**6** Typically, the air-conducted sound is heard twice as long as the bone-conducted sound. For example, if the bone-conducted sound was heard for 15 seconds, the air-conducted sound should be heard for 30 seconds.

Were your results normal? _____

What type of deafness is present if the bone-conducted sound is heard longer than the air-conducted sound?

15

## Equilibrium

A common and simple test of equilibrium is the **Romberg test**, in which the person is asked to stand still, first with the eyes open and then with the eyes closed. Under normal conditions, the vestibular apparatus should be able to maintain equilibrium in the absence of visual input. If the vestibular apparatus is impaired, however, the brain relies on visual cues to maintain balance.

### Procedure  Romberg Test

**1** Have your partner stand erect with the feet together and the arms at the sides in front of a chalkboard.

**2** Use chalk to draw lines on the board on either side of your partner's torso. These lines are for your reference in the next part.

**3** Have your partner stand in front of the chalkboard for 1 minute, staring forward with his or her eyes open. Use the lines on either side of his or her torso to note how much he or she sways as she stands. Below, record the amount of side-to-side swaying (i.e., minimal or significant):

_____

**4** Now have your partner stand in the same position for 1 minute with his or her eyes closed. Again note the amount of side-to-side swaying, using the chalk lines for reference.

Was the amount of swaying more or less with his/her eyes closed? _____

Why do you think this is so? _____

_____

_____

_____

What do you predict would be the result for a person with an impaired vestibular apparatus? Explain.

_____

_____

_____

_____

_____

_____

15

# Exercise 15-3

## Olfactory and Taste Senses

**MATERIALS**
- ❑ Head and neck models
- ❑ Tongue model

Both olfaction and taste are sometimes referred to as the **chemosenses**, because they both rely on chemoreceptors to relay information about the environment to the brain. The chemoreceptors of the olfactory sense are located in a small patch in the roof of the nasal cavity called the **olfactory epithelium** (**Figure 15.14**). The olfactory epithelium contains bipolar neurons called **olfactory receptor cells**. Their axons penetrate the holes in the cribriform plate to synapse on the olfactory bulb, which then sends the impulses down the axons of the olfactory nerves (cranial nerve I) to the olfactory cortex.

Taste receptors are located on taste buds housed on projections from the tongue called **papillae** (**Figure 15.15**). Of the four types of papillae—**filiform, fungiform, foliate**, and **circumvallate**—all but filiform papillae house taste buds. Fungiform papillae are scattered over the surface of the tongue, whereas the large circumvallate papillae are located at the posterior aspect of the tongue, arranged in a V shape.

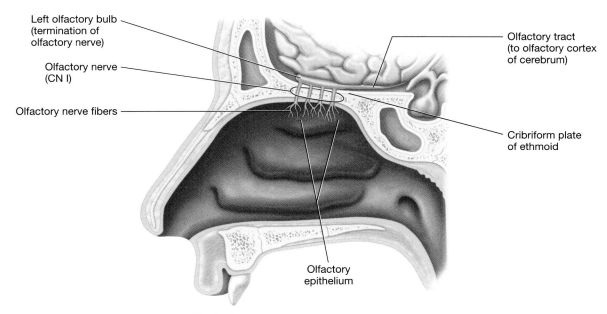

Left olfactory bulb
(termination of
olfactory nerve)

Olfactory nerve
(CN I)

Olfactory nerve fibers

Olfactory tract
(to olfactory cortex
of cerebrum)

Cribriform plate
of ethmoid

Olfactory
epithelium

FIGURE **15.14** The nasal cavity and olfactory epithelium

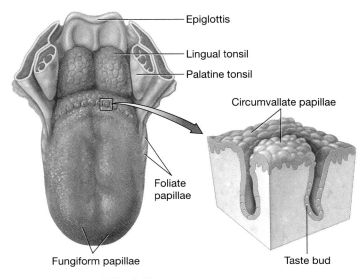

Epiglottis

Lingual tonsil

Palatine tonsil

Circumvallate papillae

Foliate
papillae

Fungiform papillae

Taste bud

FIGURE **15.15** The surface of the tongue

Identify the following structures of the olfactory and taste senses on anatomical models and charts. As you examine the anatomical models and diagrams, record on the model inventory in **Table 15.4** the name of the model and the structures you were able to identify. After you have completed the activity, answer Check Your Understanding question 6 (p. 384).

### Olfaction

1. Nasal cavity
   a. Nasal conchae
   b. Nasal septum
2. Olfactory epithelium
   a. Olfactory receptor cells
3. Olfactory bulbs
4. Olfactory nerves (cranial nerve I)
5. Cribriform plate

### Taste

1. Papillae
   a. Fungiform papillae
   b. Circumvallate papillae
   c. Foliate papillae
2. Taste buds

TABLE **15.4** Model Inventory for Olfaction and Taste

| Model/Diagram | Structures Identified |
|---|---|
|  |  |
|  |  |
|  |  |
|  |  |

15

# Exercise 15-4

## The General Senses: Cutaneous Sensation

### MATERIALS

- ❑ Water-soluble marking pens (two colors)
- ❑ Ruler
- ❑ 2 wooden applicator sticks (or toothpicks)

Sensory receptors in the skin respond to different stimuli, including temperature, touch, and pain. These receptors are not distributed throughout the skin equally but, instead, are concentrated in certain regions of the body. The following experiments will allow you to determine the relative distribution of the receptors for touch in the skin by performing two tests: the error of localization and two-point discrimination. After you complete the activities, answer Check Your Understanding question 7 (p. 384).

### Error of Localization

Every region of the skin corresponds to an area of the somatosensory association area in the cerebral cortex. Some regions are better represented than others and, therefore, are capable of localizing stimuli with greater precision than are less well-represented areas. The **error of localization** (also called tactile localization) tests the ability to determine the location of the skin touched and demonstrates how well-represented each region of the skin is in the cerebral cortex.

## **Procedure** Testing Error of Localization

**1** Have your partner sit with his or her eyes closed.

**2** Use a water-soluble marking pen to place a mark on your partner's anterior forearm.

**3** Using a different color of marker, have your partner, still with his or her eyes closed, place a mark as close as possible to where he or she believes the original spot is located.

**4** Use a ruler to measure the distance between the two points in millimeters. This is your error of localization.

**5** Repeat this procedure for each of the following locations:
- **a** Anterior thigh
- **b** Face
- **c** Palm of hand
- **d** Fingertip

**6** Record your data in **Table 15.5**.

TABLE **15.5** Error of Localization

| Location | Error of Localization (mm) |
|---|---|
| Anterior forearm | |
| Anterior thigh | |
| Face | |
| Palm of hand | |
| Fingertip | |

# Two-Point Discrimination

The **two-point discrimination test** assesses the ability to perceive the number of stimuli ("points") placed on the skin. Areas that have a higher density of touch receptors are better able to distinguish between multiple stimuli than those with fewer touch receptors.

## Procedure Testing Two-Point Discrimination

**1** Have your partner close his or her eyes.

**2** Place the ends of two wooden applicator sticks close together (they should be nearly touching) on your partner's skin on the anterior forearm. Ask your partner how many points he or she can discriminate—one or two.

**3** If he or she can sense only one point, move the sticks farther apart. Repeat this procedure until your partner can distinguish two separate points touching his or her skin.

**4** Use a ruler to measure the distance between the two sticks in millimeters. This is your two-point discrimination.

**5** Repeat this procedure for each of the following locations:

  **a** Anterior thigh

  **b** Face (around the lips and/or eyes)

  **c** Vertebral region

  **d** Fingertip

**6** Record your data in Table 15.6.

TABLE **15.6** Two-Point Discrimination

| Location | Two-Point Discrimination (mm) |
|---|---|
| Anterior forearm | |
| Anterior thigh | |
| Face | |
| Vertebral region | |
| Fingertip | |

**7** What results did you expect for each test? Explain.

**8** Did your observations agree with your expectations? Interpret your results.

UNIT
15
REVIEW

## Check Your Recall

**1** Label the following parts of the eyeball on **Figure 15.16.**

☐ Choroid
☐ Cornea
☐ Ciliary body
☐ Iris
☐ Lens

☐ Optic nerve
☐ Posterior cavity
☐ Retina
☐ Sclera
☐ Suspensory ligaments

15

FIGURE **15.16** The eyeball, sagittal section

**2** The lacrimal gland is located in the _____ and produces _____.
   a. superolateral orbit; mucus
   b. inferomedial orbit; tears
   c. superolateral orbit; tears
   d. inferomedial orbit; mucus

**3** The rods are responsible for _____, whereas the cones are responsible for _____.
   a. peripheral and dim light vision; high-acuity color vision
   b. high-acuity color vision; peripheral and dim light vision
   c. peripheral and color vision; high-acuity and dim light vision
   d. high-acuity and dim light vision; peripheral and color vision

**4** Label the following parts of the ear on **Figure 15.17**.

☐ Auricle                  ☐ Malleus                   ☐ Tympanic membrane
☐ Cochlea                  ☐ Pharyngotympanic tube     ☐ Vestibule
☐ External auditory canal  ☐ Semicircular canals       ☐ Vestibulocochlear nerve
☐ Incus                    ☐ Stapes

FIGURE **15.17** The anatomy of the ear

**5** The auditory ossicles transmit vibrations from the _____ to the _____.
   a. auricle; malleus
   b. tympanic membrane; inner ear
   c. tympanic membrane; middle ear
   d. auricle; inner ear

**6** The structures of the cochlea are responsible for _____, whereas the structures of the vestibule and semicircular canals are responsible for _____.

a. equilibrium; balance

b. equilibrium; hearing

c. hearing; equilibrium

d. hearing; audition

**7** *Mark the following statements as true (T) or false (F). If the statement is false, correct it so it is a true statement.*

_____ a. Conductive hearing loss results from damage to the inner ear or the vestibulocochlear nerve.

_____ b. The ciliary muscle controls the diameter of the iris.

_____ c. The Weber and Rinne tests assess balance and equilibrium.

_____ d. The six extraocular muscles move the eyeball.

_____ e. Smell and taste are considered chemosenses.

_____ f. The photoreceptors are located in the vascular tunic of the eyeball.

**8** The receptors for smell are located in the

a. gustatory mucosa.

b. olfactory epithelium.

c. olfactory fossa.

d. squamous epithelium.

**9** Taste buds are located on the

a. filiform, fungiform, and circumvallate papillae.

b. filiform, fungiform, and foliate papillae.

c. filiform and circumvallate papillae only.

d. fungiform, foliate, and circumvallate papillae.

**10** You would expect the error of localization and the two-point discrimination threshold to be lowest on the

a. back.

b. forearm.

c. fingertip.

d. thigh.

15

**UNIT 15 REVIEW**

# Check Your Understanding

### Critical Thinking and Application Questions

**1** The disease *macular degeneration* is characterized by a gradual loss of vision as a result of degeneration of the macula lutea. Considering the type of cells located in the macula lutea, which type of vision do you think a sufferer of macular degeneration would lose? Why?

_____

_____

_____

_____

**2** How would the signs and symptoms differ from those in question 1 in a condition that caused degeneration of rods?

_____

_____

_____

_____

**3** What signs and symptoms would you expect to see from *otitis interna* (inner ear infection)? Why?

_____

_____

_____

_____

**4** Explain why infectious *otitis media* (inflammation of the middle ear) may result in a simultaneous *pharyngitis* (inflammation of the throat).

_____

_____

_____

_____

**5** *Otosclerosis* is a condition that results in irregular ossification (bone formation) around the stapes bone. Would you expect this to result in conductive or in sensorineural hearing loss? What results would you expect from the Rinne and Weber tests in an individual with otosclerosis?

**6** The chemoreceptors of the tongue respond to monosaccharides (sugars), metal ions (salty tastes), alkaloids (bitter tastes), and acids (sour tastes). Which type of receptor do you think is most sensitive? Why? (*Hint:* One of the four types of chemicals is present in many plant poisons.)

**7** The *sensory homunculus* is a small figure of a man whose body parts are represented in proportion to the amount of the somatosensory cortex dedicated to them. Parts of the body that are well represented in the somatosensory cortex are drawn as larger, and those that are poorly represented in the somatosensory cortex are drawn as smaller. Which areas of the body do you think would be larger on the homunculus? Which areas do you think would be smaller? Explain your reasoning.

# Cardiovascular System—Part I
## The Heart

**16**

Name _____ Section _____ Date _____

# PRE-LAB EXERCISES

Complete the following exercises prior to coming to lab, using your textbook and lab manual for reference.

## Pre-Lab Exercise 16-1

 Key Terms

You should be familiar with the following terms before coming to lab.

| Term | Definition |
|------|------------|

**Layers of the Heart Wall**

Fibrous pericardium _____

_____

Pericardial cavity _____

_____

Myocardium _____

_____

Endocardium _____

_____


**Structures of the Heart**

Atria (right and left) _____

_____

Ventricles (right and left) _____

_____

Tricuspid valve_____

_____

Mitral (bicuspid) valve _____

_____

Pulmonary valve _____

_____

Aortic valve _____

_____

Chordae tendineae _____

_____

Papillary muscles _____

## Great Vessels

Superior vena cava _____

Inferior vena cava _____

Pulmonary trunk _____

Pulmonary veins _____

Aorta _____

## Structures of the Cardiac Conduction System

Sinoatrial (SA) node _____

Atrioventricular (AV) node _____

Atrioventricular bundle (bundle of His) _____

Bundle branches _____

Purkinje fibers _____

# Pre-Lab Exercise **16-2**

## Anatomy of the Thoracic Cavity

Label and color the structures in the thoracic cavity in **Figure 16.1** with the terms from Exercise 16-1 (p. 393). Use your text and Exercise 16-1 in this unit for reference.

**16**

FIGURE **16.1** The thoracic cavity

# Pre-Lab Exercise **16-3**

## Anatomy of the Heart

Label and color the three views of the heart in **Figure 16.2** with the terms from Exercise 16-1 (p. 393). Use your text and
Exercise 16-1 in this unit for reference.

**16**

FIGURE **16.2** The heart: (**A**) anterior view;
(**B**) posterior view *(continues)*

Figure **16.2** The heart *(cont.)*: (**C**) frontal section

16

# Pre-Lab Exercise 16-4

## Pathway of Blood Flow through the Heart

Answer the following questions about the pathway of blood flow through the heart. Use your textbook and Exercise 16-1 (p. 393) in this unit for reference.

1.  Regarding veins:
    a.  Where do veins carry blood? _____

    b.  Is this blood generally oxygenated or deoxygenated? _____

    c.  Does this rule have any exceptions? If yes, where? _____

2.  Regarding arteries:
    a.  Where do arteries carry blood? _____

    b.  Is this blood generally oxygenated or deoxygenated? _____

    c.  Does this rule have any exceptions? If yes, where? _____

3.  Where does each atrium pump blood when it contracts?

    a.  Right atrium: _____

    b.  Left atrium:_____

4.  Where does each ventricle pump blood when it contracts?

    a.  Right ventricle: _____

    b.  Left ventricle: _____

# EXERCISES

The cardiovascular system transports oxygen, nutrients, wastes, other solutes, and cells throughout the body. In this unit we begin our exploration of the cardiovascular system with the pump that drives it—the heart. The heart is a remarkable organ, tirelessly beating more than 100,000 times per day to pump more than 8,000 liters of blood around the body. In this unit we examine the anatomy of this remarkable organ, including the blood flow through the heart and the histology of cardiac muscle. We also examine the electrical events of cardiac action potentials and the group of cells called the *cardiac conduction system* that allows the heart to function as a unit.

## Exercise **16-1**

### Anatomy of the Heart

> **MATERIALS**
> ❏ Heart models
> ❏ Preserved heart
> ❏ Dissection equipment
> ❏ Dissection tray
> ❏ Water-soluble marking pen
> ❏ Laminated outline of the heart and lungs

The heart lies in the mediastinum and is, on average, about the size of a fist (**Figure 16.3**). The heart is surrounded by a double-layered membrane called the **pericardium**. The outermost layer of the pericardium, called the **fibrous pericardium**, anchors the heart to surrounding structures. It is made of dense irregular collagenous connective tissue that is not very distensible, which helps to prevent the heart from overfilling.

The inner layer, called the **serous pericardium**, is itself composed of two layers. The outer portion, called the **parietal pericardium**, is functionally fused to the fibrous pericardium. The inner portion is attached to the heart muscle and is called the **visceral pericardium**, also known as the **epicardium**. Between the parietal and visceral layers we find a thin layer of serous fluid that occupies a narrow space called the **pericardial cavity**. The fluid within the pericardial cavity helps the heart to beat without friction.

The heart itself is an organ that consists of three tissue layers:

1. **Epicardium**. The epicardium or visceral pericardium is considered the outermost layer of the heart wall. It consists of a layer of epithelial tissue and loose connective tissue.

**16**

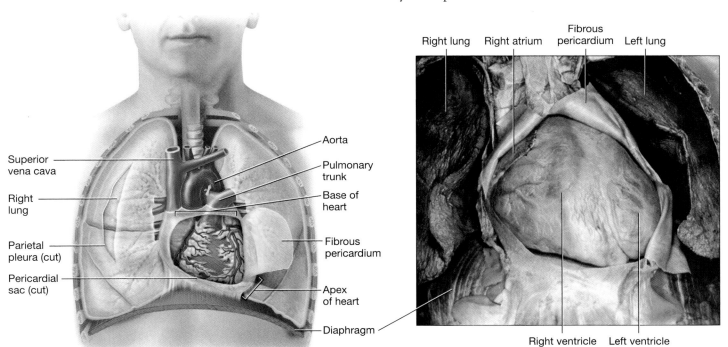

Figure **16.3** The thoracic cavity

2. **Myocardium.** The middle myocardium is the actual muscle of the heart. It consists of cardiac muscle tissue and its fibrous skeleton.

3. **Endocardium.** The innermost endocardium is a type of simple squamous epithelium called **endothelium**. It is continuous with the endothelium lining all blood vessels in the body.

As you can see in **Figure 16.4**, the heart consists of four hollow chambers: two **atria** that receive blood from the body's **veins**, and two **ventricles** that eject blood into the body's **arteries**. In between the atria is a thin wall called the **interatrial septum**. The much thicker **interventricular septum** separates the two ventricles. The heart's four chambers include the:

1. **Right atrium.** The right atrium is the superior right chamber. It receives deoxygenated blood from the body's main veins—the **superior vena cava**, the **inferior vena cava**, and the **coronary sinus**. These veins drain a series of blood vessels collectively called the **systemic circuit**, in which gases and nutrients are exchanged between the blood and the tissue cells.

2. **Right ventricle.** The right ventricle is a large chamber inferior to the right atrium, from which it receives deoxygenated blood. It ejects blood into a vessel called the **pulmonary trunk**, or the **pulmonary artery**. The pulmonary trunk branches into right and left pulmonary arteries, which deliver deoxygenated blood to the lungs through a series of vessels collectively called the **pulmonary circuit**. Within the pulmonary circuit, gases are exchanged, and the blood becomes oxygenated.

3. **Left atrium.** The superior left chamber is the left atrium. It receives oxygenated blood returning from the pulmonary circuit via four **pulmonary veins**.

4. **Left ventricle.** The left ventricle receives oxygenated blood from the left atrium and pumps it into the largest artery in the systemic circuit, the **aorta**. The aorta then branches repeatedly to deliver the oxygenated blood to the body's cells. Note that the left ventricle is considerably thicker than the right ventricle, reflecting the fact that the pressure is much higher in the systemic circuit than it is in the pulmonary circuit. The higher pressure requires the left ventricle to pump harder, and thus it is thicker.

In between the chambers of the heart and the ventricles and their vessels are **valves** that prevent the blood from flowing backward in the heart. The valves between the chambers are called **atrioventricular valves**. The three-cusped **tricuspid valve** is between the right atrium and right ventricle, and the two-cusped **mitral** or **bicuspid valve** is between the left atrium and left ventricle. Each cusp of the atrioventricular valves is attached to collagenous "strings" called **chordae tendineae**, attached to muscles within the ventricular wall called **papillary muscles**. When the ventricles contract, the papillary muscles pull the chordae tendineae taut, which puts tension on the cusps and prevents them from everting into the atria, a condition called **prolapse**.

The valves between the ventricles and their arteries are called **semilunar valves**. The **pulmonary valve** lies between the right ventricle and the pulmonary trunk, and the **aortic valve** lies between the left ventricle and the aorta. Note that there are no chordae tendineae or papillary muscles attached to the semilunar valves.

## HINTS & TIPS

### Red or Blue?

On anatomical models, vessels that carry oxygenated blood are red, whereas those that carry deoxygenated blood are blue. Systemic arteries carry oxygenated blood to the body's cells and so are red on anatomical models. Systemic veins, on the other hand, carry deoxygenated blood back to the right atrium and so are blue. But be sure to remember that the reverse is true in the pulmonary circuit: The pulmonary arteries carry deoxygenated blood to the lungs and the pulmonary veins carry oxygenated blood to the heart. So, in the pulmonary circuit, the arteries are blue and the veins are red.

The final structures you will examine in this lab period are the vessels of the **coronary circulation**. The **coronary arteries** branch off the base of the aorta and bring oxygenated blood to the cells of the myocardium, and they are drained by a set of **cardiac veins**. The first coronary artery, the **right coronary artery**, travels in the right coronary sulcus. It branches into the **marginal artery**, which serves the lateral part of the right atrium and right ventricle, and the **posterior interventricular artery**, which serves the posterior heart. The other coronary artery is the **left coronary artery**, which branches shortly after it forms into the **anterior interventricular artery** (also known as the **left anterior descending artery**), which travels along the interventricular septum to supply the anterior heart, and the **circumflex artery**, which travels in the

**A**

Superior vena cava

Ascending aorta

Branches of right pulmonary artery

Auricle of right atrium

Right pulmonary veins

Right atrium

Right coronary artery and vein

Right ventricle

Inferior vena cava

Thoracic aorta

Aortic arch

Ligamentum arteriosum

Branches of left pulmonary artery

Pulmonary trunk

Left atrium

Left pulmonary veins

Circumflex artery

Anterior interventricular artery

Anterior interventricular vein

Left ventricle

Apex of heart

**B**

Aortic arch

Left pulmonary artery

Left pulmonary veins

Left atrium

Auricle of left atrium

Great cardiac vein

Left ventricle

Apex

Middle cardiac vein

Right ventricle

Posterior interventricular artery (in posterior interventricular sulcus)

Superior vena cava

Right pulmonary artery

Right pulmonary veins

Right atrium

Coronary sinus

Inferior vena cava

Right coronary artery (in coronary sulcus)

**16**

**C**

Superior vena cava

Opening of coronary sinus

Branches of right pulmonary artery

Interatrial septum

Right atrium

Right pulmonary veins

Pectinate muscles

Tricuspid valve

Right ventricle

Myocardium

Inferior vena cava

Aortic arch

Left pulmonary artery

Pulmonary trunk

Pulmonary valve

Left atrium

Aortic valve

Mitral valve

Left ventricle

Chordae tendinae

Papillary muscle

Interventricular septum

Endocardium

Visceral pericardium

Trabeculae carneae

Figure **16.4** The heart: (**A**) anterior view; (**B**) posterior view; (**C**) frontal section left coronary sulcus to supply the left atrium and posterior left ventricle. When a coronary artery is blocked, the reduced blood flow to the myocardium can result in hypoxic injury and death to the tissue, a condition termed **myocardial infarction** (commonly called a heart attack).

The anatomy of the cardiac veins often varies from person to person, but the following three main veins generally are present:

1. **small cardiac vein**, which drains the inferolateral heart,

2. **middle cardiac vein**, which drains the posterior heart, and

3. **great cardiac vein**, which drains most of the left side of the heart.

All three veins drain into the large **coronary sinus** located on the posterior right atrium. The coronary sinus drains into the right atrium.

## Procedure Model Inventory for the Heart

Identify the following structures of the heart on models and diagrams, using your textbook and this unit for reference. As you examine the anatomical models and diagrams, record on the model inventory in **Table 16.1** the name of the model and the structures you were able to identify. When you have completed the activity, answer Check Your Understanding questions 1–3 (p. 409).

1. General structures
   a. Mediastinum
      (1) Pericardial cavity
   b. Pericardium
      (1) Fibrous pericardium
      (2) Parietal pericardium
      (3) Visceral pericardium (epicardium)
   c. Myocardium
   d. Endocardium
   e. Apex of the heart
   f. Base of the heart
   g. Anterior interventricular sulcus
   h. Posterior interventricular sulcus
   i. Coronary sulcus
2. Right atrium
   a. Opening of the superior vena cava
   b. Opening of the inferior vena cava
   c. Opening of the coronary sinus
   d. Right auricle
   e. Interatrial septum
   f. Fossa ovalis
   g. Pectinate muscles
3. Left atrium
   a. Opening of the pulmonary veins
   b. Left auricle
   c. Interatrial septum
4. Structures of the ventricles
   a. Right ventricle
   b. Left ventricle
   c. Interventricular septum
   d. Chordae tendineae
   e. Papillary muscles
   f. Trabeculae carneae

5. Atrioventricular valves
   a. Tricuspid valve
   b. Mitral valve
6. Semilunar valves
   a. Pulmonary valve
   b. Aortic valve
7. Great vessels
   a. Superior vena cava
   b. Inferior vena cava
   c. Pulmonary trunk (artery)
   d. Right and left pulmonary arteries
   e. Pulmonary veins
   f. Aorta
   g. Ligamentum arteriosum
8. Coronary arteries
   a. Right coronary artery
   b. Marginal artery
   c. Posterior interventricular artery
   d. Left coronary artery
   e. Anterior interventricular artery
   f. Circumflex artery
9. Cardiac veins
   a. Small cardiac vein
   b. Middle cardiac vein
   c. Great cardiac vein
   d. Coronary sinus

TABLE **16.1** Model Inventory for the Heart

| Model | Structures Identified |
| --- | --- |
|  |  |
|  |  |
|  |  |
|  |  |
|  |  |
|  |  |
|  |  |

## **Procedure** Heart Dissection

You will now examine a preserved heart or a fresh heart, likely from a sheep or a cow. Follow the procedure below to find the structures indicated. When you have completed the dissection, answer Check Your Understanding question 4 (p. 410).

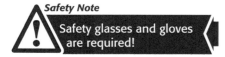

**1** Orient yourself by first determining the superior aspect and the inferior aspect of the heart. The superior aspect of the heart is the broad end, and the inferior aspect (apex) is the pointy tip. Now orient yourself to the anterior and posterior sides. The easiest way to do this is to locate the pulmonary trunk—the vessel directly in the middle of the anterior side. Find the side from which the pulmonary trunk originates, and you will be on the anterior side, which you can see in **Figure 16.5**. Structures to locate at this time are the

   a. parietal pericardium (may not be attached),
   b. visceral pericardium (shiny layer over the surface of the heart),
   c. aorta,
   d. pulmonary trunk,
   e. superior vena cava,
   f. inferior vena cava,
   g. pulmonary veins,
   h. ventricles, and
   i. atria.

Finding the coronary vessels tends to be difficult because the superficial surface of the heart is covered with adipose tissue. To see the coronary vessels, carefully dissect the adipose tissue.

**2** Locate the superior vena cava. Insert scissors or a scalpel into the superior vena cava and cut down into the right atrium. Before moving on to step 3, note the structure of the tricuspid valve, and draw it below. How many flaps do you see? What is the function of this valve?

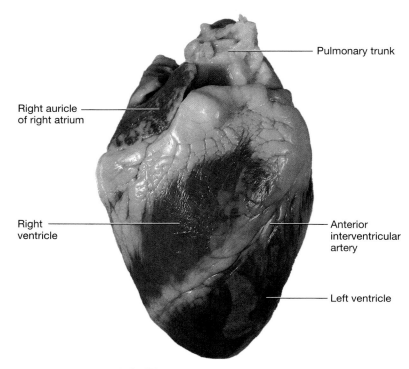

Figure **16.5** Anterior view of a sheep heart

**16**

**3** Once the right atrium is exposed, continue the cut down into the right ventricle, which is shown in **Figure 16.6**. Structures to locate at this time include the

a. tricuspid valve,

b. chordae tendineae,

c. papillary muscles,

d. myocardium, and

e. endocardium (shiny layer on the inside of the heart).

**4** Insert the scissors into the pulmonary trunk. Note the structure of the pulmonary valve, and draw it below. How does it differ structurally from the tricuspid valve? What is the function of this valve?

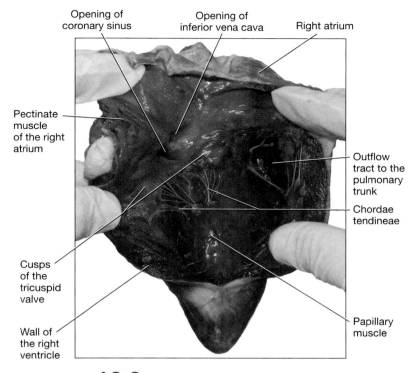

Figure **16.6** The right ventricle of a sheep heart

**16**

**5** Insert the scissors into a pulmonary vein. Cut down into the left atrium. Note the structure of the mitral valve, and draw it below. What is the function of this valve? How does its structure differ from that of the pulmonary and tricuspid valves?

**6** Continue the cut into the left ventricle. Note the thickness of the left ventricle, as shown in **Figure 16.7**. How does it compare with the thickness of the right ventricle? Why is there a difference?

_____

_____

_____

_____

_____

_____

_____

_____

_____

_____

_____

Figure **16.7** The left ventricle of a sheep heart

**7** Insert the scissors into the aorta. Extend the cut until you can see the aortic valve. Draw the aortic valve below. Is it structurally more similar to the pulmonary valve or the mitral valve? What is the function of this valve?

_____

_____

**8** Your instructor may wish you to identify other structures on the heart. List any additional structures below.

## **Procedure** Tracing Blood through the Heart

Use water-soluble markers and a laminated outline of the heart to trace the pathway of blood as it flows through the heart and pulmonary circulation. Use a blue marker to indicate areas that contain deoxygenated blood and a red marker to indicate areas that contain oxygenated blood. If no laminated outline is available, use **Figure 16.8**. When you have completed the activity, answer Check Your Understanding question 5 (p. 410).

Figure **16.8** Heart, lungs, and pulmonary circulation

## Exercise 16-2

### Cardiac Muscle Histology

**MATERIALS**

❏ Cardiac muscle tissue slide
❏ Light microscope
❏ Colored pencils

Recall from Unit 5 (p. 85) that cardiac muscle tissue is striated like skeletal muscle tissue but otherwise is quite different (**Figure 16.9**). Following are some important differences:

▌ Cells of cardiac muscle, known as **cardiac myocytes**, are shorter, wider, and branched, whereas skeletal muscle fibers are long, thin, and unbranched.

▌ Cardiac muscle cells typically are uninucleate. The nucleus is generally located in the center of the cell.

▌ These cells contain specialized adaptations called **intercalated disks** that appear as dark lines parallel to the striations. They contain desmosomes and gap junctions that hold adjacent cardiac cells tightly together to allow the cells to communicate chemically and electrically. The intercalated disks enable the heart to contract as a unit.

Figure **16.9** Cardiac muscle tissue, photomicrograph

**Procedure** Microscopy of Cardiac Muscle Tissue

Examine a prepared slide of cardiac muscle tissue on high power. Use colored pencils to draw what you see, and label the structures indicated. When you have completed the activity, answer Check Your Understanding question 6 (p. 410). Label the following structures:

1. Cardiac myocyte
2. Intercalated disks
3. Striations
4. Nucleus

# Exercise **16-3**

## The Cardiac Conduction System

The heart has two populations of cardiac myocytes. The majority of the cells (about 99%) are known as **nonpacemaker cells**, or regular, contractile myocytes. The remaining 1% are specialized cells known as **pacemaker cells**. Pacemaker cells have the unique property of spontaneously depolarizing and generating action potentials. The action potentials they initiate trigger action potentials of the non-pacemaker cells, the stimulus for these cells to contract.

The heart's pacemaker cells are located at specific places throughout the heart in clusters. Together, these clusters of pacemaker cells make up the **cardiac conduction system**. The parts of the cardiac conduction system, shown in **Figure 16.10**, include the following:

1. **Sinoatrial (SA) node.** The SA node is located in the upper right atrium, where it acts as the main pacemaker of the heart. It depolarizes spontaneously around 60 times per minute. (Note that the heart rate is generally higher than 60 because of the influence of the autonomic nervous system.)

2. **Atrioventricular (AV) node.** The AV node is located in the lower right atrium just medial to the tricuspid valve. It acts as a backup pacemaker and is capable of pacing the heart at about 40 beats per minute. Conduction through the AV node is slow, producing what is known as the **AV node delay**. This delay allows the atria to depolarize and contract before the ventricles, which allows the ventricles to fill with blood.

3. **Purkinje system.** The impulses are transmitted from the AV node to the ventricles by the group of pacemaker cells collectively called the Purkinje system. The myocytes of the Purkinje system pace the heart slowly but conduct impulses more rapidly than any part of the conduction system. The Purkinje system has three components:

    a. **Atrioventricular (AV) bundle.** The AV bundle is the small group of fibers in the lower interatrial septum and upper interventricular septum that transmits impulses from the AV node to the ventricles.

    b. **Right and left bundle branches.** Impulses are transmitted from the AV bundle down either side of the interventricular septum by the right and left bundle branches.

    c. **Purkinje fibers.** At the end of the interventricular septum, the right and left bundle branches fan out through the myocardium as the Purkinje fibers. These fibers extend about one-third of the way into the heart muscle, after which they blend with regular nonpacemaker cardiac myocytes.

16

Figure **16.10** The cardiac conduction system

The general pathway of impulse conduction through the heart starts with the cells of the SA node generating an action potential. This pacemaker action potential spreads to the nonpacemaker cells of the atria and triggers them to have action potentials and contract. The impulses are carried through the atria until they hit the AV node, where they encounter a delay. Once the AV nodal cells are depolarized, this spreads to the Purkinje system and finally to the nonpacemaker cardiac myocytes of the ventricles. These cells are then triggered to depolarize and contract. Note that while ventricular depolarization is occurring, the atria are repolarizing. As the ventricles are repolarizing, the cells of the SA node are slowly depolarizing, which begins the cycle again.

## Procedure Tracing the Electrical Events of the Heart

Let's put all of these electrophysiological events together and trace the electrical events that occur within the heart, including

- the steps of the pacemaker action potentials that initiate the cycle;
- the spread of electrical activity through the atria;
- the steps of the nonpacemaker action potentials of the atria;
- the spread of electrical activity through the ventricles;
- the steps of the nonpacemaker action potentials of the ventricles through repolarization.

Note that many of the steps of the tracing focus on ion channels and ion movements during the different stages of the pacemaker and nonpacemaker action potentials. You will want to refer to your textbook for parts of this exercise—particularly the steps of the action potentials—to make sure that you have all of the ion channels and ion movements correct. You will find hints and blank spaces throughout this tracing exercise to ensure that you stay on the right track. When you have completed the exercise, answer Check Your Understanding question 7 (p. 410).

**Start:** SA nodal cells slowly depolarize to threshold →

_____ open and the _____ depolarize →

the _____ spreads to the _____ cells →

nonpacemaker atrial cells depolarize to threshold →

_____ channels _____ and the _____ cells _____ →

_____ channels _____ and the cells begin _____ →

calcium ion channels open, calcium ions exit the cell, and the nonpacemaker atrial cells enter the plateau phase →

calcium ion channels _____ and the cells _____ →

the _____ spreads through the _____ until it reaches the _____ →

impulse delayed at the AV node →

impulse conducted through the _____ →

impulse conducted through the _____ →

impulse conducted through the _____ →

nonpacemaker ventricular cells depolarize to threshold →

_____ channels open and the _____ cells _____ →

_____ channels open and the _____ cells begin _____ →

_____ channels open and the _____ begins →

nonpacemaker ventricular cells repolarize as potassium ions exit the cells and return them to their resting membrane potential. **End**

# Check Your Recall

**1** Label the following parts of the heart on **Figure 16.11**.

☐ Anterior interventricular artery
☐ Aorta
☐ Circumflex artery
☐ Inferior vena cava
☐ Pulmonary trunk
☐ Pulmonary veins
☐ Right coronary artery
☐ Superior vena cava

Figure **16.11** Heart, anterior view

**2** Label the following parts of the heart on **Figure 16.12.**

- ☐ Aortic valve
- ☐ Chordae tendineae
- ☐ Interventricular septum
- ☐ Left atrium
- ☐ Left ventricle
- ☐ Mitral valve
- ☐ Papillary muscles
- ☐ Pulmonary valve
- ☐ Right atrium
- ☐ Right ventricle
- ☐ Tricuspid valve

Figure **16.12** Heart, frontal section

**3** *Matching:* Match the following terms with the correct definition.

_____ Myocardium

_____ Parietal pericardium

_____ Tricuspid valve

_____ Aortic valve

_____ Papillary muscles

_____ Pulmonary veins

_____ Mitral valve

_____ Visceral pericardium

_____ Pulmonary trunk

_____ Coronary sinus

_____ Chordae tendineae

_____ Aorta

A. Located between the left ventricle and the aorta
B. Located between the left atrium and left ventricle
C. Largest artery of the systemic circuit
D. Bring(s) oxygenated blood to the left atrium from the lungs
E. Layer of the heart composed of cardiac muscle tissue
F. Bring(s) deoxygenated blood to the lungs
G. Located between the right atrium and the right ventricle
H. Largest vein of the coronary circulation
I. Outer layer of the serous pericardium
J. Fibrous cord(s) that attach(es) to valves
K. Inner layer of the serous pericardium
L. Fingerlike muscular projections from the ventricles

**4** The veins of the systemic circuit carry _____ blood, and the veins of the pulmonary circuit carry _____ blood.

a. oxygenated; deoxygenated
b. oxygenated; oxygenated
c. deoxygenated; deoxygenated
d. deoxygenated; oxygenated

**5** The pulmonary and aortic valves are known as the

a. atrioventricular (AV) valves.
b. semilunar valves.
c. coronary valves.
d. chordae tendineae.

**6** The right and left coronary arteries are the first branches off the

a. aorta.
b. superior vena cava.
c. pulmonary trunk.
d. pulmonary veins.

**7** *Fill in the blanks:* Cardiac muscle cells are also known as _____. Adjacent cells are joined together by _____, which allow the heart to _____.

**8** Label the following components of the cardiac conduction system on **Figure 16.13**.

- ☐ AV bundle
- ☐ AV node
- ☐ Left bundle branch
- ☐ Purkinje fibers
- ☐ Right bundle branch
- ☐ SA node

Figure **16.13** The cardiac conduction system

**9** Which of the following is the normal pacemaker of the heart?
a. AV node
b. AV bundle
c. Purkinje fibers
d. SA node

**10** The purpose of the AV node delay is to
a. allow the atria and ventricles to depolarize and contract simultaneously.
b. allow the ventricles to depolarize and contract before the atria.
c. allow the atria to depolarize and contract before the ventricles and allow the ventricles to fill.
d. allow the right side of the heart to depolarize and contract before the left side of the heart.

# Check Your Understanding

### Critical Thinking and Application Questions

**1** When the pericardium fills with blood, it produces a condition called *cardiac tamponade*, which can be rapidly lethal. Why is this condition so dangerous? (*Hint:* Consider the structure of the fibrous pericardium.)

_____

_____

_____

_____

_____

_____

**2** High pressure in the systemic and pulmonary circuits often results in *ventricular hypertrophy*, in which the ventricle enlarges to pump against greater force. Which side(s) of the heart would be affected by high pressure in the pulmonary circuit? Which side(s) of the heart would be affected by high pressure in the systemic circuit? Explain.

_____

_____

_____

_____

_____

**16**

**3** One potential cause of valve dysfunction is rupture of the chordae tendineae. Why would this lead to valve dysfunction? Would this affect the atrioventricular valves, the semilunar valves, or both? Explain.

_____

_____

_____

_____

_____

**4** The condition known as *atrial septal defect* is characterized by the presence of a hole in the interatrial septum. How would this condition affect the normal pattern of blood flow? What effect would this have on the oxygenation of the blood?

_____

_____

_____

_____

_____

**5** In the heart dissection you performed, you saw the smooth texture of the valves and how the leaflets fit together. How do you think the function of the valves would be affected if the valves were tough and filled with calcium deposits? Explain.

_____

_____

_____

_____

_____

**6** Skeletal muscle cells exhibit a phenomenon known as *recruitment*, in which the number of muscle cells recruited to contract is proportional to the strength of muscle contraction needed. In this way, we activate a few fibers to produce a small contraction to pick up a piece of paper, and we activate many fibers to produce a larger contraction to pick up a textbook. Would you expect to see recruitment in cardiac muscle tissue? Why or why not?

_____

_____

_____

_____

**7** One of the more common electrical abnormalities seen in the heart is a *heart block*, in which conduction is blocked or delayed along some part of the cardiac conduction system. Predict what you would see if a heart block were present in each of the following locations:

**a** AV node:

_____

**b** Right or left bundle branches:

_____

**c** Purkinje fibers:

_____

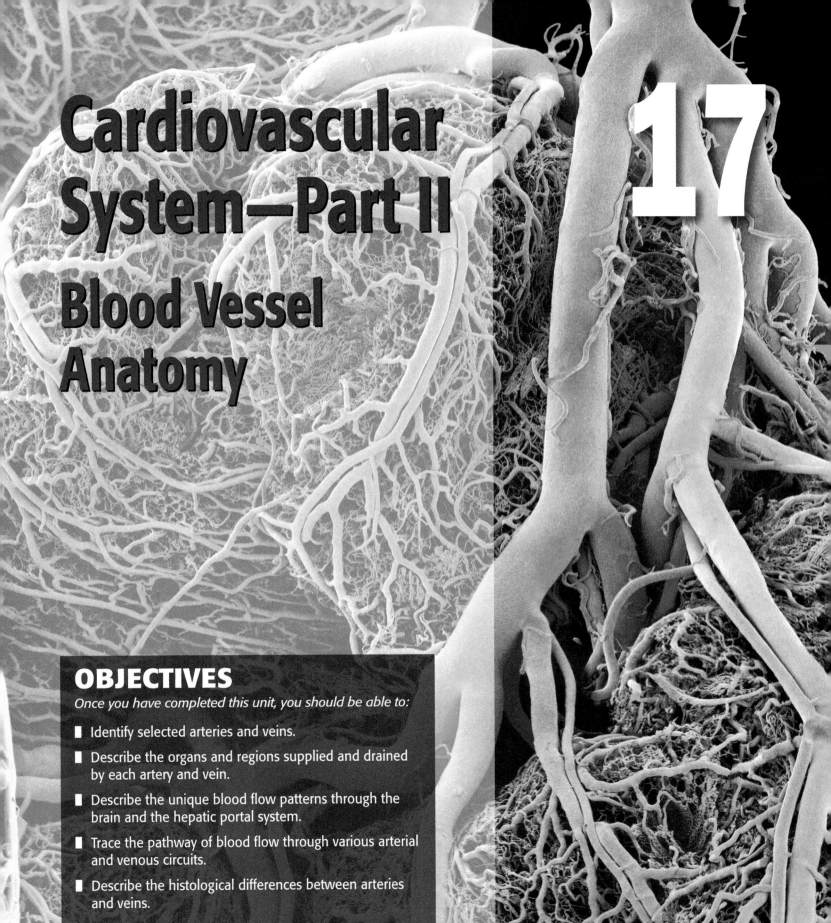

# Cardiovascular System—Part II

# Blood Vessel Anatomy

# 17

## OBJECTIVES

*Once you have completed this unit, you should be able to:*

■ Identify selected arteries and veins.

■ Describe the organs and regions supplied and drained by each artery and vein.

■ Describe the unique blood flow patterns through the brain and the hepatic portal system.

■ Trace the pathway of blood flow through various arterial and venous circuits.

■ Describe the histological differences between arteries and veins.

# PRE-LAB EXERCISES

Complete the following exercises prior to coming to lab, using your textbook and lab manual for reference.

## Pre-Lab Exercise 17-1

✎ Key Terms: Arteries

Describe the location and the organ or region of the body supplied by each of the arteries in **Table 17.1.**

TABLE **17.1** Major Arteries of the Body

| Artery | Location | Organ/Region Supplied |
|---|---|---|
| **Arteries of the Trunk** | | |
| Brachiocephalic artery | | |
| Celiac trunk | | |
| Inferior mesenteric artery | | |
| Renal artery | | |
| Splenic artery | | |
| Superior mesenteric artery | | |
| **Arteries of the Head and Neck** | | |
| Basilar artery | | |
| Circle of Willis | | |
| External carotid artery | | |

*(continues)*

17

TABLE **17.1** Major Arteries of the Body *(cont.)*

| Artery | Location | Organ/Region Supplied |
|---|---|---|
| **Arteries of the Head and Neck** | | |
| Internal carotid artery | | |
| Temporal artery | | |
| Vertebral artery | | |
| **Arteries of the Upper Limb** | | |
| Axillary artery | | |
| Brachial artery | | |
| Radial artery | | |
| Subclavian artery | | |
| Ulnar artery | | |
| **Arteries of the Lower Limb** | | |
| Dorsalis pedis artery | | |
| Femoral artery | | |
| Posterior tibial artery | | |

17

# Pre-Lab Exercise **17-2**

✎ Key Terms: Veins

Describe the location and the organ or region of the body drained by each of the veins in Table 17.2.

TABLE **17.2** Major Veins of the Body

| Vein | Location | Organ/Region Drained |
|---|---|---|
| **Veins of the Trunk** | | |
| Brachiocephalic vein | | |
| Hepatic portal vein | | |
| Inferior mesenteric vein | | |
| Renal vein | | |
| Splenic vein | | |
| Superior mesenteric vein | | |
| **Veins of the Head and Neck** | | |
| Dural sinuses | | |
| Internal jugular vein | | |
| **Veins of the Upper Limb** | | |
| Basilic vein | | |

*(continues)*

TABLE **17.2** Major Veins of the Body *(cont.)*

| Vein | Location | Organ/Region Drained |
|---|---|---|
| Brachial vein | | |
| Cephalic vein | | |
| Subclavian vein | | |
| **Veins of the Lower Limb** | | |
| Femoral vein | | |
| Greater saphenous vein | | |

17

# Pre-Lab Exercise **17-3**

## Arterial Anatomy

Label the arterial diagrams depicted in **Figure 17.1A–D** with the terms from Exercise 17-1 (p. 419).
Use your text and Exercise 17-1 in this unit for reference. Note that these diagrams are presented
in color to facilitate identification of the vessels.

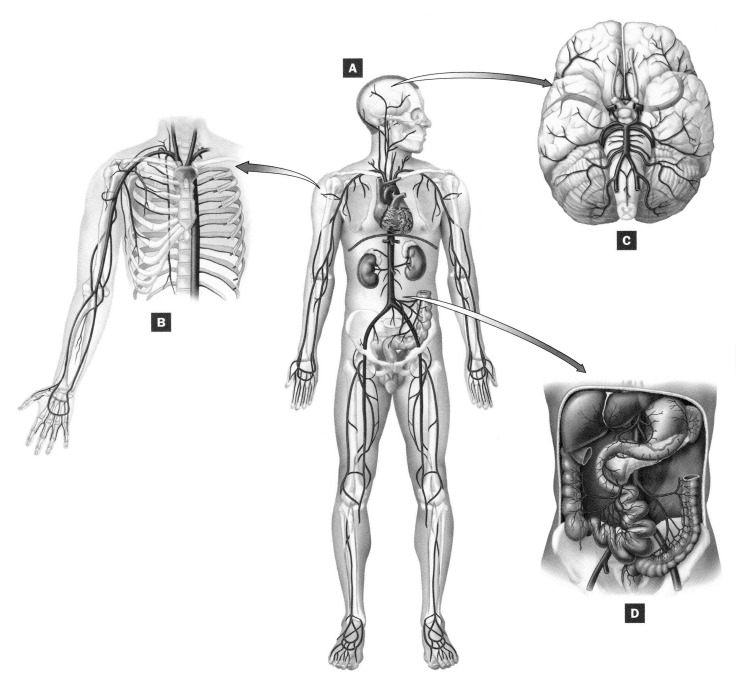

FIGURE **17.1** (**A**) Major arteries of the body: (**B**) right arm and thorax; (**C**) brain; (**D**) abdomen

# Pre-Lab Exercise 17-4

## Venous Anatomy

Label the venous diagrams depicted in **Figure 17.2A–D** with the terms from Exercise 17-2 (p. 424). Use your text and Exercise 17-2 in this unit for reference. Note that these diagrams are presented in color to facilitate identification of the vessels.

FIGURE **17.2** (**A**) Major veins of the body: (**B**) right arm and thorax; (**C**) head and neck; (**D**) abdomen

# EXERCISES

The blood vessels are a closed system of tubes that carry blood around the body. The heart pumps blood away from the heart through a series of **arteries**. Arteries branch as they pass through organs and tissues to form progressively smaller vessels until they branch into tiny **capillary beds**, where gas, nutrient, and waste exchange take place. The blood is drained from the capillaries via a series of **veins** that return the blood to the heart.

The three major circuits of blood flow in the body are

1. the **systemic circuit**, which delivers oxygenated blood to most organs and tissues in the body,

2. the **coronary circuit**, which delivers oxygenated blood to the heart, and

3. the **pulmonary circuit**, which delivers deoxygenated blood to the lungs.

In this unit we will address primarily the systemic circuit, as the coronary and pulmonary circuits were discussed in Unit 16. In the upcoming exercises, you will identify the systemic circuit's major blood vessels, and trace various pathways of blood flow through the body. In the final exercise you will investigate the histology of the blood vessel wall.

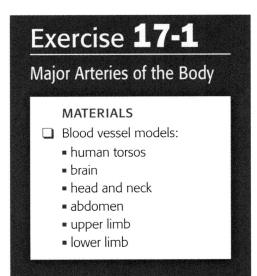

## Exercise 17-1

### Major Arteries of the Body

> **MATERIALS**
> ❏ Blood vessel models:
>   - human torsos
>   - brain
>   - head and neck
>   - abdomen
>   - upper limb
>   - lower limb

The systemic arterial circuit begins with the largest artery in the body, the **aorta**. The aorta originates from the left ventricle as the **ascending aorta**, which ascends until it curves around to form the **aortic arch**. The aortic arch has three major branches:

1. **Brachiocephalic artery.** The first branch is the brachiocephalic artery, which travels superiorly and to the right. Near the clavicle it splits into

two branches—the **right subclavian artery** and the **right common carotid artery (Figure 17.3)**. The right subclavian artery supplies the upper limb and becomes the **axillary artery** near the axilla. In the arm, the axillary artery becomes the **brachial artery**, which branches into the **radial artery** and the **ulnar artery**. The right common carotid artery travels to the neck, where it splits into the **internal carotid artery**, which supplies the brain, and the **external carotid artery**, which supplies the face and the scalp.

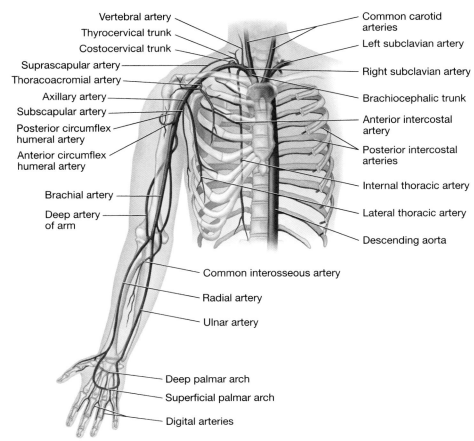

**17**

Vertebral artery
Thyrocervical trunk
Costocervical trunk
Suprascapular artery
Thoracoacromial artery
Axillary artery
Subscapular artery
Posterior circumflex humeral artery
Anterior circumflex humeral artery
Brachial artery
Deep artery of arm
Common carotid arteries
Left subclavian artery
Right subclavian artery
Brachiocephalic trunk
Anterior intercostal artery
Posterior intercostal arteries
Internal thoracic artery
Lateral thoracic artery
Descending aorta
Common interosseous artery
Radial artery
Ulnar artery
Deep palmar arch
Superficial palmar arch
Digital arteries

FIGURE **17.3** Arteries of the right arm and thorax

2. **Left common carotid artery**. The second branch is the left common carotid artery. Like the right common carotid artery, the left common carotid artery splits into internal and external carotid arteries in the neck.

3. **Left subclavian artery**. The final branch off of the aortic arch is the left subclavian artery, which supplies the left upper limb. The branches and names of the left subclavian artery are the same as those of the right subclavian artery.

The arterial supply of the brain comes primarily from the internal carotid arteries and the **basilar artery (Figure 17.4)**. The basilar artery is located on the base of the brainstem and is derived from the fusion of two branches off of the subclavian arteries called the **vertebral arteries**. Once the internal carotid arteries and basilar artery enter the brain, they contribute to a structure called the **circle of Willis** (also called the **cerebral arterial circle**). This circle is composed of branches of these vessels and a set of **anterior** and **posterior communicating arteries**. These vessels are connected to provide alternate routes of circulation to the brain if one of the arteries supplying the brain becomes blocked.

After these three vessels branch off of the aortic arch, it turns downward to become the **thoracic aorta**. The thoracic aorta descends through the thoracic cavity posterior to the heart, after which it passes through the diaphragm to become the **abdominal aorta**. The major branches of the abdominal aorta, shown in **Figure 17.5**, include the following:

1. **Celiac trunk**. The short, stubby celiac trunk is the first branch off of the abdominal aorta. It splits almost immediately into the **common hepatic artery**, the **splenic artery**, and the **left gastric artery**.

2. **Renal arteries**. Inferior to the celiac trunk we find the two **renal arteries**, which serve the kidneys. Note that the renal arteries are not illustrated in **Figure 17.5**, but they are visible in **Figure 17.6**, p. 423.

3. **Superior mesenteric artery**. Right around the renal arteries is another branch called the superior mesenteric artery. As its name implies, it travels through the membranes of the intestines (called the *mesentery*) and supplies the small and much of the large intestine.

4. **Inferior mesenteric artery**. The last large branch off of the abdominal aorta is the inferior mesenteric artery, which supplies the remainder of the large intestine.

The abdominal aorta terminates by bifurcating into two **common iliac arteries**, which themselves bifurcate into an **internal iliac artery** and an **external iliac artery (Figure 17.6)**. The internal iliac artery supplies structures of the pelvis, and the external iliac artery passes deep to the inguinal ligament to enter the thigh, where it becomes the **femoral artery**. The femoral artery continues to the area around the popliteal fossa (the posterior knee), where it becomes the **popliteal artery**. Shortly thereafter, the popliteal artery divides into its two main branches: the **anterior tibial artery**, which continues in the foot as the **dorsalis pedis artery**, and the **posterior tibial artery**.

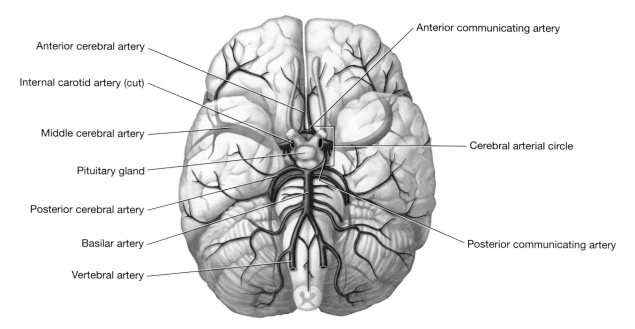

FIGURE **17.4** Arteries of the brain, inferior view

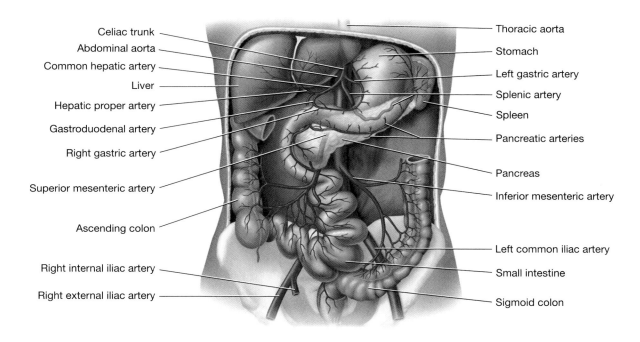

Celiac trunk
Abdominal aorta
Common hepatic artery
Liver
Hepatic proper artery
Gastroduodenal artery
Right gastric artery
Superior mesenteric artery
Ascending colon
Right internal iliac artery
Right external iliac artery

Thoracic aorta
Stomach
Left gastric artery
Splenic artery
Spleen
Pancreatic arteries
Pancreas
Inferior mesenteric artery
Left common iliac artery
Small intestine
Sigmoid colon

FIGURE **17.5** Arteries of the abdomen

## Procedure Model Inventory for Arteries

Identify the following arteries on models and diagrams, using your textbook and this unit for reference. As you examine the anatomical models and diagrams, record on the model inventory in **Table 17.3** the name of the model and the structures you were able to identify. When you have completed the activity, answer Check Your Understanding questions 1 and 2 (p. 441).

### Arteries of the Trunk

1. Aorta
   a. Ascending aorta
   b. Aortic arch
   c. Thoracic (descending) aorta
   d. Abdominal aorta
2. Brachiocephalic artery
3. Celiac trunk
   a. Splenic artery
   b. Left gastric artery
   c. Common hepatic artery

4. Superior mesenteric artery
5. Suprarenal artery
6. Renal artery
7. Inferior mesenteric artery
8. Gonadal artery
9. Common iliac artery
   a. Internal iliac artery

### Arteries of the Head and Neck

1. Common carotid artery
   a. External carotid artery
      (1) Temporal artery
   b. Internal carotid artery
      (1) Carotid sinus
2. Vertebral artery
3. Basilar artery

4. Cerebral arterial circle
   a. Anterior communicating arteries
   b. Posterior communicating arteries
5. Anterior cerebral artery
6. Middle cerebral artery
7. Posterior cerebral artery

## Arteries of the Upper Limbs

1. Subclavian artery
2. Axillary artery
3. Brachial artery
4. Radial artery
5. Ulnar artery

## Arteries of the Lower Limbs

1. External iliac artery
2. Femoral artery
3. Popliteal artery
   a. Anterior tibial artery
      (1) Dorsalis pedis artery
   b. Posterior tibial artery

TABLE **17.3** Model Inventory for Arteries

| Model/Diagram | Structures Identified |
|---|---|
|  |  |
|  |  |
|  |  |
|  |  |
|  |  |
|  |  |

17

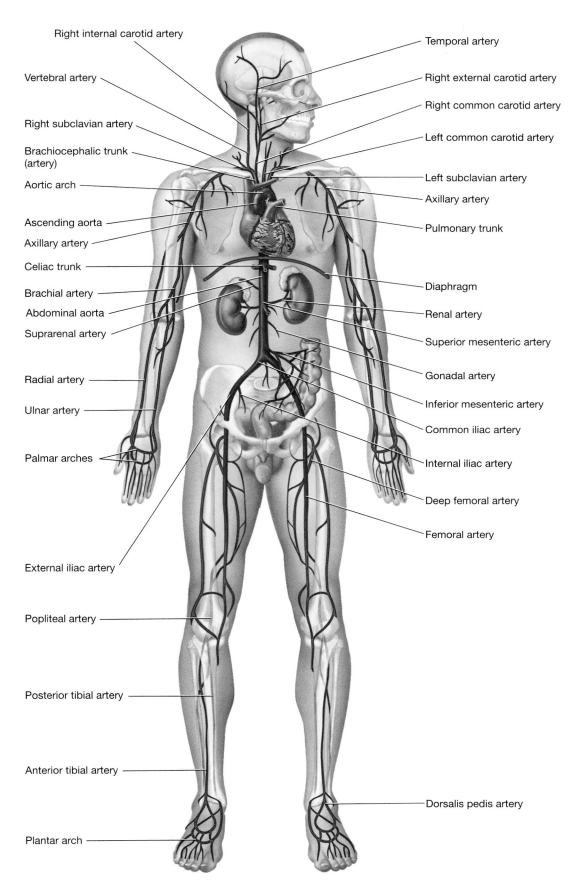

Right internal carotid artery

Vertebral artery

Right subclavian artery

Brachiocephalic trunk (artery)

Aortic arch

Ascending aorta

Axillary artery

Celiac trunk

Brachial artery

Abdominal aorta

Suprarenal artery

Radial artery

Ulnar artery

Palmar arches

External iliac artery

Popliteal artery

Posterior tibial artery

Anterior tibial artery

Plantar arch

Temporal artery

Right external carotid artery

Right common carotid artery

Left common carotid artery

Left subclavian artery

Axillary artery

Pulmonary trunk

Diaphragm

Renal artery

Superior mesenteric artery

Gonadal artery

Inferior mesenteric artery

Common iliac artery

Internal iliac artery

Deep femoral artery

Femoral artery

Dorsalis pedis artery

17

FIGURE **17.6** Major arteries of the body

# Exercise 17-2

## Major Veins of the Body

### MATERIALS

❏ Blood vessel models:
- human torsos
- brain
- head and neck
- abdomen
- upper limb
- lower limb
- dural sinuses

Arteries of the systemic circuit deliver oxygenated, nutrient-rich blood to capillary beds, where gases, nutrients, and wastes are exchanged. The deoxygenated, carbon dioxide–rich blood is then drained from the capillary beds by a series of veins. The two largest veins in the body are the **superior vena cava**, which drains the structures superior to the diaphragm, and the **inferior vena cava**, which drains the structures inferior to the diaphragm.

The head and the neck are drained primarily by the **internal** and **external jugular veins** (**Figure 17.7**). The much smaller external jugular vein drains the face and the scalp, and the larger internal jugular vein, which travels in a sheath with the common carotid artery, drains the brain. Note, however, that venous blood from the brain does not simply drain into one vein and exit the head. Instead, it drains into spaces between the two layers of the dura mater called the **dural sinuses** (**Figure 17.8**). Blood from the brain capillaries drains into the dural sinuses; the inferior dural sinus, called the **transverse sinus**, drains into the internal jugular vein.

Veins draining blood from the organs of the abdomen are named largely in parallel with the arteries that serve the organs: The **renal veins** drain the kidneys (**Figure 17.9**), the **splenic vein** drains the spleen, the **gastric veins** drain the stomach, the **superior mesenteric vein** drains the small intestine and much of the large intestine, and the **inferior mesenteric vein** drains the remainder of the large intestine. Although the renal vein empties into the inferior vena cava, the blood from the latter four veins does not drain into the inferior vena cava directly. Instead, note in **Figure 17.10** that each vein drains into a common vein called the **hepatic portal vein**. Here, the nutrient-rich blood percolates through the liver, where it is processed and detoxified. In this way, everything we ingest (except lipids, which we discuss in Unit 23) must travel through the liver before entering the systemic circulation. Once the blood has filtered through the hepatic portal system, it exits via **hepatic veins** and drains into the inferior vena cava.

FIGURE **17.7** Veins of left superficial aspect of head and neck

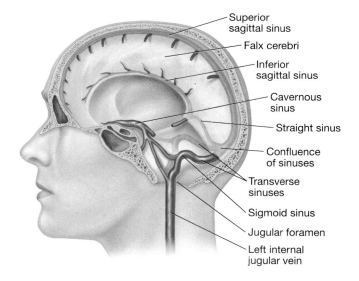

FIGURE **17.8** Veins draining dural sinuses of brain

FIGURE **17.9** Veins draining the abdominal area (excluding the hepatic portal system)

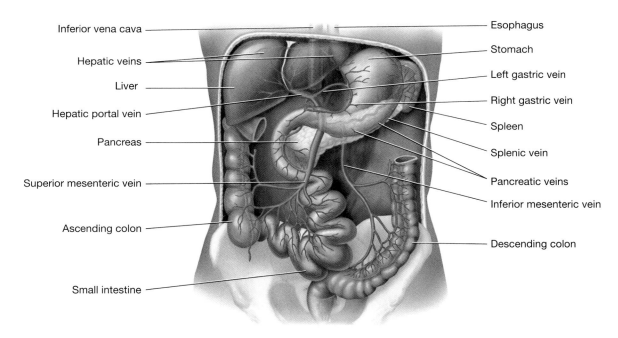

FIGURE **17.10** Veins of the abdomen and the hepatic portal system

Blood from the deep structures of the upper limb is drained by the **radial** and **ulnar veins,** both of which parallel the bones for which they are named (**Figure 17.11**). These two veins merge in the arm to form the **brachial vein,** which becomes the **axillary vein** in the axilla. Near the clavicle, the axillary vein becomes the **subclavian vein,** which drains into the **brachiocephalic vein** and finally into the superior vena cava. The superficial structures of the upper limb are drained by the **cephalic vein** on the lateral side and the **basilic vein** on the medial side. Note in **Figure 17.12** that the cephalic vein and the basilic vein are united in the antecubital fossa by the **median antecubital vein**. This is a frequent site for drawing blood with a syringe. The basilic vein joins the brachial vein to form the axillary vein, and the cephalic vein joins the axillary vein to form the subclavian vein.

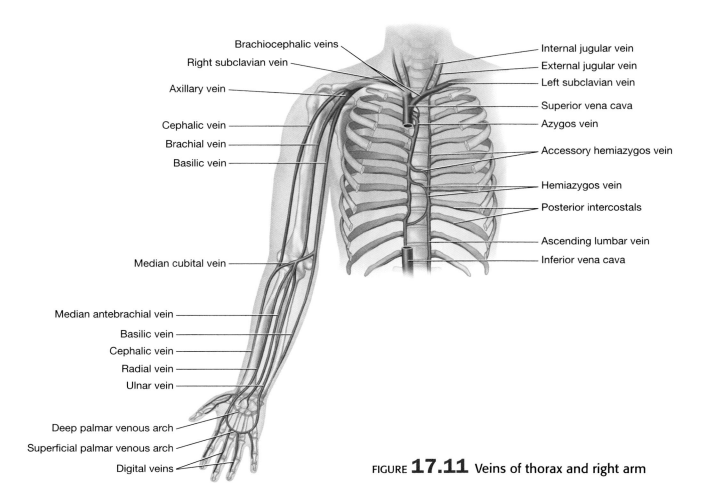

Brachiocephalic veins

Right subclavian vein

Axillary vein

Cephalic vein

Brachial vein

Basilic vein

Median cubital vein

Median antebrachial vein

Basilic vein

Cephalic vein

Radial vein

Ulnar vein

Deep palmar venous arch

Superficial palmar venous arch

Digital veins

Internal jugular vein

External jugular vein

Left subclavian vein

Superior vena cava

Azygos vein

Accessory hemiazygos vein

Hemiazygos vein

Posterior intercostals

Ascending lumbar vein

Inferior vena cava

FIGURE **17.11** Veins of thorax and right arm

The deep structures of the lower limb are drained by the **anterior** and **posterior tibial veins,** which unite in the popliteal fossa to form the **popliteal vein (Figure 17.12,** p. 428). In the distal thigh, the popliteal vein becomes the **femoral vein,** which becomes the **external iliac vein** after it passes under the inguinal ligament. The external iliac vein merges with the **internal iliac vein,** which drains pelvic structures, forming the **common iliac vein.** The two common iliac veins unite to form the inferior vena cava near the superior part of the pelvis. The largest superficial vein of the lower limb is the **greater saphenous vein,** which drains the medial leg and thigh and empties into the femoral vein.

## Procedure Model Inventory for Veins

Identify the following veins on models and diagrams, using your textbook and this unit for reference.
As you examine the anatomical models and diagrams, record on the model inventory in **Table 17.4** the name of the model and the structures you were able to identify. When you have completed the activity, answer Check Your Understanding questions 3 and 4 (pp. 441–442).

### Veins of the Trunk

1. Superior vena cava
2. Inferior vena cava
3. Brachiocephalic vein
4. Hepatic veins
5. Hepatic portal vein
6. Splenic vein
7. Superior mesenteric vein
8. Inferior mesenteric vein
9. Gastric vein
10. Renal vein
11. Suprarenal vein
12. Gonadal vein
13. Internal iliac vein
14. Common iliac vein

## Veins of the Head and Neck

1. Dural sinuses
   a. Superior sagittal sinus
   b. Cavernous sinus
   c. Transverse sinus
2. Internal jugular vein
3. External jugular vein
4. Vertebral vein

## Veins of the Upper Limbs

1. Ulnar vein
2. Radial vein
3. Median cubital vein
4. Brachial vein
5. Basilic vein
6. Cephalic vein
7. Axillary vein
8. Subclavian vein

## Veins of the Lower Limbs

1. Anterior tibial vein
2. Posterior tibial vein
3. Popliteal vein
4. Great saphenous vein
5. Femoral vein
6. External iliac vein

TABLE **17.4** Model Inventory for Veins

| Model | Structures Identified |
|---|---|
| | |
| | |
| | |
| | |
| | |
| | |
| | |
| | |
| | |

17

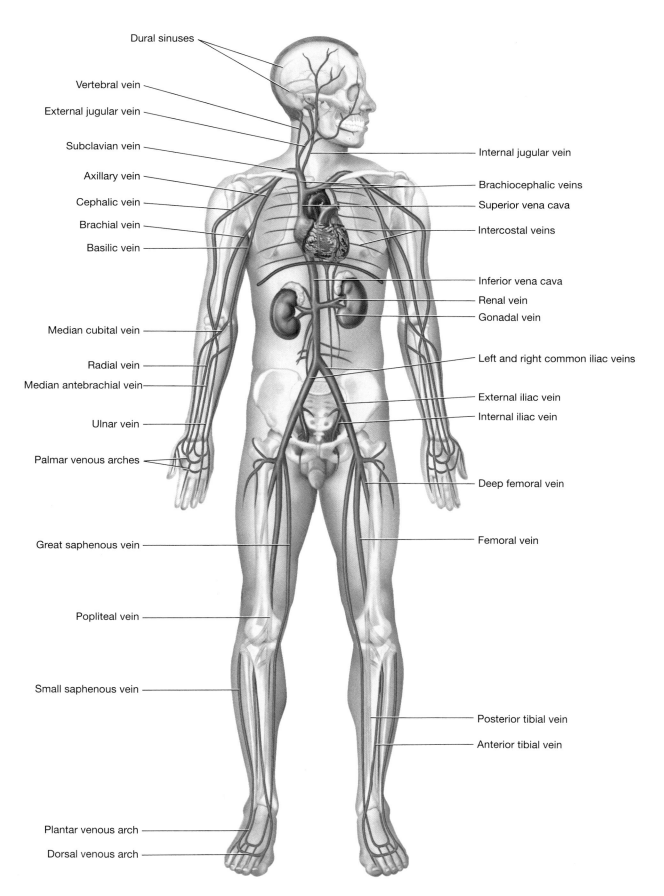

Dural sinuses

Vertebral vein

External jugular vein

Subclavian vein

Axillary vein

Cephalic vein

Brachial vein

Basilic vein

Median cubital vein

Radial vein

Median antebrachial vein

Ulnar vein

Palmar venous arches

Great saphenous vein

Popliteal vein

Small saphenous vein

Plantar venous arch

Dorsal venous arch

Internal jugular vein

Brachiocephalic veins

Superior vena cava

Intercostal veins

Inferior vena cava

Renal vein

Gonadal vein

Left and right common iliac veins

External iliac vein

Internal iliac vein

Deep femoral vein

Femoral vein

Posterior tibial vein

Anterior tibial vein

17

FIGURE **17.12** Major veins of the body

# Exercise 17-3

## Time to Trace!

### MATERIALS
❑ Laminated outline of the human body
❑ Water-soluble marking pen

In this exercise you will trace the blood flow through various places in the body. As you trace, keep the following hints in mind:

▮ Don't forget about the hepatic portal system! Remember that most venous blood coming from the abdominal organs has to go through the hepatic portal vein and the hepatic portal system before it can enter the general circulation.

▮ Don't forget that the venous blood in the brain drains first into the dural sinuses, then into a vein.

▮ If you start in a vein, you have to go through the venous system and through the heart before you can get back to the arterial system.

▮ If you start in an artery, you have to go through the arterial system and then through a *capillary bed* before you can go through the venous system. You can't go backward through the arterial system—that's cheating!

▮ If you start in an artery and end in an artery, you likely will have to go through the arterial circuit, through a capillary bed, through the venous circuit, back to the heart and lungs, and *then* reenter the arterial circuit. Whew!

▮ Sometimes there is more than one right answer, because you may take multiple paths.

▮ Following is an example in which we have started in the right popliteal vein and ended in the left internal carotid artery:

**Start:** right popliteal vein → right femoral vein → right external iliac vein → right common iliac vein → inferior vena cava → right atrium → tricuspid valve → right ventricle → pulmonary valve → pulmonary artery → lungs → pulmonary veins → left atrium → mitral valve → left ventricle → aortic valve → ascending aorta → aortic arch → left common carotid artery → left internal carotid artery → **End**

Wasn't that easy?

## Procedure Tracing Blood Flow Patterns

Trace the path of blood flow through the following circuits, using the example on page 429 for reference. It is helpful to draw the pathway out on a laminated outline of the human body, and label each vessel as you trace. If a laminated outline is not available, use **Figures 17.13–17.16** instead.

**1** *Start:* Right radial vein

   *End:* Right renal artery

FIGURE **17.13** Outline of the human body

**17**

**2** *Start:* Left coronary artery

*End:* Dorsalis pedis artery

FIGURE **17.14** Outline of the human body

**3**  *Start:* Superior mesenteric vein

   *End:* Superior mesenteric artery

FIGURE **17.15** Outline of the human body

**4**  *Start:*  Renal artery

*End:*  Internal jugular vein

FIGURE **17.16** Outline of the human body

# Exercise 17-4

## Histology of the Blood Vessel Wall

### MATERIALS

- ❏ Blood vessel slide
- ❏ Light microscope
- ❏ Colored pencils

Three distinct tissue layers make up the walls of arteries and veins. The three layers, shown in **Figure 17.17**, are the following:

1. **Tunica interna.** The innermost lining of the blood vessel is called the tunica interna. It consists of a specialized type of simple squamous epithelium called **endothelium**. It rests on top of a thin layer of connective tissue.

2. **Tunica media.** The middle layer of the blood vessel wall is called the tunica media and consists of smooth muscle and elastic fibers. The smooth muscle, innervated by the sympathetic nervous system, controls the diameter of the vessel and plays an important role in tissue perfusion and blood pressure. The elastic fibers allow the vessel to expand with changing pressure and return to its original shape and diameter.

3. **Tunica externa.** The outermost layer of the blood vessel wall is the tunica externa (or **tunica adventitia**), which consists of dense irregular collagenous connective tissue with abundant collagen fibers. The collagen fibers reinforce the blood vessel and prevent it from rupturing when the pressure in the vessel increases.

Artery

Vein

Tunica externa

Tunica media

Tunica interna

FIGURE **17.17** Photomicrograph, artery and veins

Like any other organ, the structure of each type of blood vessel follows its function. For this reason, the characteristics of the three layers of the blood vessel wall are considerably different in arteries, capillaries, and veins. They are easy to identify if you bear the following in mind:

ℹ️ Arteries have a much thicker tunica media, with prominent elastic fibers that typically appear as a wavy purple line. They appear circular in cross sections because of their thick walls. Arteries closer to the heart are under very high pressure and so have more elastic fibers in the tunica media and more collagen fibers in the tunica externa.

ℹ️ Veins have a thin tunica media with few elastic fibers. Because the wall is so much thinner, the lumen is wider, and the vein is typically collapsed on the slide.

ℹ️ Capillaries are extremely thin-walled and consist only of a thin tunica interna. The smallest capillaries are large enough for only one red blood cell to fit through at a time.

## Procedure Microscopy

Examine prepared microscope slides of an artery, a capillary, and a vein. The capillary may be on a separate slide. Use colored pencils to draw what you see, and label your diagrams with the terms indicated. When you have completed the activity, answer Check Your Understanding questions 6 and 7 (p. 442).

**1** Artery
   a. Tunica interna (endothelium)
   b. Tunica media
      (1) Smooth muscle
      (2) Elastic fibers
   c. Tunica externa
   d. Lumen

**17**

**2** Capillary
    a.  Tunica interna
    b.  Blood cell(s)

**3** Vein
    a.  Tunica interna
    b.  Tunica media
        (1) Smooth muscle
    c.  Tunica externa
    d.  Lumen

17

Name _____

Section _____ Date _____

## Check Your Recall

**1** Label the following arteries in **Figures 17.18** and **17.19**.

**Figure 17.18**

☐ Basilar artery
☐ Brachial artery
☐ Cerebral arterial circle
☐ Common iliac artery
☐ Dorsalis pedis artery
☐ Femoral artery
☐ Left common carotid artery
☐ Middle cerebral artery
☐ Posterior tibial artery
☐ Radial artery
☐ Renal artery
☐ Right common carotid artery
☐ Right subclavian artery
☐ Ulnar artery
☐ Vertebral artery

FIGURE **17.18** Major arteries of the body

## Figure 17.19

- ☐ Celiac trunk
- ☐ Common hepatic artery
- ☐ Inferior mesenteric artery
- ☐ Splenic artery
- ☐ Superior mesenteric artery

FIGURE **17.19** Arteries of the abdomen

**2** Label the following veins on **Figures 17.20** and **17.21**.

## Figure 17.20

- ☐ Brachial vein
- ☐ Brachiocephalic vein
- ☐ Cephalic vein
- ☐ Common iliac vein
- ☐ External jugular vein
- ☐ Femoral vein
- ☐ Great saphenous vein
- ☐ Internal jugular vein
- ☐ Renal vein
- ☐ Subclavian vein
- ☐ Vertebral vein

FIGURE **17.20** Major veins of the body

**Figure 17.21**

☐  Hepatic portal vein
☐  Hepatic veins
☐  Inferior mesenteric vein
☐  Left gastric vein
☐  Splenic vein
☐  Superior mesenteric vein

FIGURE **17.21** Veins of the abdomen and the hepatic portal system

**3** Which of the following is *not* a major circuit of blood flow in the body?

a.  Coronary circuit

b.  Cerebral circuit

c.  Pulmonary circuit

d.  Systemic circuit

**4** How do the right and left common carotid arteries differ in terms of the arteries from which they originate?

_____

_____

_____

_____

_____

_____

**5** The cerebral arterial circle (circle of Willis):

a.  provides alternate routes of blood flow in the brain.

b.  supplies the face and the scalp.

c.  provides alternate routes of blood flow in the liver.

d.  supplies the myocardium.

17

**6** The venous blood of the brain drains into a set of _____ before draining into a vein.

   a. coronary arteries

   b. cerebral veins

   c. dural sinuses

   d. paranasal sinuses

**7** Which of the following veins drain into the hepatic portal vein? *(Circle all that apply.)*

   a. Renal vein

   b. Splenic vein

   c. Superior mesenteric vein

   d. Inferior mesenteric vein

   e. Hepatic veins

   f. Gastric veins

**8** Which of the following describes the purpose of the hepatic portal system?

   a. It provides the liver with oxygenated blood.

   b. It allows the liver to process and detoxify nutrient-rich blood from most of the abdominal organs.

   c. It drains blood from the liver to the inferior vena cava.

   d. All of the above.

**9** From superficial to deep, the layers of the blood vessel wall include the:

   _____

   _____

   _____

**10** Which of the following correctly describes the role of the smooth muscle cells in the blood vessel wall?

   a. They play an important role in determining the amount of blood flowing to a tissue (tissue perfusion).

   b. They play an important role in blood pressure.

   c. They control the diameter of the blood vessel.

   d. All of the above.

   e. None of the above.

## Check Your Understanding

### Critical Thinking and Application Questions

**1** Would a blood clot lodged in one of the anterior or posterior communicating arteries of the cerebral arterial circle be likely to cause significant symptoms? Why or why not?

_____

_____

_____

_____

_____

_____

**2** Your patient presents with severe abdominal pain. Your team orders a CT scan and discovers a large blood clot lodged in the celiac trunk. What organs could this blood clot potentially affect? Why do you think the clot is causing abdominal pain?

_____

_____

_____

_____

_____

_____

**3** Certain drugs cannot be taken by mouth because the entire dose of the drug is destroyed in the liver before it ever reaches the general circulation. Explain why these same drugs can be given by injection, either intravenously or intramuscularly. (*Hint*: Consider the hepatic portal system.)

_____

_____

_____

_____

_____

**17**

**4** In the surgical procedure called a bypass graft, a vessel is removed from a patient and used to make a vascular bridge that bypasses one of that patient's blocked arteries. A common vessel used for this procedure is the greater saphenous vein. Why do you think a surgeon might use the greater saphenous vein instead of the femoral vein or a large artery?

**5** Predict what might happen if the elastic fibers in a blood vessel's wall were defective. Would such a condition be more problematic for arteries or for veins? Explain.

**6** The diseases known as *collagen vascular diseases* are characterized by defects in the collagen in the tunica externa. What effects would you expect such diseases to have on the structure and function of blood vessels?

**7** Collagen vascular diseases may lead to the formation of an aneurysm, or the ballooning of a blood vessel, generally in the aorta. Why do you think aneurysms are more likely to form in the aorta than in other vessels?

# Cardiovascular System— Part III

## Cardiovascular Physiology

## OBJECTIVES

*Once you have completed this unit, you should be able to:*

- Auscultate the sounds of the heart.

- Describe and demonstrate common physical examination tests of the blood vessels.

- Measure blood pressure using a stethoscope and a sphygmomanometer.

- Describe the effects of the autonomic nervous system and caffeine on blood pressure and pulse rate.

- Measure and interpret the ankle-brachial index using a Doppler ultrasound device.

- Describe the waves and intervals on an ECG, and perform basic ECG interpretation.

# PRE-LAB EXERCISES

Complete the following exercises prior to coming to lab, using your textbook and lab manual for reference.

## Pre-Lab Exercise 18-1

### ✎ Key Terms

You should be familiar with the following terms before coming to lab.

| Term | Definition |
|------|------------|

**Heart Sounds and Pulses**

Auscultation _____

_____

S1 _____

_____

S2 _____

_____

Tachycardia _____

_____

Bradycardia _____

_____

Pulse point _____

_____

**Blood Pressure**

Systolic pressure _____

_____

Diastolic pressure _____

_____

Sounds of Korotkoff _____

_____

# Pre-Lab Exercise 18-2

## Autonomic Nervous System and Cardiovascular System

As you learned in Unit 14, the autonomic nervous system (ANS) is an important body system because of its regulation of other body systems. The cardiovascular system is no exception, which you will see in this unit's exercises. Fill in Table 18.1 with the basic effects of the two divisions of the ANS to refresh your memory.

TABLE **18.1** Effects of the Divisions of the ANS

| Organ | Sympathetic Nervous System Effects | Parasympathetic Nervous System Effects |
|---|---|---|
| Heart | | |
| Blood vessels | | |
| Net effect on blood pressure | | |
| Lungs (bronchial smooth muscle) | | |
| Metabolic rate | | |
| Urinary functions | | |
| Digestive functions | | |

# EXERCISES

Every physical examination includes an assessment of the cardiovascular system in which the heart rate is counted, heart sounds are auscultated, and the blood pressure is measured. In this unit you will learn how to perform these and other common procedures along with the basis for their interpretation. You will also determine the effects of the autonomic nervous system and caffeine on these measurements.

## Exercise 18-1

### Heart Auscultation

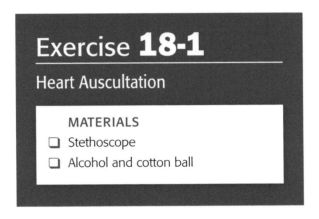

**MATERIALS**
- ❏ Stethoscope
- ❏ Alcohol and cotton ball

The two main sounds you hear when you listen to the heart are produced by the closing of valves at certain points during the cardiac cycle. The first heart sound is known as **S1**, and it is caused by closure of the mitral and tricuspid valves at the beginning of ventricular systole and isovolumetric contraction. The second heart sound, **S2**, is caused by closure of the aortic and pulmonary valves at the beginning of ventricular diastole and isovolumetric relaxation.

The process of listening to heart sounds is known as **auscultation**. Heart sounds are typically auscultated in four areas, each of which is named for the valve best heard at that specific location. The position of each area is described relative to the sternum and the spaces between the ribs, known as *intercostal spaces*. The first intercostal space is located between the first and second rib, roughly below the clavicle. From the clavicle you can count down to consecutive spaces to auscultate in the appropriate areas. The four areas are as follows (**Figure 18.1**):

1. **Aortic area.** This is the area where the sounds of the aortic valve are best heard. It is located in the second intercostal space at the right sternal border.

2. **Pulmonic area.** The pulmonic valve is best heard over the second intercostal space at the left sternal border.

3. **Tricuspid area.** The sounds produced by the tricuspid valve are best heard over the fourth intercostal space at the left sternal border.

4. **Mitral area.** The mitral area is located in the fifth intercostal space at the left midclavicular line (draw an imaginary line down the middle of the clavicle, roughly in line with the nipple).

The following variables are checked during heart auscultation:

▌ **Heart rate.** The heart rate refers to the number of heartbeats per minute. If the rate is more than 100, it is termed **tachycardia**. If the rate is below 60 beats per minute, it is termed **bradycardia**.

▌ **Heart rhythm.** The heart's rhythm refers to the pattern and regularity with which it beats. Some rhythms are *regularly irregular*, in which the rhythm is irregular but still follows a defined pattern. Others are *irregularly irregular*, in which the rhythm follows no set pattern.

▌ **Additional heart sounds.** Sometimes sounds in addition to S1 and S2 are heard that could be a sign of pathology. These sounds are called **S3**, which occurs just after S2, and **S4**, which occurs just prior to S1.

▌ **Heart murmur.** A heart murmur is a clicking or "swooshing" noise heard between the heart sounds. Murmurs are caused by a valve leaking, called **regurgitation**, or by a valve that has lost its pliability, called **stenosis**.

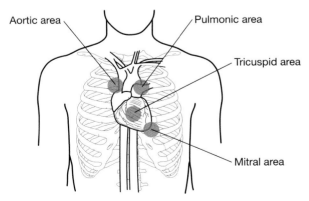

FIGURE **18.1** Areas of auscultation

Heart sounds are auscultated with a stethoscope, which you will use in this unit. Most stethoscopes contain the following parts (Figure 18.2):

■ **Earpieces** are gently inserted into the external auditory canal and allow you to auscultate the heart sounds.

■ The **diaphragm** is the broad, flat side of the end of the stethoscope. It is used to auscultate higher-pitched sounds and is the side used most often in auscultation of heart sounds.

■ The **bell** is the concave, smaller side of the end of the stethoscope. It is used to auscultate lower-pitched sounds.

Note that sounds are not audible through both the bell and the diaphragm at the same time. Typically, the end can be flipped from one side to the next. Before auscultating with either side, lightly tap the end to ensure that you can hear sound through it. Note also that the ends of some stethoscopes have only one side—the diaphragm. In this case, placing light pressure on the end as you are auscultating yields sounds associated with the diaphragm, whereas placing heavier pressure yields sounds associated with the bell.

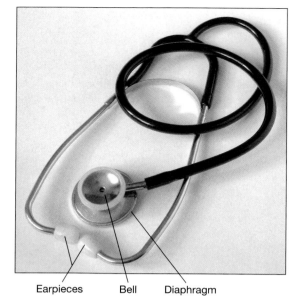

Earpieces    Bell    Diaphragm

FIGURE **18.2** A stethoscope

 **Procedure** Heart Auscultation

**1** Obtain a stethoscope, and clean the earpieces and diaphragm with alcohol and cotton balls.

**2** Place the earpieces in your ears, and gently tap the diaphragm to ensure that it is on the proper side. If it is not, flip it to the other side.

**3** Lightly place the diaphragm on your partner's chest in the aortic area. (Note: You may wish to have your lab partner place the stethoscope on his or her chest under the shirt, because the sounds are heard best on bare skin.)

**4** Auscultate several cardiac cycles, and determine which sound is S1 and which sound is S2. Measure your partner's heart rate, and determine if the rhythm is regular. In addition, check for heart murmurs and extra heart sounds.

**5** Record your results in **Table 18.2**.

**6** Move on to the next area, and repeat.

TABLE **18.2** Heart Auscultation Results

| Area | Rate | Rhythm | Extra Heart Sounds? (yes/no) | Murmurs Present? (yes/no) |
|------|------|--------|------------------------------|----------------------------|
| Aortic area | | | | |
| Pulmonic area | | | | |
| Tricuspid area | | | | |
| Mitral area | | | | |

# Exercise 18-2

## Vascular Examination

### MATERIALS
❏ Blood vessel model or diagram
❏ Stethoscope
❏ Alcohol and cotton swab

A vascular examination is the portion of a physical examination that assesses the health of the arteries and veins. Commonly performed exams include *pulse palpation*, auscultation of vessels to check for noises called *bruits*, and measuring the time it takes for capillary beds to refill. Each of those procedures is outlined in the sections that follow.

Please note before you begin that you should never assess both of your lab partner's carotid arteries at the same time. This might initiate the **baroreceptor reflex**, in which the parasympathetic nervous system triggers a reflexive and often dramatic drop in blood pressure and heart rate. This could cause your lab partner to momentarily lose consciousness.

## Procedure Pulse Palpation

**Pulse palpation** is the process of feeling the pulse with the fingertips. It is performed to assess rate, rhythm, and regularity of the heartbeat and to assess the arterial circulation to different parts of the body. The pulses commonly measured are those found at the radial, ulnar, brachial, carotid, temporal, femoral, popliteal, posterior tibial, and dorsalis pedis arteries, shown in **Figure 18.3**.

When pulses are palpated, they are **graded** according to a standard scale. This allows health-care professionals to communicate about a patient unambiguously and to assess the progress or deterioration of a patient's condition. The scale utilizes the following four grades:

Grade **0/4**: The pulse is absent.
Grade **1/4**: The pulse is barely or only lightly palpable.
Grade **2/4**: The pulse is normal.
Grade **3/4**: The pulse is quite strong.
Grade **4/4**: The pulse is bounding and visible through the skin.

Note that this scale has no negative numbers or decimal numbers (e.g., you would not use −1/4 or 2.5/4). In a healthy person most pulses are grade 2/4, although occasionally a pulse is weak or absent. This is simply normal anatomical variation and does not signify pathology. Students often mistakenly grade any strong pulse (such as the carotid pulse) as 4/4. If a pulse were truly 4/4, however, this would be a sign of extremely high blood pressure in that artery.

The following procedure asks you to palpate certain pulses on your lab partner. Note that we will not be palpating each pulse illustrated in **Figure 18.3**. When you have completed this activity, answer Check Your Understanding questions 1–3 (p. 461).

**1** Wash your hands prior to palpating your lab partner's pulses.

**2** On a model or diagram, locate the artery you are palpating.

**3** Lightly place your index finger and middle finger over the artery. You may increase the pressure slightly, but be careful not to press too hard, because you could cut off blood flow through the artery and also could mistake the pulse in your

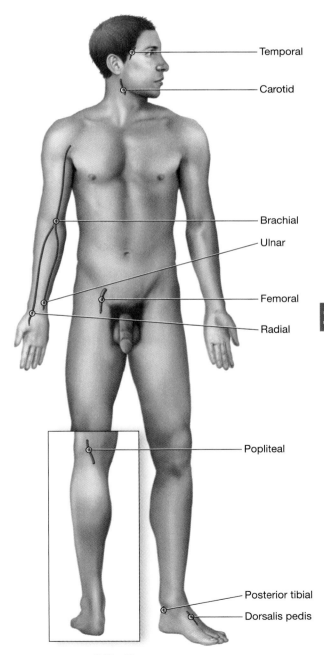

FIGURE **18.3** Common pulse points

Labels: Temporal, Carotid, Brachial, Ulnar, Femoral, Radial, Popliteal, Posterior tibial, Dorsalis pedis

18

fingertips for your partner's pulse. If you are unsure if the pulse is yours or your partner's, feel the lab table. If the lab table "has a pulse," you are feeling the pulse in your own fingertips.

**4** Palpate *only* one side (right or left) at a time, especially in the carotid artery.

**5** Grade your partner's pulses according to the 0/4 to 4/4 scale, and record the results in **Table 18.3**.

TABLE **18.3** Pulse Points Grades

| Artery | Right-Side Grade | Left-Side Grade |
|---|---|---|
| Carotid | | |
| Temporal | | |
| Brachial | | |
| Radial | | |
| Ulnar | | |
| Dorsalis pedis | | |
| Posterior tibial | | |

**18**

## Procedure Auscultating for Carotid Bruits

Vascular disease in large vessels may lead to turbulent blood flow through the vessels. This produces a sound you can auscultate with a stethoscope called a **bruit**. The following procedure outlines the process for auscultating for bruits in the carotid arteries.

**1** Obtain a stethoscope, and switch from the diaphragm to the bell. Clean the earpieces and bell with alcohol and cotton balls.

**2** Place the earpieces in your ears, and gently tap the bell to ensure that it is on the proper side. If it is not, flip it to the other side. If your stethoscope has only one side, remember that you will have to use slightly heavier pressure to hear the sounds associated with the bell.

**3** Palpate the carotid pulse (remember to palpate only one carotid artery at a time), and place the bell over it. Auscultate for bruits, which sound like a "swooshing" sound with each heartbeat.

**4** Repeat the process for the other carotid artery.

**5** Were bruits present in either side? _____

## **Procedure** Measuring Capillary Refill Time

The **capillary refill time** is the time it takes for capillary beds to refill after they have been forcibly emptied by pressure. This standard physical examination test is done to evaluate a patient for a variety of vascular diseases, particularly arterial disease. A normal capillary refill time measures 1–3 seconds; a value greater than 3 seconds may signify some sort of pathology. Measure the capillary refill time for each of your lab partner's digits in the procedure outlined below. (Note that you cannot perform this procedure if your lab partner is wearing nail polish.) When you have completed the activity, answer Check Your Understanding question 4 (p. 462).

**1** Squeeze the end of the fingertip over the fingernail until the fingernail bed blanches (turns white).

**2** Release the fingernail, and count the number of seconds it takes for the fingernail bed to return to a pinkish color (note that this may occur immediately, in which case simply record it as 1 second).

**3** Repeat for each fingertip of each hand, and record your results in **Table 18.4**.

TABLE **18.4** Capillary Refill Times

| Digits: Right Hand | Refill Time | Digits: Left Hand | Refill Time |
|---|---|---|---|
| 1 (thumb) | | 1 (thumb) | |
| 2 | | 2 | |
| 3 | | 3 | |
| 4 | | 4 | |
| 5 | | 5 | |

18

# Exercise 18-3

## Blood Pressure

### MATERIALS

- ❏ Sphygmomanometer
- ❏ Stethoscope
- ❏ Alcohol and cotton ball
- ❏ Bucket of ice water
- ❏ Caffeinated beverage
- ❏ Caffeine-free beverage

**Blood pressure** is defined as the pressure exerted by the blood on the walls of the blood vessels. It is determined by the following three factors:

1. **Cardiac output.** Cardiac output is the amount of blood each ventricle pumps in 1 minute. It is a product of **stroke volume**, or the amount pumped with each beat, and heart rate.

2. **Peripheral resistance.** Resistance is defined as any impedance to blood flow encountered in the blood vessels. It is determined largely by the degree of **vasoconstriction** or **vasodilation** in the systemic circulation. Vasoconstriction increases peripheral resistance, and vasodilation has the opposite effect. Other factors that influence resistance include obstructions, such as atheromatous plaques, within the arteries.

3. **Blood volume.** The amount of blood found in the blood vessels at any given time is known as the blood volume. It is greatly influenced by overall fluid volume and is largely controlled by the kidneys and hormones of the endocrine system.

Note that cardiac output and peripheral resistance are factors that can be altered quickly to change blood pressure. Alterations to blood volume, however, occur relatively slowly and generally require two to three days to have a noticeable effect.

Arterial blood pressure is measured clinically and experimentally using instruments called a **sphygmomanometer** and a stethoscope. This procedure yields two pressure readings:

1. **Systolic pressure.** The pressure in the arteries during ventricular systole is known as the systolic pressure. This is the larger of the two readings, averaging between 100 and 120 mmHg.

2. **Diastolic pressure.** The pressure in the arteries during ventricular diastole is the diastolic pressure. This is the smaller of the two readings, averaging between 60 and 80 mmHg.

Arterial blood pressure is measured by placing the cuff of the sphygmomanometer around the upper arm. When the cuff is inflated, it compresses the brachial artery and cuts off blood flow. When the pressure is released to the level of the systolic arterial pressure, blood flow through the brachial artery resumes but becomes turbulent. This results in sounds known as **sounds of Korotkoff**, which may be auscultated with a stethoscope.

## Procedure  Measuring Blood Pressure

Practice is necessary to develop the skills to accurately measure arterial blood pressure. Following are the steps you may use to practice using the sphygmomanometer and stethoscope together. All readings should be taken with your lab partner seated and relaxed. When you have completed the activity, answer Check Your Understanding question 5 (p. 462).

**1** Obtain a stethoscope and sphygmomanometer of the appropriate size (about 80% of the circumference of the arm).

**2** Clean the earpieces and diaphragm as in Exercise 18-1.

**3** Wrap the cuff around your partner's arm. It should not be noticeably tight, but it should stay in place when you are not holding it. It should be about 1½ inches proximal to the antecubital fossa.

**4** Place the diaphragm of your stethoscope over the brachial artery. You should *not* hear anything at this point.

**5** Support your partner's arm by cradling it in your arm, or have your partner rest his or her arm on the lab table.

**6** Locate the screw of the sphygmomanometer near the bulb, and close it by turning it clockwise. Inflate the cuff by squeezing the bulb several times. Pay attention to the level of pressure you are applying by watching the pressure gauge. You should not inflate it beyond about 30 mmHg above your partner's normal systolic pressure (for most people, this is no higher than 180 mmHg). Your lab partner will not likely be happy with you if you inflate it above about 200 mmHg, because this can be uncomfortable.

**7** Slowly open the screw by turning it counterclockwise. Watch the pressure gauge, and listen to the brachial artery with your stethoscope.

**8** Eventually you will see the needle on the pressure gauge begin to bounce; at about the same time, you will begin to hear the pulse in the brachial artery. Record the pressure at which this first happens as the *systolic pressure*.

**9** Continue to listen, and watch the gauge until you can no longer hear the pulse. At this point, the needle on the gauge will stop rhythmically bouncing. Record the pressure at which this happens as the *diastolic pressure*. The numbers should be recorded as a fraction (e.g., 110/70, where 110 is the systolic pressure and 70 is the diastolic pressure).

Practice reading 1: _____

Practice reading 2: _____

## Procedure Measuring the Effects of the Autonomic Nervous System on Blood Pressure and Heart Rate

The autonomic nervous system (ANS) exerts a great deal of control over blood pressure through its influence on cardiac output, peripheral resistance, and blood volume. This exercise limits us to measuring only its effects on cardiac output and peripheral resistance (unless you want to stay in lab for the next two days and measure urine output—but I'm guessing you don't want to do that). When you have completed this activity, answer Check Your Understanding question 6 (p. 462).

**1** Have your lab partner remain seated and relaxed for 3 minutes. After 3 minutes, measure your partner's blood pressure and heart rate.

*Safety Note*
⚠️ Students with health problems or known cardiovascular disorders should not engage in this activity.

**2** Have your partner place one of his or her hands into a bucket of ice water.

**3** Repeat the blood pressure and pulse measurements with your partner's hand still in the ice water. (Note: Be kind to your partner and do this quickly!)

**4** Have your partner remove his or her hand from the ice water.

**5** Wait 5 minutes, then repeat the blood pressure and pulse measurements. Record your results in Table 18.5.

TABLE **18.5** Blood Pressure and Pulse Readings: Ice Water Experiment

| Test Situation | Blood Pressure | Pulse Rate |
|---|---|---|
| At rest | | |
| After immersing in ice water | | |
| 5 minutes after removing from ice water | | |

**6** Interpret your results:

**a** Which situation(s) represented activation of the sympathetic nervous system?

_____

**b** Which situation(s) represented activation of the parasympathetic nervous system?

_____

## Procedure Measuring the Effects of Caffeine on the Blood Pressure and Heart Rate

You are likely already familiar with caffeine's stimulant effects on the central nervous system, but caffeine also has stimulant effects on the cardiovascular system. This exercise will allow you to examine and measure these effects.

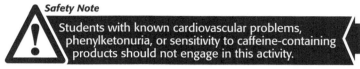

**Safety Note**

Students with known cardiovascular problems, phenylketonuria, or sensitivity to caffeine-containing products should not engage in this activity.

**1** Choose two classmates from your lab table who have not consumed caffeine for a minimum of 2 hours prior to lab to be your test subjects.

**2** Have each subject sit quietly for 5 minutes. After 5 minutes, measure each subject's blood pressure and heart rate. Record the results in **Table 18.6**.

**3** Give subject 1 a diet caffeine-free soda, and give subject 2 a diet caffeine-containing soda. Have each subject drink his or her entire can of soda as quickly as possible, ideally in 2 minutes or less.

**4** Have each subject remain seated for 15 minutes, then measure each subject's blood pressure and pulse rate. Record the results in **Table 18.6**.

TABLE **18.6** Blood Pressure and Pulse Readings: Caffeine Experiment

| Blood Pressure and Heart Rate Measurements | Subject 1: Caffeine-Free Soda | Subject 2: Caffeinated Soda |
|---|---|---|
| Blood pressure before soda | | |
| Blood pressure after soda | | |
| Heart rate before soda | | |
| Heart rate after soda | | |

**5** Report these data to your lab instructor, who will pool the data for the entire class. Calculate the average blood pressure and heart rate of all subjects before and after they consumed the sodas, and record these data in Table 18.7.

TABLE **18.7** Average Blood Pressure and Pulse Readings: Caffeine Experiment

| Blood Pressure and Heart Rate Measurements | Average of All Subject 1s: Caffeine-Free Soda | Average of All Subject 2s: Caffeinated Soda |
|---|---|---|
| Average blood pressure before soda | | |
| Average blood pressure after soda | | |
| Average heart rate before soda | | |
| Average heart rate after soda | | |

**6** Interpret your results:

**a** How did the average blood pressure and heart rate change before and after consuming the soda?

**b** What can you conclude about the effects of caffeine on blood pressure and heart rate from this experiment?

18

# Exercise 18-4

## The Ankle-Brachial Index

### MATERIALS

- ❑ Doppler ultrasound device
- ❑ Sphygmomanometer
- ❑ Stethoscope
- ❑ Alcohol and cotton ball
- ❑ Ultrasound gel

**Peripheral vascular disease**, also called *peripheral artery disease*, is any disease of the arteries outside of the brain and coronary circuit—in particular the arteries of the legs. It can be caused by diabetes, atherosclerosis, and numerous other conditions. One commonly performed test to assess the severity of peripheral vascular disease is the **ankle-brachial index** (**ABI**). The ABI compares the blood pressure in the legs (the "ankle" portion) with the blood pressure in the arms (the "brachial" portion). Generally, the blood pressure in the ankles declines with worsening disease in the arteries of the legs.

The ABI is obtained by dividing the systolic pressure in the ankle by the systolic pressure in the arm. Typically, the ABI is a decimal number because ankle pressure is slightly lower than the brachial pressure. This is because the legs are farther from the heart than the arms, which causes a slight decline in blood pressure. Interpretation of the ABI is as follows:

| | | |
|---|---|---|
| 0.9–1.0 | = | normal |
| 0.5–0.9 | = | peripheral vascular disease |
| 0.2–0.5 | = | intermittent claudication (temporary blockages to blood flow) |
| <0.2 | = | severe disease causing death of tissue |

Note that in younger or more athletic persons, the ABI is often greater than 1. This is simply a result of more muscle mass in the lower limb, which is difficult to completely compress with a sphygmomanometer cuff. This falsely elevates the ankle pressure and makes it appear higher than the brachial pressure.

## Procedure Ankle-Brachial Index

The brachial pressure is measured in this test in the standard way, with a sphygmomanometer and a stethoscope. You cannot easily auscultate either of the pedal pulses, however, so a Doppler ultrasound device is used to hear the blood flow instead, as shown in **Figure 18.4**. A Doppler ultrasound device uses sound waves transmitted through a liquid medium (ultrasound gel) to produce audible sounds of the blood flowing through the vessel. Following is the procedure for measuring the ankle pressure.

**1** Take the brachial pressure in the standard way with a sphygmomanometer and a stethoscope. Record *only* the systolic pressure below.

**2** Wrap the sphygmomanometer cuff around the ankle.

**3** Palpate for either the dorsalis pedis or the posterior tibial pulse.

**4** Place a small amount of ultrasound gel over the pulse. Place the Doppler probe in the gel so it is lightly touching the skin. (Note: If you press too hard with the probe, it is possible to cut off blood flow. If you do this, you will hear no sound.)

**5** Once you can hear sounds of the blood flowing, hold the probe in place as your lab partner inflates the sphygmomanometer cuff. Continue to inflate the cuff until you no longer hear any sound coming from the Doppler.

FIGURE **18.4** Performing the ankle-brachial index

**6** Have your lab partner slowly deflate the cuff while you continue to hold the probe in place. Record the pressure at which you first hear the blood flowing again. Note that this is a systolic pressure only; you will not obtain a diastolic pressure for the ankle. Remember that we record the diastolic pressure as the number when we can no longer hear blood flowing. But with the Doppler, you will continue to hear blood flow; therefore, there will be no diastolic reading.

**7** Calculate the ABI by dividing the systolic pressure of the ankle by the systolic pressure of the arm. You most likely will get a decimal number, and this is your ABI.

Systolic pressure of the ankle: _____

Systolic pressure of the arm (brachial): _____

Ankle pressure/Brachial pressure: _____

18

# Exercise 18-5

## The Electrocardiogram

The **electrocardiogram**, or **ECG**, is a recording of the electrical activity of the heart. Electrical activity is recorded by placing *electrodes* on the surface of the skin that record the changes in electrical activity. The changes in electrical activity are visible on the ECG as **waves**. Note that if there is no *net* change in electrical activity, the line on the ECG is flat. However, even when the ECG is flat between waves, the cells of the heart are in some phase of an action potential.

A standard ECG recording is shown in **Figure 18.5**. Notice that it consists of five waves, each of which represents the depolarization or repolarization of different parts of the heart.

- **P wave.** The initial **P wave** shows the depolarization of the cells of the right and left atria. It is fairly small because of the small number of cells in the atria.

- **QRS complex.** The next set of waves is called collectively the **QRS complex**, which represents the depolarization of the right and left ventricles. The first wave is the **Q wave**, a downward deflection, the next is the **R wave**, a large upward deflection, and the last is the **S wave**, the final downward deflection. The large size of the QRS complex compared with the P wave results from the large number of ventricular cells.

- **T wave.** The small **T wave** is usually the final wave, and it shows the repolarization of the right and left ventricles.

The waves aren't the only important landmarks of the ECG. In addition, we consider the periods between the waves, which show the spread of electrical activity through the heart and phases of the myocardial cells' action potentials. One important period is the **R-R interval**, or the period of time between two R waves. This interval represents the duration of the generation and spread of an action potential through the heart. It can also be used to determine the heart rate, which we discuss shortly. Another period we examine is the **P-R interval**, defined as the period from the beginning of the P wave to the beginning of the R wave. During the P-R interval, the depolarization from the SA node spreads through the atria to the ventricles via the AV node.

The **S-T segment** is between the end of the S wave and the beginning of the T wave. This period is recorded during the ventricles' plateau phase. Recall that there is no net change in electrical activity during this phase, and for this reason, the S-T segment is generally flat. The final period we discuss is the **Q-T interval**, the time from the beginning of the Q wave to the end of the T wave. During the Q-T interval, the ventricular cells are depolarizing and repolarizing.

The normal pattern seen on an ECG is known as a **normal sinus rhythm**, which means that the SA node is pacing the heart at a rate between 60 and 100 beats per minute. Any deviation from the normal sinus rhythm is known as a **dysrhythmia** or **arrhythmia**.

An electrocardiograph records the tracing at a standard speed of 25 mm/second. This allows us to determine precisely the heart rate and the duration of the intervals we discussed. As you can see in **Figure 18.5**, each small box on the ECG tracing measures 0.04s, and each large box measures 0.20s. Five large boxes together measure 1 second. Determining the duration of most intervals is simple—just count the small or large boxes, and add the seconds together. Calculating the heart rate is equally simple: count the number of large boxes, and divide 300 by this number. For example, if you count 4.2 boxes: 300/4.2 = 71 beats per minute. The normal values for the periods we discussed are given in **Table 18.8**.

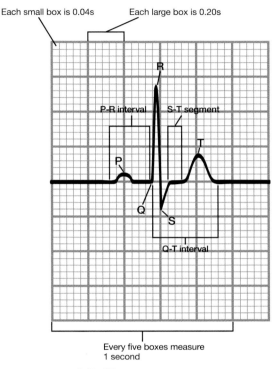

Each small box is 0.04s    Each large box is 0.20s

Every five boxes measure 1 second

FIGURE **18.5** Standard ECG recording

TABLE **18.8** Normal Values for ECG Periods

| Period | Normal Value |
|---|---|
| Heart rate | 60–100 beats per minute |
| R-R interval | 0.60–1.0s |
| P-R interval | 0.12–0.20s |
| Q-T interval | 0.42–0.44s |
| QRS complex duration | Less than or equal to 0.12s |

18

## Procedure Interpreting an ECG

You will now perform some basic ECG interpretation. Following are two tracings for which you will calculate the heart rate, and determine the duration of key periods of the ECG. When you have completed the activity, answer Check Your Understanding question 7 (p. 462).

**1** Identify and label the P wave, QRS complex, T wave, P-R interval, R-R interval, and Q-T interval on Tracings 1 and 2 in **Figure 18.6**.

**2** Calculate the heart rate for each tracing. Are the values normal or abnormal?

Heart Rate Tracing 1: _____

Heart Rate Tracing 2: _____

**3** Determine the R-R interval, QRS duration, P-R interval, and Q-T interval for each tracing, and record the values in **Table 18.9**.

**Tracing 1**

**Tracing 2**

FIGURE **18.6** ECG tracings

TABLE **18.9** Values for ECG Periods

| Value | Tracing 1 | Tracing 2 |
|-------|-----------|-----------|
| R-R interval | | |
| QRS duration | | |
| P-R interval | | |
| Q-T interval | | |

## Check Your Recall

**1** *Fill in the blanks:* The first heart sound is known as _____ and is caused by closure of the

_____. The second heart sound is known as _____ and is caused by

closure of the _____.

**2** *Matching:* Match the following terms with the correct definition.

| | | | |
|---|---|---|---|
| _____ | Bruit | A. | A noise heard in the heart because of defective valves |
| _____ | Murmur | B. | Extra heart sounds that may signify pathology |
| _____ | Tachycardia | C. | A decreased heart rate below 60 beats per minute |
| _____ | Rhythm | D. | A noise heard in an artery as a result of turbulent blood flow |
| _____ | S3 and S4 | E. | The pattern and regularity with which the heart beats |
| _____ | Bradycardia | F. | An elevated heart rate above 100 beats per minute |

**3** Which of the following statements is *false*?
a. The diaphragm of a stethoscope is used to hear higher-pitched sounds.
b. The bell of a stethoscope is used to hear lower-pitched sounds.
c. The aortic area is the area where the aortic valve is best heard.
d. Pulse palpation is the process of listening to an artery with a stethoscope.

**4** What is the baroreceptor reflex?

_____

_____

_____

**5** Which of the following is *not* a factor that determines blood pressure?
a. Cardiac output
b. Ankle-brachial index
c. Peripheral resistance
d. Blood volume

**6** *Fill in the blanks:* The _____ is the pressure in the arteries during ventricular systole and

averages about _____. The _____ is the pressure in the arteries during

ventricular diastole and averages about _____.

18

**7** The sounds heard in an artery when measuring the blood pressure are called

a. bruits.

b. sounds of Korotkoff.

c. Doppler waves.

d. S3 and S4.

**8** The actions of the sympathetic nervous system _____ blood pressure, and the actions of the parasympathetic nervous system _____ blood pressure.

a. increase; decrease

b. increase; increase

c. decrease; increase

d. decrease; decrease

**9** The ankle-brachial index compares the

a. diastolic pressure in the arm with the diastolic pressure in the ankle.

b. systolic pressure in the arm with the diastolic pressure in the ankle.

c. diastolic pressure in the arm with the systolic pressure in the ankle.

d. systolic pressure in the arm with the systolic pressure in the ankle.

**10** The QRS complex on an ECG represents:

a. depolarization of the atria.

b. repolarization of the atria.

c. depolarization of the ventricles.

d. repolarization of the ventricles.

18

Name _____

Section _____ Date _____

# Check Your Understanding

### Critical Thinking and Application Questions

**1** What might it mean if a pulse were normal (grade 2/4) on the right limb and absent (grade 0/4) on the left limb?

_____

_____

_____

_____

_____

**2** Palpation of both carotid arteries at the same time initiates a reflex known as the *baroreceptor reflex*. If this reflex is initiated in a healthy person with normal blood pressure, the person may temporarily lose consciousness. Explain why this happens, considering the effects of the baroreceptor reflex.

_____

_____

_____

_____

_____

18

**3** The baroreceptor reflex can be used to our benefit in a procedure known as a *carotid sinus massage*, performed when the blood pressure or heart rate is pathologically high. Why would a "massage" of the carotid sinus help treat these conditions?

_____

_____

_____

_____

_____

**4** What might it mean if the capillary refill time measures 10 seconds in the digits of one hand and 2 seconds in the digits of the other hand? What might it mean if the capillary refill time measures 10 seconds in all digits?

**5** Your patient has been admitted to the emergency room with an occupational injury from an industrial saw. He has lost a significant volume of blood. What effect has this blood loss likely had on his blood pressure, and why?

**6** Predict the effects of a drug that stimulates the sympathetic nervous system on blood pressure and heart rate. How would this drug affect the three factors that determine blood pressure?

**7** The condition known as an *AV block* is characterized by slow conduction of action potentials through the AV node to the ventricles. Which segment of the ECG would this be likely to affect and how would it affect it? Explain.

18

# Blood

## OBJECTIVES

*Once you have completed this unit, you should be able to:*

■ Identify the formed elements of blood.

■ Determine the differential white blood cell count of a blood smear.

■ Perform blood typing of the ABO and Rh blood groups using simulated blood.

■ Explain the basis for blood typing and matching for blood donation.

■ Determine appropriate blood donors for a given recipient.

■ Determine your own blood type and the hemoglobin content of your blood.

# PRE-LAB EXERCISES

Complete the following exercises prior to coming to lab, using your textbook and lab manual for reference.

## Pre-Lab Exercise **19-1**

✎ Key Terms

You should be familiar with the following terms before coming to lab.

| Term | Definition |
|------|------------|

**Formed Elements**

Erythrocyte _____

_____

Leukocyte _____

_____

Granulocyte _____

_____

Neutrophil _____

_____

Eosinophil _____

_____

Basophil _____

_____

Agranulocyte _____

_____

Lymphocyte _____

_____

Monocyte _____

_____

Platelets _____

_____

**19**

## Blood Typing

Antigen _____

_____

Antiserum _____

_____

Antibody _____

_____

## Blood Donation

Universal donor _____

_____

Universal recipient _____

_____

**19**

# Pre-Lab Exercise 19-2

## Formed Elements

In this unit we will identify the formed elements of blood on a peripheral blood smear. Each formed element has unique morphological characteristics and functions. Use your text and Exercise 19-1 (p. 467) in this unit to fill in **Table 19.1** with these functions and characteristics.

TABLE **19.1** Properties of Formed Elements

| Formed Element | Nucleus Shape | Cytoplasm and/or Granule Color | Function | Prevalence |
|---|---|---|---|---|
| Erythrocyte | | | | |
| Neutrophil | | | | |
| Eosinophil | | | | |
| Basophil | | | | |
| Lymphocyte | | | | |
| Monocyte | | | | |
| Platelet | | | | |

# EXERCISES

Safety concerns often preclude the use of real blood in the laboratory, but we can still demonstrate important principles of blood by viewing prepared microscope slides of blood cells and by using simulated blood to learn about blood typing. There is no real concern over blood-borne diseases with simulated blood, but do keep in mind that the simulated blood contains chemicals that may be hazardous. Therefore, use appropriate safety protocols when handling all materials in this lab.

In this unit you will identify the formed elements of blood on microscope slides. You will also play a murder mystery game in which you use simulated blood to apply blood-typing techniques, after which you use the same cast of characters to examine blood donation. If your lab has the appropriate protocols for the use of real blood, you will also type your own blood and determine its hemoglobin content.

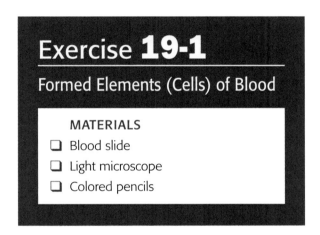

## Exercise 19-1

### Formed Elements (Cells) of Blood

**MATERIALS**
- ☐ Blood slide
- ☐ Light microscope
- ☐ Colored pencils

Whole blood consists of two main components: **plasma,** the fluid portion of blood, and **formed elements,** or the cellular portion of blood. Plasma accounts for about 55% of the volume of whole blood and consists primarily of water, proteins, and other solutes such as nutrients and ions. Formed elements account for about 45% of the volume of whole blood. Formed elements can be divided into three classes, each of which is shown in **Figure 19.1.**

1. **Erythrocytes.** Erythrocytes, also known as **red blood cells,** are the most numerous blood cells. These cells carry oxygen around the body on an iron-containing molecule called **hemoglobin.** Erythrocytes are easily distinguished from the other formed elements by their red color and the fact that mature erythrocytes lack nuclei and most organelles.

2. **Leukocytes.** Leukocytes, also known as **white blood cells,** play a role in the immune system. The two subclasses of leukocytes are based upon the presence or absence of granules in their cytoplasm.

   a. **Granulocytes.** As implied by their name, granulocytes are cells containing cytoplasmic granules that are visible when stained. The three types of granulocytes stain differently when treated with the dyes hematoxylin and eosin and are named for the type of stain with which they interact.

      i. **Neutrophils** do not interact strongly with either type of dye, and they stain a light violet color. They are the most numerous type of leukocyte and have multilobed nuclei. Their nuclei often vary in appearance, and for this reason they are often called *polymorphonucleocytes.*

      ii. **Eosinophils** interact strongly with the red dye eosin, and their granules stain bright red. They are far less numerous than neutrophils and tend to have bilobed nuclei.

      iii. **Basophils** take up the dark purple stain hematoxylin (it is a basic dye, hence their name *baso*phil), and their granules appear dark blue-purple. They tend to have bilobed nuclei, but their nuclei are often obscured by their dark granules. They are the least numerous of the leukocytes and will likely be the most difficult to find on your slide.

   b. **Agranulocytes.** The cells known as agranulocytes lack cytoplasmic granules. The two types of agranulocytes are the following:

      i. **Lymphocytes** tend to be smaller than granulocytes and have large, spherical nuclei. They are the second most numerous type of leukocyte.

      ii. **Monocytes** are the largest of the leukocytes and have "U"-shaped or horseshoe-shaped nuclei. They are the third most numerous type of leukocyte.

3. **Platelets.** Note in **Figure 19.1** that platelets aren't actually cells at all but are instead just small cellular fragments. As such, they lack nuclei and most organelles and are much smaller than the other formed elements. Platelets are involved in blood clotting.

**19**

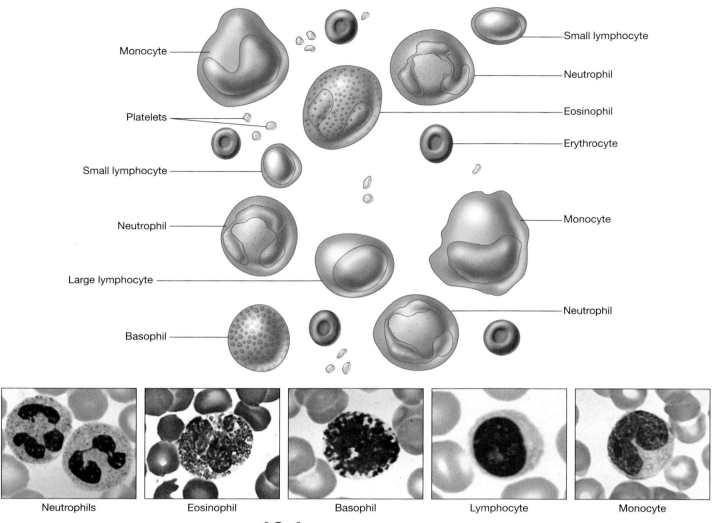

Monocyte

Platelets

Small lymphocyte

Neutrophil

Large lymphocyte

Basophil

Small lymphocyte

Neutrophil

Eosinophil

Erythrocyte

Monocyte

Neutrophil

Neutrophils   Eosinophil   Basophil   Lymphocyte   Monocyte

FIGURE **19.1** Formed elements of the blood

**19**

## Procedure  Microscopy of a Peripheral Blood Smear

In this exercise, you will examine a blood slide called
a **peripheral blood smear**. Examine the peripheral
blood smear on high power, and scroll through to find each of
the formed elements. Note that you may have to find a second
slide to locate certain cells, because some types are more difficult
to find (in particular the eosinophils and basophils). In the spaces
provided, use colored pencils to draw and describe each formed
element you locate. When you have finished, answer Check
Your Understanding questions 1–3 (pp. 485–486).

**1** Erythrocyte:

**2** Neutrophil:

**3** Eosinophil:

**4** Basophil:

**5** Monocyte:

**6** Lymphocyte:

**7** Platelets:

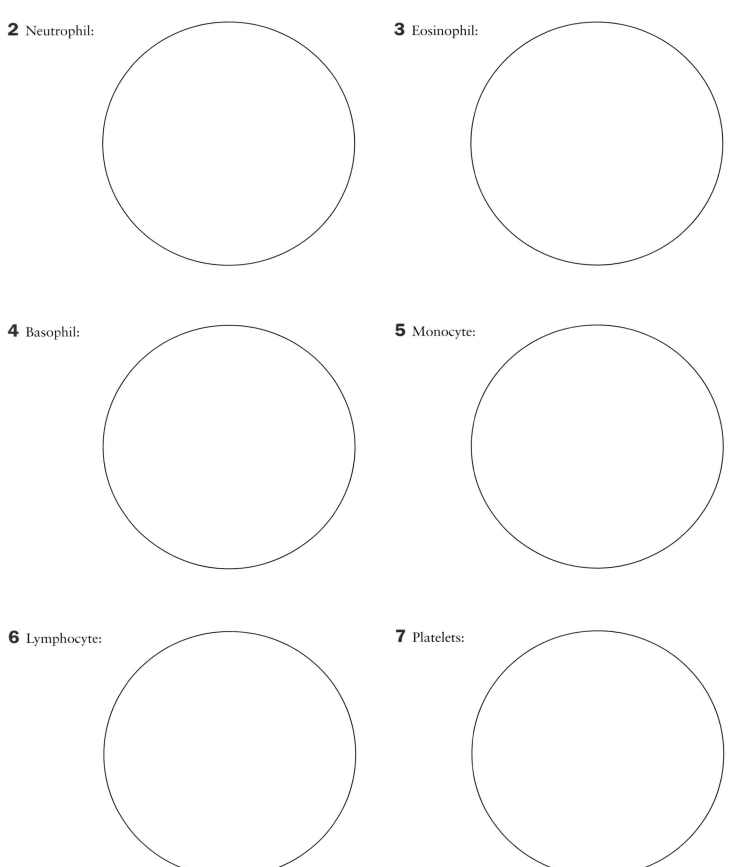

## Procedure Performing a Differential White Blood Cell Count

One of the most common blood analyses performed by a laboratory is a *differential white blood cell count*. In this procedure, the number and types of leukocytes in a blood sample are counted to determine the relative frequency of each type. This can give important information about the cause of a patient's illness, because different conditions will cause an elevation of different types of leukocytes. Differential counts were previously performed manually by a laboratory technician, as you will do here. Now, however, they are generally done by a machine, but the technician will still manually examine the cells if any abnormalities are noted.

**1** Obtain a blood smear slide, and scan it on low power to find an area where the cells appear to be evenly distributed.

**2** Advance the microscope objective to a power high enough for you to distinguish between the different types of leukocytes. For some microscopes, medium power will suffice, and for others, high power or even oil immersion might be necessary.

**3** Scan the slide for leukocytes, and create a running tally of the number of each type you find below. Do this until you have counted 100 total leukocytes.

| *Neutrophils* | *Eosinophils* | *Basophils* | *Lymphocytes* | *Monocytes* |
|---|---|---|---|---|
| _____ | _____ | _____ | _____ | _____ |
| _____ | _____ | _____ | _____ | _____ |
| _____ | _____ | _____ | _____ | _____ |
| _____ | _____ | _____ | _____ | _____ |

**4** Total the number of each leukocyte you counted, and record these data in **Table 19.2**. Refer to **Table 19.1** in the Pre-Lab Exercises (p. 466) for the predicted percentage of each type of leukocyte in the blood. How do the totals you found compare with the predicted percentages?

TABLE **19.2** Differential White Blood Cell Totals

| Leukocyte | Number Counted | Percentage (of 100) | Predicted Percentage |
|---|---|---|---|
| Neutrophils | | | |
| Eosinophils | | | |
| Basophils | | | |
| Lymphocytes | | | |
| Monocytes | | | |

# Exercise 19-2

## ABO and Rh Blood Groups

### MATERIALS
- ❑ Well plates
- ❑ Simulated blood types A−, B+, AB−, and O+
- ❑ Simulated antisera: anti-A, anti-B, anti-Rh

Blood typing is done by checking the blood for the presence or absence of specific glycoproteins called **antigens** found on the cell surface. Two clinically relevant antigens are the **A antigen** and the **B antigen**. The blood type is named based upon which of the antigens is present.

- ▌ **Type A** blood has A antigens on the cell surface.
- ▌ **Type B** blood has B antigens on the cell surface.
- ▌ **Type AB** blood has both A and B antigens on the cell surface.
- ▌ **Type O** blood has neither A nor B antigens on the cell surface.

An additional clinically relevant antigen is the Rh antigen.

- ▌ Blood that has the Rh antigen is denoted as **Rh positive** (e.g., A+).
- ▌ Blood that lacks the Rh antigen is denoted as **Rh negative** (e.g., A−).

The prevalence of different blood types in the United States varies with different ethnic groups. In general, we can say that the most common type is O+, followed by A+ and then B+.

The antigens present on the surface of an erythrocyte can be determined by combining it with a solution called an **antiserum**. An antiserum is a solution that contains proteins from the immune system called **antibodies** that bind to specific antigens. When antibodies bind to antigens on erythrocytes, they cause **agglutination,** or clumping of the erythrocytes, in the sample. In this exercise we are using simulated blood, so you won't see agglutination unless your instructor sets up a demonstration or permits you to type your blood (which is in a later, optional procedure). The antisera used to determine the blood type of a sample are named according to the antibody they bind:

- ▌ **Anti-A antiserum** contains anti-A antibodies that bind to erythrocytes with A antigens.
- ▌ **Anti-B antiserum** contains anti-B antibodies that bind to erythrocytes with B antigens.
- ▌ **Anti-Rh antiserum** contains anti-Rh antibodies that bind to erythrocytes with Rh antigens.

## Antigen-Antibody Reactions

This exercise allows you to examine the antigen-antibody reactions of known blood types. Each table should take one well plate and one set of dropper bottles. The bottles are labeled A−, B+, AB−, and O+ to represent each of those blood types, and anti-A, anti-B, and anti-Rh to represent the different antisera. You can see what a positive reaction looks like in **Figure 19.2**. Notice that a positive reaction is indicated by the formation of a white, cloudy precipitate. So, for type A+, a positive reaction is seen in the wells to which anti-A and anti-Rh antisera were added, but there is no reaction in the well containing anti-B antiserum.

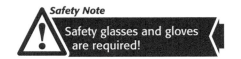

*Safety Note*

Safety glasses and gloves are required!

Positive reactions

Negative reaction

Type A+        Type B−        Type AB+        Type O−

FIGURE **19.2** Reactions of simulated blood with simulated antisera

## Procedure Testing Simulated Blood

Use **Figure 19.3** as a guide to placement of samples in the wells.

**1** Label wells on the well plate as wells 1–12.

**2** Drop two drops of type A− blood in well 1, well 2, and well 3.

**3** Drop two drops of type B+ blood in well 4, well 5, and well 6.

**4** Drop two drops of type AB− blood in well 7, well 8, and well 9.

**5** Drop two drops of type O+ blood in well 10, well 11, and well 12.

**6** Add two drops of the anti-A antiserum to wells 1, 4, 7, and 10.

**7** Add two drops of the anti-B antiserum to wells 2, 5, 8, and 11.

**8** Add two drops of the anti-Rh antiserum to wells 3, 6, 9, and 12.

**9** Observe the samples for changes in color symbolizing the agglutination or clumping that would normally occur between antisera and specific blood types. Record your results in **Table 19.3**.

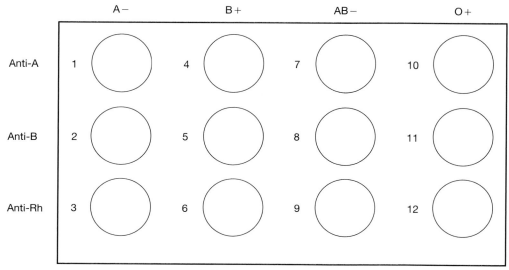

FIGURE **19.3** Well plate diagram

**19**

TABLE **19.3** Blood Typing Results

| Blood Type | Reacted with anti-A? (yes/no) | Reacted with anti-B? (yes/no) | Reacted with anti-Rh? (yes/no) | Antigens Present on Cell Surface |
|---|---|---|---|---|
| A− | | | | |
| B+ | | | | |
| AB− | | | | |
| O+ | | | | |

# Exercise 19-3

## Murder Mystery Game

### MATERIALS
- ❑ Well plate
- ❑ Simulated antisera: anti-A, anti-B, anti-Rh
- ❑ Murder mystery game

In this game you will be applying the blood-typing techniques you learned in Exercise 19-2 (p. 471) to solve a series of murder mysteries. Each of the following cases presents a victim, a murderer, three suspects, three possible murder rooms, and three possible murder weapons. Your job is to play the role of detective, and determine the identity of the murderer, which weapon he or she used, and in which room the crime was committed.

**Safety Note** Safety glasses and gloves are required!

## Procedure Solving the Murder Mysteries

For each case a unique set of bottles is marked with a number that corresponds to the specific cases (i.e., the bottles are marked 1 for Case 1, 2 for Case 2, and 3 for Case 3). The murderer, rooms, and murder weapons are different for each case, but the cast of characters remains the same.

**1** To begin the game, assemble groups of two or three students. Obtain a well plate, and choose one set of samples to test (e.g., the rooms from Case 1, the suspects from Case 2, or the weapons from Case 3).

**2** Test the samples by placing drops of the sample in three separate wells. Add two drops of anti-A antiserum to the first well, add two drops of anti-B antiserum to the second well, and add two drops of anti-Rh antiserum to the third well. After you have tested each of the samples, return them to their proper places in the front of the lab.

**3** Read and record the blood type by watching for a reaction with the antisera. Remember that this is simulated blood, just as in Exercise 19-2. A positive reaction is denoted by a color change.

**4** To determine the
- ❚ *Murderer*: Match the blood type of one of the *suspects* to that of the murderer.
- ❚ *Weapon and room*: Match the blood type of the *victim* to the blood types found in the rooms and on the weapon.

## Case 1: Ms. Magenta

We enter the scene to find the dearly departed Ms. Magenta. Forensic analysis determines that there are two types of blood on the body. One blood type is Ms. Magenta's, and the other blood type is a trace amount left behind from the murderer.

**Ms. Magenta's blood type:** _____        **Murderer's blood type:** _____

We have three suspects:

1. *Mrs. Blanc* was being blackmailed by Ms. Magenta and Col. Lemon. They had discovered that Mrs. Blanc had murdered her late husband. Mrs. Blanc knew this would ruin her reputation at the country club.
2. *Col. Lemon* wanted to keep the blackmail money for himself and wanted Ms. Magenta out of the way.
3. *Mr. Olive* had been secretly in love with Ms. Magenta for years, and when he told her of his feelings, she rejected him harshly.

**Mrs. Blanc's blood type:** _____        **Col. Lemon's blood type:** _____

**Mr. Olive's blood type:** _____

We have three possible murder rooms:

**Ballroom blood type:** _____        **Library blood type:** _____

**Den blood type:** _____

We have three possible murder weapons:

Candlestick blood type: _____     Noose blood type: _____

Knife blood type: _____

## Case 1: Conclusion

Ms. Magenta was killed by _____, in the _____,

with the _____.


## Case 2: Col. Lemon

Our next victim is poor Col. Lemon. On his body we find his blood and also trace amounts of the blood of another person, presumably the murderer.

Col. Lemon's blood type: _____     Murderer's blood type: _____

We have three potential suspects:

1. *Mrs. Blanc.* Now, with Ms. Magenta out of the way, Mrs. Blanc could easily rid herself of her problem by disposing of the only other person who knows her secret—Col. Lemon.
2. *Professor Purple* believed the colonel had stolen his groundbreaking research into collagen lip injections.
3. *Mr. Olive* couldn't stand the colonel because of his close relationship with Ms. Magenta.

Mrs. Blanc's blood type: _____     Professor Purple's blood type: _____

Mr. Olive's blood type: _____

We have blood in three different rooms:

Hall blood type: _____     Kitchen blood type: _____

Billiards room blood type: _____

Forensics found blood on three different weapons:

Copper pipe blood type: _____     Hammer blood type: _____

Revolver blood type: _____

## Case 2: Conclusion

Col. Lemon was killed by _____, in the _____,

with the _____.


## Case 3: Mr. Olive

Our next (and hopefully last) victim is Mr. Olive. Analysis demonstrates two blood types: one belonging to Mr. Olive, and trace amounts of another belonging to the murderer.

Mr. Olive's blood type: _____     Murderer's blood type: _____

We have three potential suspects:

1. *Ms. Feather* had always secretly loved Mr. Olive, but he spurned her advances in favor of Ms. Magenta.

2. *Mrs. Blanc* was worried that Mr. Olive knew her secret and wanted him out of the way.

3. *Professor Purple* discovered that Mr. Olive—not Col. Lemon—had actually stolen the collagen lip implant research. Whoops!

**Ms. Feather's blood type:** _____  **Mrs. Blanc's blood type:** _____

**Professor Purple's blood type:** _____

Blood was found in three rooms:

**Lounge blood type:** _____  **Dining room blood type:** _____

**Greenhouse blood type:** _____

We have three potential murder weapons:

**Noose blood type:** _____  **Hammer blood type:** _____

**Revolver blood type:** _____

## Case 3: Conclusion

Mr. Olive was killed by _____, in the _____,

with the _____.

# Exercise 19-4

## Blood Donation

**Blood transfusion**, the infusion of a recipient with a donor's blood cells, is a commonly performed medical procedure. Before a recipient is given a blood transfusion, the medical team must first learn the patient's blood type and then find a suitable, or "matching," donor. This is necessary because of the A, B, and Rh antigens on the surface of the donor erythrocytes and the presence of preformed antibodies in the recipient's blood. If a donor's blood has antigens the recipient's immune system recognizes as foreign, the recipient's antibodies will agglutinate the foreign erythrocytes, and the immune system will destroy them, a process known as **hemolysis**. This is called a **transfusion reaction**, and it is a medical emergency that can lead to kidney failure and death.

To ensure that a transfusion reaction does not occur, we must make sure the donor blood does not have antigens the recipient's immune system will recognize as foreign. Our immune systems produce antibodies to any antigen *not* present on the surface of our own cells.

- People with type A blood have A antigens and so produce anti-B antibodies.
- People with type B blood have B antigens and so produce anti-A antibodies.
- People with type O blood have neither A nor B antigens and so produce anti-A and anti-B antibodies.
- People with type AB blood have both A and B antigens and so produce neither anti-A nor anti-B antibodies.

If you're wondering about the Rh factor, wait just a moment—we're getting there. Let's do an example with the ABO blood groups first:

Patient 1 has type B blood, which means that he has anti-A antibodies. What will happen if we give him blood from a donor with:

- **Type A blood?** There are A antigens on these erythrocytes, and his anti-A antibodies would agglutinate them. ✘
- **Type B blood?** Patient 1's anti-A antibodies would have no effect on the B antigens on these erythrocytes, so this blood is safe. ✔
- **Type O blood?** There are no antigens on these donor erythrocytes, so Patient 1's anti-A antibodies would have no effect on them, and this blood is safe. ✔
- **Type AB blood?** There are both A and B antigens on these erythrocytes, and Patient 1's anti-A antibodies would agglutinate the erythrocytes. ✘

Now that's easy, isn't it?

Next let's address the Rh factor. The blood of an Rh-negative person does *not* contain preformed antibodies to the Rh antigen. If an Rh-negative person is exposed to the Rh antigen, however, he or she *does* produce anti-Rh antibodies. In an emergency setting, it is generally not possible to determine if an Rh-negative person has been exposed to the Rh antigen, so health-care professionals err on the side of caution and assume that the person has anti-Rh antibodies. For the sake of simplicity, we will assume the same thing in this exercise. So, for our purposes:

- People with Rh-positive blood *do not* produce anti-Rh antibodies.
- People with Rh-negative blood *do* produce anti-Rh antibodies.

Let's do one more example, taking into account the Rh factor this time:

Patient 2 has A − blood, which means that she has anti-B and anti-Rh antibodies. What will happen if we give her blood from a donor with:

- **Type A + blood?** There are A and Rh antigens on these erythrocytes, and her anti-Rh antibodies would agglutinate them. ✘
- **Type B − blood?** There are B antigens on these erythrocytes, and her anti-B antibodies would agglutinate them. ✘
- **Type O − blood?** There are no antigens on these erythrocytes, so Patient 2's anti-B and anti-Rh antibodies would have no effect on them, and this blood is safe. ✔
- **Type AB + blood?** There are A, B, and Rh antigens on these erythrocytes, and Patient 2's anti-B and anti-Rh antibodies would agglutinate them. ✘

# **Procedure** Blood-Type Matching Practice

Use the information above and your text to fill in **Table 19.4**. After you have filled in the table, answer Check Your Understanding questions 4 and 5 (p. 486).

TABLE **19.4** Blood Donation

| Blood Type | Antigens Present | Antibodies Present | Can Donate Safely to Which Blood Types? | Can Receive Safely from Which Blood Types? |
|---|---|---|---|---|
| A+ | | | | |
| A− | | | | |
| B+ | | | | |
| B− | | | | |
| AB+ | | | | |
| AB− | | | | |
| O+ | | | | |
| O− | | | | |

You should notice something from **Table 19.4**: AB+ blood can receive from any blood type, and O− blood can donate to any blood type. This is because AB+ blood has all three antigens but no antibodies to these antigens, which is why it is often called the **universal recipient**. But Type O− has none of the three antigens for a recipient's antibodies to bind, so it can be donated to any blood type. For this reason, type O− is often called the **universal donor**.

## Procedure Type Matching for Transfusions

Gasp! It turns out that Ms. Magenta, Col. Lemon, and Mr. Olive all survived their injuries! But they have lost blood and are in need of blood transfusions. All of the suspects have had a sudden change of heart and have offered to help the three victims by donating blood. Your job is to determine who among the suspects could safely donate blood to whom. You do not need to retest each person's blood type, as you may use your results from Exercise 19-3. After you have finished this activity, answer Check Your Understanding questions 6 and 7 (p. 486).

**HINTS & TIPS**

Remember—when trying to work out who can donate blood to whom, you are concerned with the recipient's antibodies and the donor's antigens. So first work out which antibodies the recipient has, then make sure the recipient's antibodies won't bind any antigens on the donor's erythrocytes.

## Recipient 1: Ms. Magenta

Ms. Magenta's blood type: _____

**Donors:**

Ms. Feather's blood type: _____

Mrs. Blanc's blood type: _____

Professor Purple's blood type: _____

Who could safely donate blood to Ms. Magenta? _____

Who could not safely donate blood to Ms. Magenta? _____

## Recipient 2: Col. Lemon

Col. Lemon's blood type: _____

**Donors:**

Ms. Feather's blood type: _____

Mrs. Blanc's blood type: _____

Professor Purple's blood type: _____

Who could safely donate blood to Col. Lemon? _____

Who could not safely donate blood to Col. Lemon? _____

## Recipient 3: Mr. Olive

Mr. Olive's blood type: _____

**Donors:**

Ms. Feather's blood type: _____

Mrs. Blanc's blood type: _____

Professor Purple's blood type: _____

Who could safely donate blood to Mr. Olive? _____

Who could not safely donate blood to Mr. Olive? _____

# Exercise 19-5

## Typing and Examining Your Own Blood

### MATERIALS

- ❏ Lancet
- ❏ Alcohol wipes
- ❏ Anti-A, anti-B, and anti-Rh antisera
- ❏ Blank microscope slide
- ❏ Sharpie pen
- ❏ Toothpicks
- ❏ Tallquist paper and scale
- ❏ Dissection microscope or magnifying glass

If your lab has the appropriate equipment and biohazard disposal means, your instructor may permit you to type and examine your own blood in this optional exercise. Working with blood is actually very safe provided you follow some basic procedures outlined here:

**1** Wash your hands with soap and water before starting and after completing the procedures.

**2** Prepare your work area by placing a disposable absorbent liner on the lab table.

**3** Wear gloves and safety glasses when handling all materials for this lab.

**4** Handle only your own lab materials to avoid coming into contact with blood other than your own.

**5** Dispose of all lancets and microscope slides in the designated sharps container only.

**6** Dispose of all nonsharp materials (disposable pads, gloves, alcohol wipes, and any other materials you have used) in the designated red biohazard bag.

**7** When you are finished with the procedures, clean your work area with the disinfectant solution provided by your lab instructor.

In the following procedures, you will determine your blood type and measure the approximate hemoglobin content of your blood. Note that these exercises involve lancing your own finger to stimulate bleeding. If you have any medical conditions that render this activity unsafe, discuss this with your lab instructor.

## Procedure Determining Your Blood Type

The process of determining the blood type of real blood is similar to the process you performed earlier with simulated blood: You use three blood samples, and apply three different antisera to look for a reaction. The reaction between real blood antigens and antibodies is more subtle than in the simulated blood; in particular, the reaction with Rh antigens and the anti-Rh antibodies can be difficult to see. You may wish to examine your samples under a magnifying glass to more clearly see the reaction.

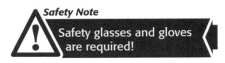

*Safety Note*

⚠ Safety glasses and gloves are required!

**1** Obtain a blank microscope slide, and draw three circles on it with a Sharpie pen. Label the circles A, B, and Rh.

**2** Wash your hands, and prepare your work area with a disposable absorbent liner.

**3** Place a glove on one hand, and prepare one finger of your ungloved hand by cleaning it with an alcohol wipe.

**4** Use a fresh lancet to lance the finger you just cleaned. Dispose of the used lancet in the sharps bin.

**5** Squeeze your finger to stimulate bleeding, and squeeze a small drop of blood onto each of the three circles on the slide.

**6** Place a drop of anti-A antiserum in the circle marked A, anti-B in the circle marked B, and anti-Rh in the circle marked Rh.

**7** Use three toothpicks, one for each circle, to gently mix the blood with the antisera. You must use a different toothpick for each circle to avoid cross-contamination. Watch carefully for a reaction, using a dissection microscope or magnifying glass if needed (be sure to hold a magnifying glass in your gloved hand to avoid coming into contact with blood from your classmates).

**8** Determine your blood type based upon the reactions with the antisera. Remember, if the blood reacts with an antiserum, that antigen is present. For example, if your blood reacts with anti-B antiserum and anti-Rh antiserum, the B and Rh antigens are present, and the blood type is B +. Record your blood type here: _____

**9** Dispose of your microscope slide in the sharps bin when you have completed the blood-typing procedure.

## Procedure Determining the Hemoglobin Content of Your Blood

As you learned in Exercise 19-1, erythrocytes are filled with the protein hemoglobin, which binds and transports oxygen through the blood. The average amount of hemoglobin in erythrocytes averages about 12–16 g/dl in females and about 14–18 g/dl in males. Medically, this is an important value, because a decreased amount of hemoglobin can indicate conditions such as *iron-deficiency anemia*. In a hospital lab, the amount of hemoglobin in the blood is measured with an instrument known as a *hemoglobinometer*. However, in this lab we will employ an older (and less expensive) method using *Tallquist paper*, which provides an estimate of the blood's hemoglobin content.

**1** Squeeze your finger to continue to stimulate bleeding. If you need to lance your finger again, obtain a new lancet, and follow the same steps from the first procedure.

**2** Obtain a piece of Tallquist paper. Roll your bleeding finger on the paper.

**3** Compare the color the paper turns with the scale on the container while the paper is still slightly wet—do not let the paper dry, or you will need to repeat the procedure.

Estimated hemoglobin value: _____

**4** Dispose of any remaining sharps in the sharps container and all other blood-containing materials, including the Tallquist paper, in the red biohazard bag. Clean your work area with the disinfectant provided by your lab instructor.

## Procedure Calculating the Average Hemoglobin and Most Common Blood Type for Your Class

Once you have completed the first two procedures and cleaned up your work area, you may pool the results for your class to find the most common blood types and the average values for hemoglobin content.

**1** Record your blood type on the whiteboard.

**2** When everyone in the class has recorded his or her blood type, pool the results, and complete **Table 19.5**. Calculate the percentage of your class that has each blood type.

**3** Your instructor will have two columns for hemoglobin values on the whiteboard, one for female and one for male members of the class. Write your estimated hemoglobin value on the whiteboard under the appropriate column.

**4** When all of your classmates have recorded their number on the board, calculate the average value for the females and males of your class. Record these data in **Table 19.6**. How do the values for your class compare with the averages?

Table **19.5** Blood Types for Your Class

| Blood Type | Number in Class | Percentage of Total |
|---|---|---|
| A+ | | |
| A− | | |
| B+ | | |
| B− | | |
| O+ | | |
| O− | | |
| AB+ | | |
| AB− | | |

TABLE **19.6** Average Hemoglobin Values for Your Class

| Group | Average Hemoglobin Value |
|---|---|
| Females | |
| Males | |

19

## Check Your Recall

**1** *Mark the following statements as true (T) or false (F).* If the statement is false, correct it to make it a true statement.

_____ a. Red blood cells are also known as leukocytes.

_____ b. White blood cells with granules in their cytoplasm are known as agranulocytes.

_____ c. The granulocytes include the neutrophils, eosinophils, and basophils.

_____ d. Erythrocytes carry oxygen through the body on the protein hemoglobin.

_____ e. Both lymphocytes and monocytes are granulocytes.

_____ f. Platelets are fully formed cells involved in blood clotting.

**2** Label the formed elements on the peripheral blood smear in **Figure 19.4.**

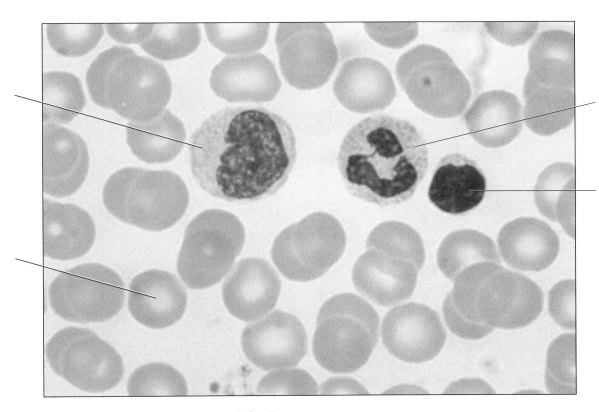

FIGURE **19.4** Peripheral blood smear

19

**3** Which of the following is not an antigen that may be found on the surface of an erythrocyte?

    a. A antigen

    b. B antigen

    c. O antigen

    d. Rh antigen

**4** State which antigens are present on the surface of erythrocytes of the following blood types:

    a. B− _____

    _____

    b. O+ _____

    _____

    c. AB− _____

    _____

    d. A+ _____

    _____

**5** Antibodies cause _____ of the antigens on erythrocytes.

    a. agglutination

    b. aggregization

    c. neutralization

    d. They have no effect on erythrocytes.

**6** A person with type A blood has

    a. anti-A antibodies.

    b. anti-B antibodies.

    c. anti-O antibodies.

    d. no antibodies.

**7** A person with type B− blood has which of the following antibodies? (Assume the person has been exposed to Rh antigens. Circle all that apply.)

    a. anti-A antibodies

    b. anti-B antibodies

    c. anti-Rh antibodies

    d. no antibodies

**8** List all of the blood types to which the following people could donate:

a. Person 1: Type A − _____

_____

b. Person 2: Type O + _____

_____

c. Person 3: Type AB − _____

_____

d. Person 4: Type B + _____

_____

**9** Which blood type is considered the universal donor? Why can this blood type be given to all other blood types?

_____

_____

_____

_____

_____

_____

_____

_____

**10** Which blood type is considered the universal recipient? Why can people with this blood type receive blood from all other blood types?

_____

_____

_____

_____

_____

_____

_____

19

# Check Your Understanding

### Critical Thinking and Application Questions

**1** When someone is admitted to the hospital, one of the first procedures health-care professionals perform is a blood draw. One lab value checked is a white blood cell (WBC) count. If the WBC count is elevated (normal range = 4,000–11,000/mm$^3$), this could indicate the presence of inflammation. Further analysis (called a differential) is performed to determine the relative prevalence of different types of white blood cells and the potential cause of the inflammation. What might be causing the inflammation if . . . (use your textbook for reference):

**a** neutrophils were elevated? Explain your reasoning.

_____

_____

_____

_____

**b** lymphocytes were elevated? Explain your reasoning.

_____

_____

_____

_____

**c** eosinophils were elevated? Explain your reasoning.

_____

_____

_____

_____

**2** Another lab value monitored routinely is the total red blood cell (RBC) count. A normal RBC count is typically 4.2–5.9 million/mm$^3$. Having a low RBC count is called *anemia*. Predict the possible effects of having anemia, considering the functions of erythrocytes.

_____

_____

_____

_____

**3** The poison ricin, derived from the castor bean plant, interferes with protein synthesis. Which formed elements of blood would ricin affect directly? Are there any formed elements in the blood that ricin would not affect? Explain.

**4** The disease *erythroblastosis fetalis* (also called *hemolytic disease of the newborn*) develops in a fetus or a newborn infant with Rh-positive blood and an Rh-negative mother. Symptoms result when maternal anti-Rh antibodies cross the placenta and interact with the fetus' erythrocytes. Why are the children of Rh-positive mothers not at risk for this disease? Why are Rh-negative fetuses not at risk for this disease?

**5** Explain why a person who is blood type AB+ can receive blood from any blood type but can only donate to individuals who are also blood type AB+.

**6** When Col. Lemon arrived at the hospital, the staff determined that his blood had been mistyped, and he was in fact blood type AB−. Which of our three suspects (Mrs. Blanc, Ms. Feather, and/or Professor Purple) could safely donate blood to Col. Lemon now? (Refer to p. 478 for their blood types.)

**7** What would have happened to Col. Lemon if he had received blood from an incompatible donor?

# Lymphatics and Immunity

# 20

## OBJECTIVES

*Once you have completed this unit, you should be able to:*

▮ Identify structures of the lymphatic system.

▮ Describe the role of the immune system in processing lymph as it is returned to the heart.

▮ Trace the pathway of lymph as it is returned to the cardiovascular system.

▮ Identify histological features of lymphoid organs on microscope slides.

▮ Trace the response of the immune system to an invading pathogen.

▮ Describe the theory behind the ELISA procedure.

Name _____ Section _____ Date _____

# PRE-LAB EXERCISES

Complete the following exercises prior to coming to lab, using your textbook and lab manual for reference.

## Pre-Lab Exercise **20-1**

 Key Terms

You should be familiar with the following terms before coming to lab.

| Term | Definition |
|---|---|

**Lymphatic Structures**

Lymphatic capillary _____

_____

Lymph _____

_____

Lymph-collecting vessel _____

_____

Lymph trunk _____

_____

Lymph duct _____

_____

Spleen _____

_____

Thymus _____

_____

Tonsil _____

_____

Mucosal-associated lymphatic tissue _____

_____

Lymph node _____

_____

**Components of the Immune System**

Nonspecific response _____

_____

Inflammation _____

_____

Specific response _____

_____

Antigen _____

_____

Humoral immunity _____

_____

Antibody _____

_____

Cell-mediated immunity _____

_____

# Pre-Lab Exercise 20-2

## Anatomy of the Lymphatic System

Label and color the structures of the lymphatic system depicted in **Figure 20.1** with the terms from Exercise 20-1 (p. 491). Use your text and Exercise 20-1 in this unit for reference.

20

FIGURE **20.1** Lymphoid organs: (**A**) overview of the lymphatic organs and vessels; (**B**) the tonsils

# EXERCISES

The **lymphatic system** serves numerous homeostatic functions in the body. For one, it is an important part of the immune system and combats harmful agents in the internal and external environments. It also works with the cardiovascular system to maintain fluid homeostasis in the extracellular fluid and with the gastrointestinal system to absorb fats.

The exercises in this unit introduce you to this important system. In the first exercise, you will study the lymphatic system's anatomy, and trace the flow of lymph throughout the body. In the second exercise, you will examine three types of lymph tissue on microscope slides. Finally, in the last two exercises you will explore the immune system response.

## Exercise **20-1**

### Lymphatic System Anatomy

#### MATERIALS
- ❏ Human torso models
- ❏ Head and neck models
- ❏ Intestinal villus model
- ❏ Laminated outline of the human body
- ❏ Water-soluble marking pens

As you have learned, the lymphatic system consists of a diverse group of organs that have three primary functions:

1. **Transporting excess interstitial fluid back to the heart.** Hydrostatic pressure is stronger than colloid osmotic pressure in blood capillaries, which forces fluid out of the blood capillaries and into the interstitial space. Approximately 1.5 ml/min. of fluid is lost from the circulation in this manner. This may not sound like a lot, but if this fluid were not returned to the blood vessels, we could lose our entire plasma volume in about a day! Fortunately, the lymphatic system picks up this lost fluid, carries it through lymphatic vessels, and returns it to the cardiovascular system. The fluid is picked up first by small, blind-ended **lymph capillaries** that surround blood capillary beds. These lymph capillaries are distinct from blood capillaries and contain highly permeable walls that allow large substances and large volumes of fluid to enter and exit. Once inside the lymph capillaries, the fluid is called **lymph** and is delivered to larger **lymph-collecting vessels**. Lymph-collecting vessels drain the fluid into larger **lymph trunks**, and these drain into **lymph ducts** that return the lymph to the cardiovascular system (**Figure 20.2A**). Note in **Figure 20.2A** that there are two lymph ducts—the **right lymphatic duct**, which drains the right arm and the right side of the head, neck, and thorax, and the **thoracic duct**, which drains lymph from the remainder of the body. The right lymphatic and thoracic ducts drain lymph into the blood at the junctions of the right and left subclavian and internal jugular veins, respectively.

2. **Activating the immune system.** Several of the lymphatic organs activate the immune system. These include the **thymus**, the organ in which T lymphocytes mature; the **spleen**, which filters the blood and houses phagocytes; and the **tonsils**, aggregates of unencapsulated lymphoid tissue found in the oropharynx and nasopharynx (**Figure 20.2B**). As you can see in **Figure 20.2B**, there are three main sets of tonsils: the **pharyngeal tonsils**, located in the posterior nasopharynx; the **palatine tonsils**, located in the posterior oropharynx; and the **lingual tonsil**, located at the base of the tongue. Clusters of lymphatic tissue similar in structure to the tonsils are scattered throughout the gastrointestinal tract, where it is called **mucosal-associated lymphoid tissue**, or **MALT**. The lymphoid organs called **lymph nodes** are similar to tonsils and MALT, but they are surrounded by a connective tissue capsule (note that lymph nodes are often called "lymph glands," but this is a misnomer because they do not secrete any products). Lymph nodes are found along lymphatic vessels, where they filter lymph and remove pathogens, toxins, and cells (such as cancer or virally infected cells).

3. **Absorbing dietary fats.** Fats are not absorbed from the small intestine directly into the blood capillaries because they are too large to enter these small vessels. Instead, fats enter a lymphatic capillary called a **lacteal**. The lacteal delivers the fats to the lymph in a large lymphatic vessel called the **cisterna chyli**, which then drains into the thoracic duct.

20

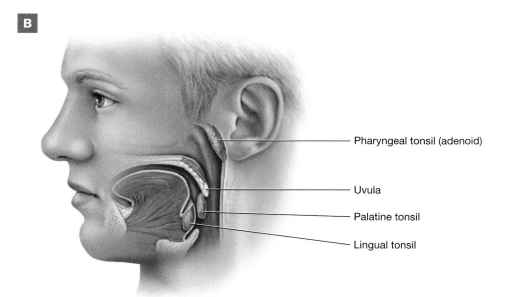

Thymus

Spleen

Lymph node

Tonsil

Cervical lymph nodes

Right lymphatic duct

Right subclavian vein

Axillary lymph nodes

Lymph trunks

Intestinal lymph nodes

Inguinal lymph nodes

Lymph-collecting vessels

Left subclavian vein

Axillary lymph nodes

Thoracic duct

Cubital lymph nodes

Cisterna chyli

Lymphatic nodules

Pharyngeal tonsil (adenoid)

Uvula

Palatine tonsil

Lingual tonsil

FIGURE **20.2** Lymphoid organs: (**A**) overview of the lymphatic organs and vessels; (**B**) the tonsils

In the following activities you will identify structures of the lymphatic system, and then trace the flow of lymph through the vessels on its way back to the cardiovascular system. Your instructor may also wish you to dissect a preserved small mammal to identify some of the structures difficult to see on models, such as the thymus. If so, follow the procedure outlined in Unit 1 (p. 11) to open the animal and identify the required structures.

## Procedure  Model Inventory for the Lymphatic System

Identify the following structures of the lymphatic system on models and diagrams, using your textbook and this unit for reference. As you examine the anatomical models and diagrams, record on the model inventory in **Table 20.1** the name of the model and the structures you were able to identify. When you have completed this activity, answer Check Your Understanding questions 1–3 (p. 507).

1. Lymph vessels
   a. Thoracic duct
   b. Right lymphatic duct
   c. Lymph trunks
   d. Lymph-collecting vessels
   e. Lacteal
   f. Cisterna chyli
2. Lymph nodes
   a. Cervical lymph nodes
   b. Axillary lymph nodes
   c. Inguinal lymph nodes
   d. Intestinal lymph nodes

3. Spleen
4. Thymus (this is best viewed on a fetal pig)
5. Mucosal-associated lymphoid tissue (MALT)
6. Tonsils
   a. Palatine tonsil
   b. Pharyngeal tonsil
   c. Lingual tonsil

TABLE **20.1** Model Inventory for the Lymphatic System

| Model | Structures Identified |
|---|---|
|  |  |
|  |  |
|  |  |
|  |  |
|  |  |

20

## Procedure Tracing the Flow of Lymph through the Body

In this exercise you will trace the pathway of lymph flow from the starting point to the point at which the lymph is delivered to the cardiovascular system. You will trace the flow through the major lymph-collecting vessels, trunks, and ducts, and highlight clusters of lymph nodes through which the lymph passes as it travels.

**1** Write the sequence of the flow.

**2** Then use differently colored water-soluble markers to draw the pathway on a laminated outline of the human body. If no outline is available, use colored pencils and **Figure 20.3**.

Trace the flow from the following locations:

**Start:** Right foot _____

_____

_____

_____

_____

_____

_____

_____

**Start:** Right arm _____

_____

_____

_____

_____

_____

_____

_____

_____

_____

**Start:** Intestines (with fat) _____

_____

_____

_____

_____

_____

_____

FIGURE **20.3** Outline of human body, anterior view

# Exercise 20-2

## Lymphatic Organ Histology

### MATERIALS
- ❏ Spleen section slide
- ❏ Lymph node slide
- ❏ Peyer's patch slide
- ❏ Light microscope
- ❏ Colored pencils

In this exercise, you will examine the microscopic anatomy of three different lymphoid organs: the spleen, a lymph node, and a special type of MALT found in the terminal portion of the small intestine called a **Peyer's patch**. Following are some hints regarding what to look for on each slide.

1. **Spleen**: Note in **Figure 20.4A** that the spleen consists of two types of tissue:

   a. **red pulp**, which is involved in the destruction of old and worn-out red blood cells, and

   b. **white pulp**, which contains phagocytes and lymphocytes that play a role in the immune system.

   Be careful here in identifying the red and white pulp. Notice that the red pulp is reddish in color, but the white pulp doesn't show up on the slide as white; instead, it appears as purplish "islands" within the red pulp. The purple color is caused by the presence of white blood cells, which have prominent nuclei that stain purple with commonly used stains.

2. **Lymph nodes**: Lymph nodes consist of an outer **cortex** that contains spherical clusters of cells (primarily B lymphocytes) called **lymphatic nodules** or **germinal centers** (**Figure 20.4B**). Deep to the lymphatic nodules we find an area of the cortex that houses primarily T lymphocytes. The innermost region of the node, called the **medulla**, houses macrophages, highly active phagocytes. The entire lymph node is surrounded by a capsule made of dense irregular collagenous connective tissue.

3. **Peyer's patch**: Peyer's patches somewhat resemble lymph nodes, although their lymphatic nodules are less well defined, and their capsules are incomplete (**Figure 20.4C**). To have the best chance of finding them, examine your section on low power, and look for the epithelial lining of the small intestine. Scroll downward (deep to the epithelium) into the connective tissue, where you will note large, oval or teardrop-shaped, purplish clusters. These are Peyer's patches.

White pulp

Red pulp

FIGURE **20.4** Lymphatic organs, photomicrographs: (**A**) spleen; *(continues)*

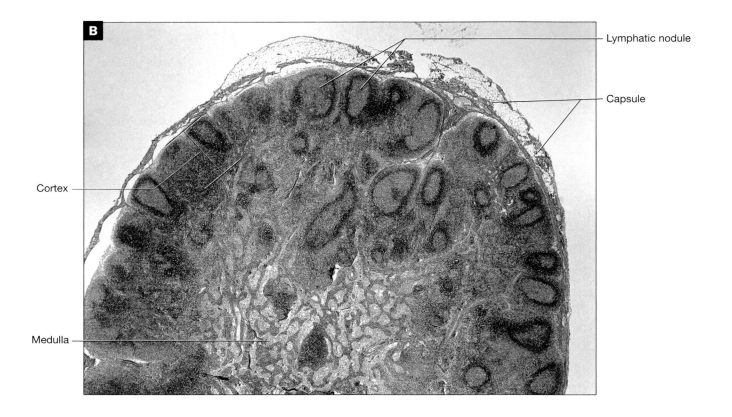

Lymphatic nodule

Capsule

Cortex

Medulla

Peyer's patches

FIGURE **20.4** Lymphatic organs, photomicrographs *(cont.)*: (**B**) lymph node; (**C**) Peyer's patches in the ileum

20

## Procedure Microscopy

Obtain prepared slides of the spleen, a lymph node, and the small intestine (the *ileum*). Use your colored pencils to draw what you see in the field of view, and label your drawing with the terms below.

**Spleen**

1. Red pulp
2. White pulp

**Lymph Node**

1. Capsule
2. Cortex
3. Medulla
4. Lymphatic nodule

**Peyer's Patch**

1. Mucosa (simple columnar epithelial tissue)
2. Peyer's patch

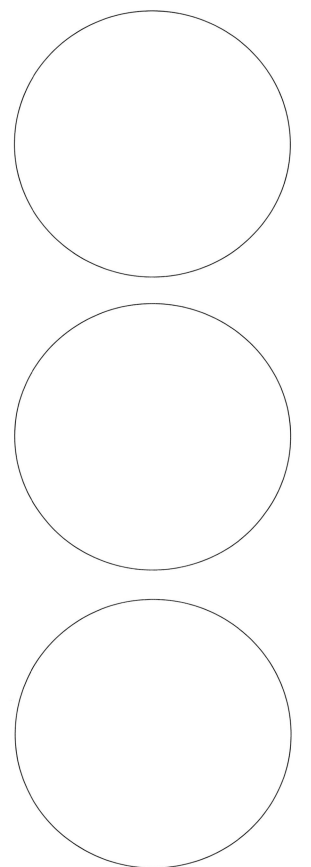

20

# Exercise 20-3

## The Immune Response

The immune system may be divided into two major types of responses:

1. the **nonspecific response**, which reacts the same way to any invading organism or cellular injury, and

2. the **specific response**, which responds to foreign cells and molecules with a unique set of cells and molecules for each individual threat.

The nonspecific responses are sometimes called **innate immunity**, because we are born with these responses, and they do not require exposure to a foreign cell or molecule to be produced. Specific responses, in contrast, do require exposure to a foreign agent to be produced, and for this reason they are often referred to as **acquired immunity**.

The nonspecific responses may be subdivided into the following:

1. **Surface barriers.** Surface barriers include the skin and mucous membranes. Both provide continuous, avascular surfaces that secrete substances that either deter the growth of microorganisms or trap foreign particles.

2. **Inflammation and fever.** When cells are damaged by microorganisms, foreign cells, toxins, or trauma, injured cells (primarily macrophages) release chemicals called **inflammatory mediators**. The inflammatory mediators attract other immune cells to the area, especially neutrophils and other macrophages. They also cause local *vasodilation*, which causes the area to become red and warm, *increased capillary permeability*, which causes the area to become swollen, and *increased sensitivity to pain*. Certain inflammatory mediators called **pyrogens** act on the hypothalamus to raise the body temperature "set point," which results in the elevated temperature of **fever**.

3. **Granulocytes and macrophages.** The body's main phagocytic cells, the **granulocytes** (**neutrophils, basophils,** and **eosinophils**) and the **macrophages** (mature monocytes), will nonspecifically ingest and destroy any foreign or damaged cell. Macrophages are generally the first cells to respond to tissue damage and release inflammatory mediators. These mediators attract other phagocytes, particularly neutrophils, to the damaged area in large numbers. Macrophages can ingest and destroy foreign and damaged cells repeatedly; however, neutrophils can ingest a cell only once, after which they die as well. When tissue damage is extensive, many neutrophils in one area will die and accumulate, leaving a whitish liquid called **pus**.

4. **Antimicrobial proteins.** Several proteins in the blood will nonspecifically attack and lyse invading organisms when activated. The largest such system is the **complement system**, a collection of more than 20 proteins, most of which are made by the liver. Complement proteins lyse foreign cells, enable phagocytes to function better, and enhance the inflammatory response.

The specific immune response involves primarily B and T lymphocytes, which respond specifically to unique **antigens**, or chemical markers. Antigens are present in many molecules, such as toxins and allergens, and on the surfaces of all cells, including our own cells. There are two types of specific immunity:

1. **Humoral immunity.** Humoral immunity is mediated by B lymphocytes. The humoral response is initiated when a macrophage or T lymphocyte activates a B lymphocyte. This causes the B lymphocyte to differentiate into two cell types: **plasma cells**, which secrete molecules called **antibodies** that bind to antigens, and **memory cells**, which will "remember" the antigen and permit a more rapid antibody response upon subsequent exposures. Antibodies are proteins that bind and **agglutinate** antigens and facilitate their destruction by phagocytes and other components of the immune system. Antibodies also activate the nonspecific response, particularly complement, and neutralize toxins such as venoms.

2. **Cell-mediated immunity.** As its name implies, cell-mediated immunity is mediated by cells rather than by antibodies, specifically T lymphocytes. Of the several different kinds of T lymphocytes, we will consider two here—the **CD4**, or **helper T cells**, and the **CD8**, or **cytotoxic T cells**. The cell-mediated response begins when a macrophage ingests an antigen and presents it to a helper T cell. The helper T cell then secretes chemicals called **cytokines**, which activate cells of the nonspecific response such as macrophages, encourage inflammation, activate B lymphocytes, and activate cytotoxic T cells. The cytotoxic T cells locate cells infected with viruses or intracellular bacteria, cancer cells, and transplanted cells and lyse and destroy them. Both helper and cytotoxic T cells may become **memory T cells**, which help prevent reinfection with the same agent.

**20**

Note that all types of cell damage and/or invading organisms are potentially met with multiple facets of the nonspecific *and* specific responses. For example, cancer cells are prevented from spreading by surface barriers, are phagocytized by granulocytes and macrophages in the blood and in the lymph nodes, are lysed by antimicrobial proteins released when healthy cells are damaged, cause local and systemic inflammation and fever, are met with an antibody response from B lymphocytes, and are located and lysed by helper T and cytotoxic T cells. This same cascade of responses is seen with viral infections, bacterial infections, envenomations, severe burns, major trauma, organ and tissue transplantations, fungal infections, parasitic infections, and more.

The immune response is . . . well, *complicated*, to say the least, and you could nearly fill this entire book tracing through everything that happens when you get a cold. The concept can be simplified somewhat because most types of invading organisms or cellular damage will trigger one type of response slightly more than the others. For example, bacterial infections tend to lead to elevated levels of neutrophils that phagocytize the bacterial cells. Similarly, viral infections and cancer cells tend to trigger a predominantly cell-mediated response and lead to elevated levels of cytotoxic T lymphocytes that lyse affected cells.

## Procedure  Time to Trace

In this exercise, you will be tracing two invaders: bacteria in the extracellular fluid that enter the lymphatic vessels and virus-infected cells in the blood. To keep things simple, you will trace the *main* pathway and the *main* immune response to the invading organism. Do note, however, that the actual response is much more complex, and your instructor may wish you to include more information. I offer hints and the number of steps you should aim to include for each organism, but you will likely want to consult your book for more help. After you have completed the activity, answer Check Your Understanding questions 4 and 5 (p. 508).

**Start:** bacteria in the extracellular fluid  →  bacteria damage local macrophages and damaged macrophages release

_____  →  _____ attract _____ to the area

→  neutrophils ingest _____  → bacteria enter lymph capillaries

→ bacteria travel through _____ → then enter _____

→ bacteria encounter macrophages in lymph node → macrophages _____ bacteria

→ macrophages present antigens to _____ , which differentiate into _____

→ _____ proliferate and release _____ ,

which _____ **End**

**Start:** virus-infected cells in the blood → viruses damage tissue and damaged cells release _____

→ _____ attract _____

→ _____ release additional _____ and _____

→ inflammation and fever begin → virus-infected cells enter _____ → virus-infected cells encounter

_____ in the _____ → _____ ingest infected cells

→ macrophages present antigens to → _____ → _____

proliferate and secrete _____ , which activate _____

→ these cells _____ and _____ infected cells **End**

20

# Exercise 20-4

## ELISA Test for Antigens

### MATERIALS

- ❏ ELISA well plate
- ❏ Micropipettor and tips
- ❏ Microcentrifuge tube with "blood" sample
- ❏ Empty microcentrifuge tube
- ❏ Antibody solution
- ❏ Colorant solution
- ❏ Washing solution
- ❏ Sharpie markers

One of the most commonly performed diagnostic laboratory tests is **ELISA**, which stands for **enzyme-linked immunosorbent assay**. ELISA allows you to detect the presence of an antigen from a sample using a reaction with a specific antibody. Laboratories commonly use ELISA to determine if an individual has been exposed to certain infectious agents such as influenza or HIV.

The basic procedure for the type of ELISA we use in this exercise is outlined on page 501. As you can see, solutions are added to wells, and any antigens present in the solution bind to the base of the wells. The solutions are allowed to sit for 5 minutes, after which a washing solution is applied to remove any unbound antigens. In the next step, a solution containing antibodies is added to all three wells. Again, the solutions are allowed to sit undisturbed for 5 minutes, after which a second washing solution is applied. Antibodies that have bound to their specific antigens will remain in the wells, whereas those that did not bind to antigens will be washed away.

In the final step, a colorant that reacts with an enzyme associated with the antibody is added to the solution. Notice in **Figure 20.5** that the colorants in wells 1 and 2 react with the enzyme on the antibody and change the color of the solution. This is read as a positive result, meaning that the antigen for which you were initially testing was present in the solution. Well 3, on the other hand, has no antibodies with which the colorant can react, and so it does not change color. This is interpreted as a negative reaction, meaning your test solution did not contain the antigen in question.

In this exercise, you will be demonstrating the spread of infection of the "influenza virus" through your class (note that no samples have the actual influenza virus—the antigens in the samples are harmless). Each student will be given a sample that represents his or her own "blood." One or two random members of your class will receive a sample that has the antigen; these samples represent the "infected" individuals. During the first part of the procedure (steps 1–4), you will exchange and combine portions of your sample with three classmates. For the second part (step 5, see **Figure 20.6**), you will run the ELISA and determine who among your classmates has become infected with the flu! When you have completed the procedure, answer Check Your Understanding question 6 (p. 508).

Three samples are placed in wells. Note that the sample placed in wells 1 and 2 have antigen; the sample placed in well 3 has no antigen.

An antibody solution is added to all three wells. The antibodies bind to the antigens in wells 1 and 2 and are not removed in the washing step. The antibodies in well 3 do not bind to antigens and so are washed away in the washing step.

A colorant is added that binds to an enzyme associated with the antibody. This leads to a colored solution in wells 1 and 2 and no color in well 3.

FIGURE **20.5** Interpretation of ELISA results

20

## Procedure Who's Got the Flu?

**1** Obtain two microcentrifuge tubes from your instructor: one empty and one containing a fluid to be your "blood" sample. Write your initials on both tubes with a Sharpie.

**2** Use a micropipettor to withdraw 200 μl of your sample, and transfer it to the empty tube. Set this tube aside for now.

**3** At your instructor's prompt, begin the swapping portion of the lab. Note that you should choose partners from the entire lab, not just your immediate lab group.

   **a** Randomly choose a partner in your lab. Each of you will use a micropipettor to withdraw 100 μl of your sample, and place it into one another's microcentrifuge tube.

   **b** Use the tip of your micropipettor to mix your remaining solution with your partner's solution so that they are thoroughly mixed.

   **c** Choose another partner in your lab, and repeat the procedure. Your tube should now contain your original sample and portions of two of your classmates' samples.

   **d** Choose your third and final partner, and repeat the procedure. At the end, your centrifuge tube will contain four total samples: yours and three of your classmates'. Record the names of each of your partners here:

     Partner 1: _____

     Partner 2: _____

     Partner 3: _____

**4** Mix your sample thoroughly, and then transfer 100 μl into three wells on your group's ELISA plate using a micropipettor (300 μl total). You are running three tests to ensure that your results are accurate.

**5** When everyone has added their samples to the wells, follow the ELISA procedural steps illustrated in **Figure 20.6**:

   **a** Allow the plate to sit undisturbed on your lab table for 5 minutes.

   **b** After 5 minutes have passed, empty the contents of the plate into a sink or designated container. If antigens were present in your solution, they will be bound to the bottom of the well.

**Step 5a:** When everyone has added samples to the wells, allow the plate to sit undisturbed on your lab table for 5 minutes.

**Step 5b:** After 5 minutes have passed, empty the contents of the plate into a sink or designated container. If antigens were present in your solution, they will be bound to the bottom of the well.

**Step 5c:** Rinse the wells with washing solution, and discard the fluid in the sink or designated container. Repeat this procedure three times.

**Step 5d:** Add about three drops (100 μl) of antibody solution to each well. Allow the plate to sit undisturbed on your lab table for 5 minutes.

**Step 5e:** Discard the fluid in the sink or designated container. Rinse the wells with washing solution three times.

**Step 5f:** Add about three drops (100 μl) of colorant solution to each well. Allow the plate to sit undisturbed on your lab table for 5 minutes.

FIGURE **20.6** ELISA procedure

c Rinse the wells with washing solution, and discard the fluid in the sink or designated container. Repeat this procedure three times.

d Add about three drops (100 µl) of antibody solution to each well. Allow the plate to sit undisturbed on your lab table for 5 minutes.

e Discard the fluid in the sink or designated container, and rinse the wells with washing solution three times.

f Add about three drops (100 µl) of colorant solution to each well. Allow the plate to sit undisturbed on your lab table for 5 minutes.

6 Interpret your results. Who among your group was infected with the flu?

_____

_____

_____

_____

7 Compile the results for your class.

a How many were infected? _____

b Determine as a class who you believe had the initial infected sample(s). _____

8 Determine to whom the initial infected sample belonged by running an ELISA on the microcentrifuge tube you set aside at the beginning of the procedure. Repeat step 5 with this uncontaminated sample.

9 Who in your class had the initial infected sample(s), according to these results? _____

20

## Check Your Recall

**1** Label **Figure 20.7** with the terms below.

### Figure 20.7A:

- ☐ Cisterna chyli
- ☐ Lymph nodes: cervical, axillary, intestinal, and inguinal
- ☐ Spleen
- ☐ Thoracic duct
- ☐ Thymus

FIGURE **20.7** Lymphoid organs: (**A**) overview of the lymphatic organs and vessels *(continues)*

20

**Figure 20.7B:**

☐ Lingual tonsil
☐ Palatine tonsil
☐ Pharyngeal tonsil

FIGURE **20.7** Lymphoid organs *(cont.)*: (**B**) the tonsils

**2** Which of the following is not a function of the lymphatic system?
  a. Maintaining blood pressure
  b. Absorbing dietary fats
  c. Activating the immune system
  d. Transporting excess interstitial fluid back to the heart

**3** Lymph nodes filter _____, and the spleen filters _____.
  a. lymph; lymph
  b. blood; blood
  c. lymph; blood
  d. blood; lymph

**4** The vessels into which fats are absorbed are called
  a. blood capillaries.
  b. lymph ducts.
  c. lacteals.
  d. Fats are not absorbed.

**5** Describe the structure and location of the tonsils, lymph nodes, and MALT.

20

**6** The cortex of a lymph node contains spherical clusters called _____ that contain primarily _____.
  a.  lymphatic nodules; T lymphocytes
  b.  lymphatic nodules; B lymphocytes
  c.  medullary centers; T lymphocytes
  d.  medullary centers; B lymphocytes

**7** Explain how each of the following sets of terms differ.
  a.  Specific immune response and nonspecific immune response:

_____

_____

_____

_____

_____

  b.  Cell-mediated immunity and humoral immunity:

_____

_____

_____

_____

_____

**8** Which of the following is not a type of nonspecific immune response?
  a.  Inflammation and fever
  b.  Antimicrobial proteins
  c.  Cell-mediated immunity
  d.  Surface barriers
  e.  Neutrophils and macrophages

**9** *Fill in the blanks:* When tissues are damaged, macrophages release inflammatory mediators that cause

_____, _____, and _____.

Certain inflammatory mediators called _____ increase the hypothalamus's "set point"
for temperature and cause fever.

**10** _____ *True or false:* The immune system's response to an infection involves either the nonspecific response or the
specific response, but not both.

20

**UNIT
20
REVIEW**

# Check Your Understanding

### Critical Thinking and Application Questions

**1** Predict some potential consequences of removing the spleen.

_____

_____

_____

_____

_____

_____

**2** In a condition called *DiGeorge syndrome*, infants are born with either an absent thymus or a thymus that isn't functional. Predict the consequences of this disease.

_____

_____

_____

_____

_____

_____

**3** Explain why blockage or removal of the lymphatic vessels can result in significant *edema* (accumulation of fluid in a limb or body part).

_____

_____

_____

_____

_____

20

**4** A common symptom of *pharyngitis* is swelling of the anterior cervical lymph nodes. Why might lymph nodes swell in the presence of an infection? What else may cause swollen lymph nodes? (*Hint:* What part of the immune response causes swelling?)

_____

_____

_____

_____

_____

_____

**5** Explain why conditions that cause destruction of B lymphocytes and/or T lymphocytes cause malfunctioning of both the specific and nonspecific immune responses.

_____

_____

_____

_____

_____

_____

**6** You are running an ELISA on a sample to test for the presence of antigens from the *influenza A virus*. When it comes time to add the antibody, you realize that you only have antibodies for the *influenza B virus*. Would these antibodies give you accurate results? Why or why not?

_____

_____

_____

_____

_____

20

# Respiratory System Anatomy

# 21

## OBJECTIVES

*Once you have completed this unit, you should be able to:*

■ Identify structures of the respiratory system.

■ Trace the pathway of gases through the respiratory system.

■ Identify histological structures of the respiratory system on microscope slides.

■ Describe the anatomical changes associated with inflation of the lungs.

# PRE-LAB EXERCISES

Complete the following exercises prior to coming to lab, using your textbook and lab manual for reference.

## Pre-Lab Exercise **21-1**

✎ Key Terms

You should be familiar with the following terms before coming to lab.

| Term | Definition |
|------|------------|

**General Structures of the Respiratory System**

Respiratory tract _____

_____

Parietal pleura _____

_____

Visceral pleura _____

_____

Pleural cavity _____

_____

Lungs and lobes _____

_____

**Structures of the Respiratory Tract**

Nasal cavity _____

_____

Pharynx _____

_____

Larynx _____

_____

Trachea _____

_____

Primary bronchi _____

_____

Secondary bronchi _____

_____

21

Terminal bronchioles _____

_____

Respiratory bronchioles _____

_____

Alveolar ducts _____

_____

Alveoli _____

_____

21

Label and color the diagrams of the structures of the respiratory system in **Figure 21.1** with the terms from Exercise 21-1 (p. 515). Use your text and Exercise 21-1 in this unit for reference.

**21**

FIGURE **21.1** Structures of the respiratory system: (**A**) the lungs and respiratory tract *(continues)*

B

C

FIGURE **21.1** Structures of the respiratory system (*cont.*): (**B**) the larynx, anterior view; (**C**) the larynx, midsagittal section (*continues*)

21

FIGURE **21.1** Structures of the respiratory system *(cont.)*: (**D**) bronchiole and alveolar sac

21

# EXERCISES

Cells require oxygen in the reactions that synthesize ATP, and these reactions produce carbon dioxide as a waste product. The respiratory system and the cardiovascular system work together to supply the cells the oxygen they need and to rid them of carbon dioxide.

The exercises in this unit will familiarize you with the organs and histology of the respiratory system, including the paired **lungs** and the collection of airway passages known as the **respiratory tract**. In the final exercise you will observe the changes in the texture, shape, and size of the lungs as they inflate and deflate with the process known as **ventilation**.

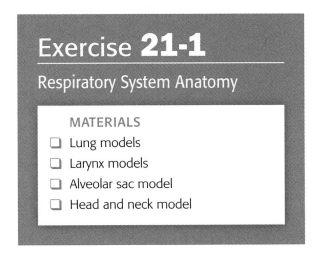

## Exercise 21-1

### Respiratory System Anatomy

> **MATERIALS**
> ❑ Lung models
> ❑ Larynx models
> ❑ Alveolar sac model
> ❑ Head and neck model

The lungs are composed of elastic connective tissue and tiny air sacs called **alveoli**, where gas exchange takes place. Each lung is divided into smaller structures called **lobes**. The right lung has three lobes (upper, middle, and lower), and the left lung has two lobes (upper and lower) (**Figure 21.2**). The lobes are separated from one another by deep indentations called **fissures**. The **horizontal fissure** separates the right upper and right middle lobes; the **right oblique fissure** separates the right middle and right lower lobes; and the **left oblique fissure** separates the left upper and left lower lobes.

Each lung is surrounded by serous membranes similar to the pericardial membranes called the **pleural membranes**. There are two layers of the pleural membranes:

1. **Parietal pleura.** The outer parietal pleura lines the interior of the thoracic cavity and the superior surface of the diaphragm.

2. **Visceral pleura.** When the parietal pleura reaches the structures of the mediastinum, it folds inward to become the visceral pleura. The visceral pleura adheres tightly to the surface of the lung.

There is a very thin potential space between the parietal and visceral pleurae called the **pleural cavity**. The space is only a "potential" space because it is filled with a thin layer of serous fluid that reduces friction as the lungs change in shape and size during ventilation.

Air is delivered to the lungs' alveoli through the passageways of the respiratory tract. The respiratory tract may be divided into two regions according to both structure and function. Structurally, it is divided into:

1. The **upper respiratory tract**, which consists of the passages superior to the thoracic cavity, and

2. the **lower respiratory tract**, which consists of the passages within the thoracic cavity.

Functionally, it is divided into:

1. The **conducting zone**, which consists of passages that carry, or "conduct," air to the lower passages where gas exchange takes place. The majority of the passages of the respiratory tract belong to the conducting zone.

2. The **respiratory zone**, which consists of the passages in which gas exchange takes place, located in the terminal respiratory tract.

The conducting zone begins with the **nasal cavity** and the **paranasal sinuses**, which filter, warm, and humidify the inhaled air (**Figure 21.3**). Air from the nasal cavity next enters the **pharynx** (pronounced "fair-inks"), also known as the throat, which has the following three divisions:

1. **Nasopharynx.** The nasopharynx is the region posterior to the nasal cavity. The muscles of the soft palate move superiorly to close off the nasopharynx during swallowing to prevent food from entering the passage. Sometimes this mechanism fails (such as when a person is laughing and swallowing simultaneously), and the unfortunate result is that food or liquid comes out of the nose. The nasopharynx is lined with pseudostratified ciliated columnar epithelium with copious mucus-secreting **goblet cells**, a type of tissue known as respiratory epithelium.

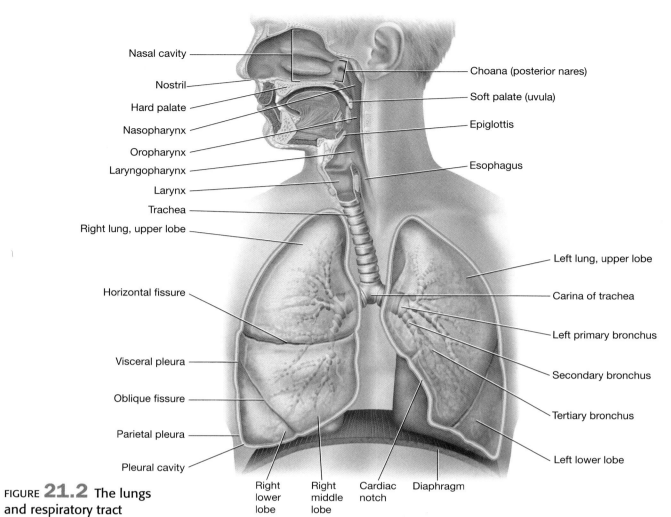

Nasal cavity

Nostril

Hard palate

Nasopharynx

Oropharynx

Laryngopharynx

Larynx

Trachea

Right lung, upper lobe

Horizontal fissure

Visceral pleura

Oblique fissure

Parietal pleura

Pleural cavity

Choana (posterior nares)

Soft palate (uvula)

Epiglottis

Esophagus

Left lung, upper lobe

Carina of trachea

Left primary bronchus

Secondary bronchus

Tertiary bronchus

Left lower lobe

Right lower lobe

Right middle lobe

Cardiac notch

Diaphragm

FIGURE **21.2** The lungs and respiratory tract

Frontal sinus

Nasal conchae

Epiglottis

Vocal folds

Larynx

Sphenoid sinus

Nasopharynx

Oropharynx

Laryngopharynx

FIGURE **21.3** Midsagittal section of the head and neck

21

2. **Oropharynx.** The oropharynx is the region posterior to the oral cavity. Both food and air pass through the oropharynx, and it is therefore lined with stratified squamous epithelium. This tissue provides more resistance to mechanical and thermal stresses.

3. **Laryngopharynx.** The laryngopharynx is the intermediate region between the larynx and the esophagus. As with the oropharynx, both food and air pass through the laryngopharynx, and it is lined with stratified squamous epithelium.

Air passes from the pharynx to the **larynx** (pronounced "lair-inks"), a short passage framed by nine cartilages (**Figure 21.4**). The "lid" of the larynx is a piece of elastic cartilage called the **epiglottis**. During swallowing, muscles of the pharynx and larynx move the larynx superiorly, and the epiglottis seals off the larynx from food and liquids. The largest cartilage of the larynx is the shield-like **thyroid cartilage**, which forms the larynx's anterior and lateral walls. Inferior to the thyroid cartilage is a smaller cartilage called the **cricoid cartilage**. Between the two is a soft piece of connective tissue called the **cricothyroid ligament**.

As its common name "voice box" implies, the larynx is the structure where sound is produced. It contains two sets of elastic ligaments known as the **vocal folds** or **vocal cords**. The superior set of vocal folds, the **false vocal cords** (also called the *vestibular folds*), plays no role in sound production. They do, however, serve an important sphincter function and can constrict to close off the larynx. The inferior set of vocal folds, called the **true vocal cords**, vibrates as air passes over it to produce sound.

Inspired air passes from the larynx into a tube supported by C-shaped rings of hyaline cartilage called the **trachea**. The trachea bifurcates in the mediastinum into two **primary bronchi** that begin the large and branching **bronchial tree**. The right primary bronchus is short, fairly straight, and wide, and the left primary bronchus is long, more horizontal, and narrow because of the position of the heart.

Each primary bronchus divides into smaller **secondary bronchi**, each of which serves one lobe of the lung. The two left secondary bronchi serve the two lobes of the left lung, and the three right secondary bronchi serve the three lobes of the right lung. The bronchi continue to branch and become tertiary bronchi, quaternary bronchi, and so on until the air reaches tiny air passages smaller than 1 millimeter in diameter called **bronchioles** (**Figure 21.5**).

Bronchioles smaller than 0.5 mm in diameter are called the **terminal bronchioles**, and these mark the end of the conducting zone. The respiratory zone begins with small branches called **respiratory bronchioles** that have thin-walled sacs called **alveoli** in their walls. As the respiratory bronchioles progressively branch, the number of alveoli in the wall increases

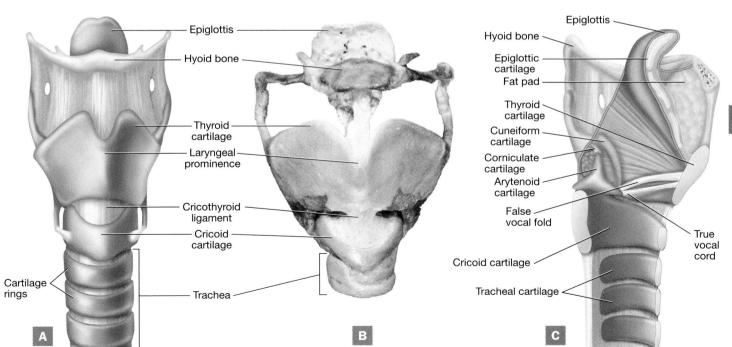

FIGURE **21.4** Larynx: (**A**) anterior view; (**B**) anterior view; (**C**) midsagittal section

until the wall is made up exclusively of alveoli, at which point it is termed an **alveolar duct**. The terminal portions of the respiratory zone, called **alveolar sacs**, are grape-like clusters of alveoli surrounded by **pulmonary capillaries**. This junction is where pulmonary gas exchange takes place: Oxygen from the alveoli diffuses into the blood, and carbon dioxide in the blood diffuses into the alveoli to be exhaled. The structure of the aveolar sacs creates a huge surface area (around 1,000 square feet on average). This large surface area is necessary for gas exchange to take place rapidly and efficiently.

FIGURE **21.5** Bronchiole and alveolar sac

Identify the following structures of the respiratory system on models and diagrams, using your textbook and this unit for reference. As you examine the anatomical models and diagrams, record on the model inventory in **Table 21.1** the name of the model and the structures you were able to identify. After you have completed the activity, answer Check Your Understanding questions 1–3 (p. 531).

1. Lungs
   a. Parietal pleura
   b. Visceral pleura
   c. Pleural cavity
2. Right lung
   a. Upper, middle, lower lobes
   b. Horizontal fissure
   c. Oblique fissure
3. Left lung
   a. Upper and lower lobes
   b. Oblique fissure
   c. Cardiac notch
4. Nasal cavity
   a. Nares
   b. Vestibule
   c. Nasal septum
   d. Nasal conchae
      (1) Superior nasal conchae
      (2) Middle nasal conchae
      (3) Inferior nasal conchae

5. Paranasal sinuses
   a. Sphenoid sinus
   b. Ethmoid sinus
   c. Frontal sinus
   d. Maxillary sinus
6. Hard palate
7. Soft palate and uvula
8. Pharynx
   a. Nasopharynx
   b. Oropharynx
   c. Laryngopharynx
9. Larynx
   a. Epiglottis
   b. Thyroid cartilage
   c. Cricoid cartilage
   d. Cricothyroid ligament
   e. False vocal cords
   f. True vocal cords
10. Trachea
    a. Hyaline cartilage rings
    b. Carina

11. Bronchi
    a. Right and left primary bronchi
    b. Secondary bronchi
    c. Tertiary bronchi
12. Bronchioles
    a. Terminal bronchioles
    b. Respiratory bronchioles
    c. Alveolar duct
13. Alveoli and alveolar sacs
14. Muscles
    a. Diaphragm
    b. External intercostals
    c. Internal intercostals
15. Vascular structures
    a. Pulmonary arteries
    b. Pulmonary arterioles
    c. Pulmonary capillaries
    d. Pulmonary venules
    e. Pulmonary veins

TABLE **21.1** Model Inventory for the Respiratory System

| Model/Diagram | Structures Identified |
|---|---|
|  |  |
|  |  |
|  |  |
|  |  |
|  |  |

**21**

## Procedure Time to Trace

### Part 1

You are a molecule of oxygen floating happily through the atmosphere when all of a sudden you are inhaled by Ms. Magenta.

**1** Trace your pathway through Ms. Magenta's respiratory tract beginning in her nasal cavity to the point where you enter the pulmonary capillaries.

**Start:** _____

_____

_____

_____

_____

_____ **End**

**2** Trace your pathway from Ms. Magenta's pulmonary capillaries through her lungs and heart and to your final destination in a hepatocyte (liver cell). You likely will have to review Unit 17 (p. 411) for some help with this.

**Start:** _____

_____

_____

_____

_____

_____ **End**

**21**

**3** Explain what happens to you in Ms. Magenta's hepatocyte. What do you do, and what do you become? (You can be somewhat general here in your answer unless your lab instructor says otherwise. You may have to refer to your textbook, and read about the basics of oxidative catabolism and the electron transfer system.)

_____

_____

_____

_____

_____

_____

_____

## Part 2

While in her hepatocyte, you notice that a molecule of carbon dioxide has just been produced.

**1** How was the molecule of carbon dioxide produced in Ms. Magenta's hepatocyte? (Again, you can be somewhat general in your answer unless your lab instructor says otherwise. You may have to refer to your textbook, and read about the basics of glycolysis and the citric acid cycle.)

_____

_____

_____

_____

_____

_____

_____

**2** Trace the carbon dioxide's pathway from Ms. Magenta's hepatocyte through her heart and to her pulmonary capillaries. You will likely need to review Unit 17 (p. 411) for some help with this.

**Start:** _____

_____

_____

_____

_____

_____

_____

_____

_____

_____ **End**

**3** Trace the pathway of the carbon dioxide from the pulmonary capillaries through the respiratory tract to the point where it exits from Ms. Magenta's body through her nares.

**Start:** _____

_____

_____

_____

_____

_____

_____

_____ **End**

# Exercise 21-2

## Histology of the Respiratory Tract

### MATERIALS

- ❏ Trachea slide
- ❏ Lung tissue slide
- ❏ Bronchiole slide
- ❏ Light microscope
- ❏ Colored pencils

As you learned in Exercise 21-1, the respiratory tract is a series of branching tubes that conduct air to and from the respiratory zone for gas exchange. The "tubes" of the respiratory tract are similar in structure histologically to other hollow organs of the body and consist of the following three tissue layers (**Figure 21.6A**):

1. **Mucosa.** Mucous membranes, or **mucosae,** line all passageways that open to the outside of the body, including the passages of the respiratory tract. Mucosae consist of epithelial tissue overlying the basement membrane. As we have seen elsewhere in the body, there is a clear relationship of structure and function in the respiratory tract, because the epithelial tissue of the mucosae is adapted for each region's function. Some examples of the epithelia found in different regions include:

   a. the nasopharynx, larynx (inferior to the vocal cords), trachea, and bronchi are lined with pseudostratified ciliated columnar epithelium with mucus-secreting cells called **goblet cells** (sometimes called **respiratory epithelium**) (**Figure 21.6B**). The cilia sweep dust and other particulates that become trapped in the mucus out of the respiratory tract. This helps to prevent this material from reaching the deeper parts of the respiratory tract.

   b. the oropharynx, laryngopharynx, and larynx (superior to the vocal cords) are lined with stratified squamous epithelium

   c. the bronchioles: simple cuboidal epithelium (**Figure 21.6C**)

   d. alveoli are lined with simple squamous epithelium. Note in **Figure 21.6D** that the alveoli consist of a mucosa only and have no other tissue layers. This minimizes the distances gases must diffuse across the alveoli and into the pulmonary capillaries. The structure through which gases must diffuse, called the **respiratory membrane,** consists of the squamous epithelial cells, the endothelial cells of the pulmonary capillaries, and their shared basal lamina.

2. **Submucosa.** Deep to the mucosa we find the submucosa, a layer of loose connective tissue. The submucosa contains specialized **seromucous glands** that secrete watery mucus. The submucosa of larger passages contains hyaline cartilage for support; it also contains smooth muscle that controls the diameter of the airways.

21

FIGURE **21.6** Photomicrograph of: (**A**) the trachea; (**B**) nasal mucosa with goblet cell; *(continues)*

**Pulmonary blood vessel**

**Lumen of bronchiole**

**Simple cuboidal epithelium of bronchiole**

**Bronchiolar smooth muscle**

**Walls of the alveoli (simple squamous epithelium)**

**Walls of alveoli (simple squamous epithelium)**

**Alveoli**

**Alveolar ducts**

**Blood vessel**

FIGURE **21.6** Photomicrograph of *(cont.)*: (**C**) bronchiole; (**D**) lung tissue

3. **Adventitia.** The outermost layer of the respiratory tract, the adventitia, consists of dense irregular collagenous connective tissue with elastic fibers for support.

As you examine different regions of the respiratory tract, you will note a few trends:

▌ The epithelial tissue changes from taller (pseudostratified) in the upper passages to shorter (cuboidal) in the lower passages and finally to flat (squamous) in the alveoli.

▌ The amount of hyaline cartilage gradually decreases (it is absent in bronchioles), as does the number of goblet cells.

▌ The amount of smooth muscle and elastic fibers increases.

These trends can help you to determine the portion of the respiratory tract from which the section on your slide was taken. In the following exercise you will be able to see these trends as you examine photomicrographs of the trachea, a bronchiole, and the alveoli.

View prepared slides of a section of the trachea, a bronchiole, and alveoli. Note that the bronchiole may be on the same slide as the alveoli. Begin your examination of the slides on low power, and advance to medium and high power to observe details. Use colored pencils to draw what you see under the microscope, and label your drawing with the terms indicated. When you have completed the activity, answer Check Your Understanding questions 4–6 (p. 532).

### Trachea

Label the following on your drawing:

1. Pseudostratified columnar epithelium
2. Cilia
3. Goblet cells
4. Submucosa
5. Seromucous glands
6. Hyaline cartilage

### Bronchioles

Label the following on your drawing:

1. Simple columnar or cuboidal epithelium
2. Smooth muscle
3. Pulmonary arteriole

### 21 Alveoli

Label the following on your drawing:

1. Simple squamous epithelium
2. Alveolar duct
3. Alveolus

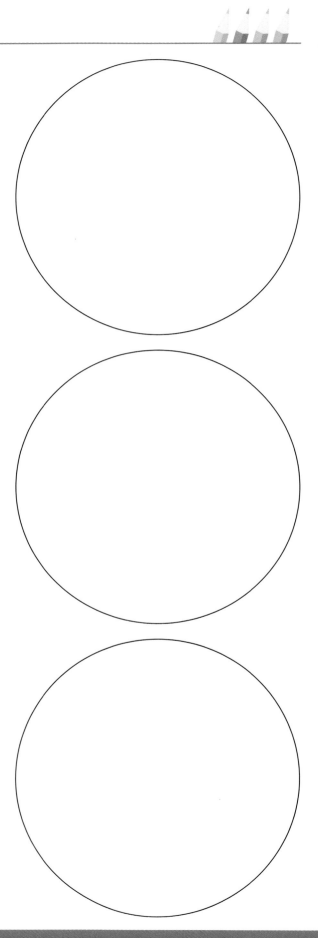

# Exercise 21-3

## Lung Inflation

### MATERIALS

- ❑ Preserved small mammal
- ❑ Dissection equipment
- ❑ Dissection trays
- ❑ Straw
- ❑ Air hose
- ❑ Fresh lungs

In movies or television programs you may have seen a hero rescuing a choking victim by poking a hole in the victim's neck (usually with a pocketknife) and inserting a straw (or pen) into the hole to restore the airway and save the victim's life. This is actually a somewhat crude version of a legitimate procedure called a **cricothyroidotomy**—but don't try this at home!

A cricothyroidotomy is performed when the upper respiratory tract is blocked, and air is prevented from moving into the lungs. The first step in the procedure is to place an incision (hopefully with a scalpel, not a pocketknife!) in the cricothyroid ligament, the soft spot between the thyroid and cricoid cartilages. A tube is then inserted into the opening, and the patient is ventilated artificially. This restores the patient's airway by bypassing the upper respiratory tract.

In the following procedure you will be able to see the effects of a cricothyroidotomy firsthand by performing the procedure on a fetal pig (or other preserved small mammal) and inflating its lungs. You will also inflate fresh lungs from a sheep, a pig, or a cow. As you perform both procedures, note the difference in the textures and appearances of the preserved lungs and fresh lungs. After you have completed both procedures, answer Check Your Understanding question 7 (p. 532).

## Procedure  Cricothyroidotomy and Inflating Preserved Lungs

**Safety Note**

⚠ Safety glasses and gloves are required!

**1**  Obtain a fetal pig or other preserved small mammal and dissection equipment.

**2**  Carefully dissect the animal's neck, and remove tissue so the larynx is clearly visible.

**3**  Locate the cricothyroid ligament, and make a small incision with a scalpel into its anterior surface.

**4**  Insert a small straw into the hole you have just made.

**5**  Attach the straw to a small hose, and attach the hose to an air outlet.

**6**  Turn on the air slowly, and watch the lungs inflate. Crimp the hose to cut off the airflow, and watch the lungs deflate.

21

## Procedure Inflating Fresh Lungs

**1** Obtain a fresh specimen and a large air hose.

**2** Examine the specimen for structures covered in Exercise 21-1 (p. 515), in particular the epiglottis and vocal folds, the other laryngeal cartilages, the trachea and its hyaline cartilage rings, and the pleural membranes.

**3** Squeeze the deflated lungs between your fingertips, and note their texture. Deflated lungs are shown in **Figure 21.7A**. Compare their texture with that of the lungs of the preserved mammal:

_____

_____

_____

**4** Insert the air hose into the larynx, and feed it down into the trachea. Take care not to get the hose stuck in one of the primary bronchi.

**5** Attach the hose to the air outlet, and turn it on slowly. You may have to squeeze the trachea and the hose to prevent air leakage.

**6** Observe the lungs as they inflate, shown in **Figure 21.7B**. You may inflate the lungs quite fully. Don't worry—they're very unlikely to pop.

**7** Squeeze the inflated lungs between your fingers, and note their texture:

_____

_____

**8** Crimp the air hose, and watch the lungs deflate. Again feel the lungs, and note changes in texture:

_____

_____

FIGURE **21.7** Fresh lungs: (**A**) deflated; (**B**) inflated

## Check Your Recall

**1** Label **Figure 21.8A** and **21.8B** with the terms below. Note that 21.8B is on the next page.

☐ Laryngopharynx      ☐ Nasal cavity      ☐ Pleural cavity

☐ Larynx      ☐ Nasopharynx      ☐ Right primary bronchus

☐ Left primary bronchus      ☐ Oropharynx      ☐ Secondary bronchi

FIGURE **21.8** Respiratory structures: (**A**) the lungs and respiratory tract *(continues)*

21

☐ Cricoid cartilage
☐ Epiglottis
☐ Hyoid bone
☐ Thyroid cartilage

FIGURE **21.8** Respiratory structures *(cont.)*:
(**B**) the larynx, anterior view

**2** Label **Figure 21.9** with the terms below.
☐ Alveolar duct
☐ Alveolar sac
☐ Alveolus
☐ Pulmonary arteriole
☐ Pulmonary capillaries
☐ Pulmonary venule
☐ Respiratory bronchiole
☐ Terminal bronchiole

FIGURE **21.9** Bronchiole and alveolar sac

**3** *Fill in the blanks:* The _____ is the lungs' outer membrane, which adheres to the inner wall of the thoracic cavity. At the lungs' root, it folds inward on itself to become the inner membrane called the

_____, which adheres to the lungs' surface. Between the two layers of membrane is the

_____.

**4** How do the respiratory zone and the conducting zone differ?

_____

_____

_____

**5** Number the following structures of the respiratory tract in the proper order. The structure that comes into contact with oxygenated air first should be number 1, and the structures where gas exchange takes place should be number 11.

_____ Oropharynx

_____ Trachea

_____ Nasal cavity

_____ Bronchiole

_____ Alveolar sac

_____ Larynx

_____ Nasopharynx

_____ Bronchi

_____ Terminal bronchiole

_____ Laryngopharynx

_____ Respiratory bronchiole

**6** The piece of elastic cartilage that seals off the larynx during swallowing is called the
a. uvula.
b. false vocal cord.
c. true vocal cord.
d. epiglottis.

**7** Passages in the respiratory tract smaller than 1 mm in diameter are called
a. bronchi.
b. bronchioles.
c. alveolar ducts.
d. paranasal sinuses.

**8** Which of the following is *not* one of the tissue layers of the respiratory tract?

    a. Muscularis

    b. Mucosa

    c. Adventitia

    d. Submucosa

**9** What three structures constitute the respiratory membrane?

    a. _____

    b. _____

    c. _____

**10** Which of the following is *not* a trend we find within the respiratory tract?

    a. The epithelium gets progressively shorter in the lower passages.

    b. The amount of hyaline cartilage gradually increases as we move to smaller passages.

    c. The amount of smooth muscle increases in the smaller passages.

    d. The amount of elastic fibers increases in the smaller passages.

21

Name _____

Section _____ Date _____

# Check Your Understanding

### Critical Thinking and Application Questions

**1** Why do you think the hyaline cartilage rings of the trachea are "C"-shaped rather than "O"-shaped? (*Hint:* Think about the structure behind the trachea.)

_____

_____

_____

_____

_____

**2** Conditions such as pneumonia and lung cancer can result in what is known as a *pleural effusion*, in which the pleural cavity becomes filled with a large amount of fluid. What effects do you think a pleural effusion would have on ventilation? Explain.

_____

_____

_____

_____

_____

**3** Diseases such as emphysema cause the destruction of the walls between the alveoli, reducing the alveolar surface area. What effect might this have on gas exchange? Explain.

_____

_____

_____

_____

_____

21

**4** Explain how the epithelium in each region of the respiratory tract is adapted so its form follows its function.

**5** Why are goblet cells more numerous in the upper respiratory passages than in the lower respiratory passages?

**6** Cigarette smoke has a number of harmful effects on the lungs, one of which is the destruction of the cilia lining the respiratory tract. Predict how this will affect the functions of the respiratory epithelium.

**21**

**7** The most common site for obstruction in a choking victim is the right primary bronchus. Explain why the obstruction tends to lodge in the right primary bronchus rather than the left primary bronchus. (*Hint:* Consider the structure of the right versus the left primary bronchus.)

# Respiratory System Physiology

# 22

## OBJECTIVES

*Once you have completed this unit, you should be able to:*

- Describe the pressure-volume relationships in the lungs.

- Describe and measure respiratory volumes and capacities.

- Explain the effects of carbon dioxide concentration on blood pH.

Name _____ Section _____ Date _____

# PRE-LAB EXERCISES

Complete the following exercises prior to coming to lab, using your textbook and lab manual for reference.

## Pre-Lab Exercise **22-1**

### ✎ Key Terms

You should be familiar with the following terms before coming to lab. Please note that the respiratory volumes and capacities are covered in Pre-Lab Exercises 22-2 and 22-3.

| Term | Definition |
|---|---|

**Basic Respiratory Processes**

Pulmonary ventilation _____

_____

Pulmonary gas exchange _____

_____

Gas transport _____

_____

Tissue gas exchange _____

_____


**Pulmonary Ventilation Terms**

Inspiration _____

_____

Expiration _____

_____

Boyle's law _____

_____

Intrapulmonary pressure _____

_____

Atmospheric pressure _____

_____

Spirometer _____

_____

Respiratory volume _____

_____

Respiratory capacity _____

_____

**22**

## Gas Transport and External Respiration Terms

Carbonic acid–bicarbonate buffer system _____

_____

Carbonic anhydrase _____

_____

Hyperventilation _____

_____

Hypoventilation _____

_____

# Pre-Lab Exercise **22-2**

## Defining Pulmonary Volumes and Capacities

Define and give the normal value for males and females for each of the following respiratory volumes and capacities listed in Table 22.1.

TABLE **22.1** Respiratory Volumes and Capacities

| Volume/Capacity | Definition | Normal Value |
|---|---|---|
| Tidal volume | | Male: _____ ml <br> Female: _____ ml |
| Inspiratory reserve volume | | Male: _____ ml <br> Female: _____ ml |
| Expiratory reserve volume | | Male: _____ ml <br> Female: _____ ml |
| Residual volume | | Male: _____ ml <br> Female: _____ ml |
| Inspiratory capacity | | Male: _____ ml <br> Female: _____ ml |
| Functional residual capacity | | Male: _____ ml <br> Female: _____ ml |
| Vital capacity | | Male: _____ ml <br> Female: _____ ml |
| Total lung capacity | | Male: _____ ml <br> Female: _____ ml |

**22**

Labeling Pulmonary Volumes and Capacities

Label and color the respiratory volumes and capacities depicted in **Figure 22.1** using the terms from Exercise 22-2 (p. 539). Use your text and Exercise 22-2 in this unit for reference.

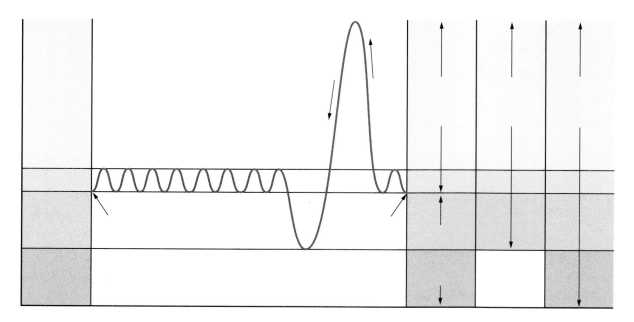

FIGURE **22.1** Respiratory volumes and capacities

# EXERCISES

In the previous unit we covered respiratory system anatomy, and now we turn our attention to respiratory system physiology. Respiration consists of four basic physiological processes:

1. **pulmonary ventilation**, the physical movement of air into and out of the lungs,

2. **pulmonary gas exchange**, the movement of gases across the respiratory membrane,

3. **gas transport**, the movement of gases through the blood, and

4. **tissue gas exchange**, the exchange of gases between the blood and the tissues.

In this unit we examine the first three processes by studying a model of the lungs, measuring respiratory volumes, and measuring the effect of exercise on carbon dioxide exchange.

## Exercise 22-1

### Pressure-Volume Relationships in the Lungs

**MATERIALS**
❑ Bell-jar model of the lungs

**Pulmonary ventilation**, the physical movement of air into and out of the lungs, consists of two phases:

1. **inspiration**, during which air is brought into the lungs, and

2. **expiration**, during which air is expelled from the lungs.

The movement of air during inspiration and expiration is driven by changes in the lungs' volume and pressure (**Figure 22.2**). The relationship of gas pressure and volume is expressed in what is known as **Boyle's law**, expressed mathematically as:

$$P_1V_1 = P_2V_2 \text{ or } P = 1/V$$

Stated simply, this means that pressure and volume are inversely proportional: As the volume of a container increases, the pressure decreases, and as the volume of a container decreases, the pressure increases.

The changes in volume during the phases of ventilation are driven by the inspiratory muscles. The main inspiratory muscle is the **diaphragm**, and it is assisted by the **external intercostal muscles**. During forced inspiration, several other muscles, termed *accessory muscles of inspiration*, assist the diaphragm and external intercostals. When the inspiratory

**Inspiration:** Lung volume increases and intrapulmonary pressure decreases.

**Expiration:** Lung volume decreases and intrapulmonary pressure increases.

FIGURE **22.2** Pressure-volume relationships in the lungs during inspiration and expiration muscles contract, they increase both the height and the diameter of the thoracic cavity, which increases its volume. Recall that the lungs are attached to the thoracic cavity directly by the pleural membranes. Therefore, the lungs increase in volume when the thoracic cavity increases in volume. As the lungs' volume increases, their pressure, called the **intrapulmonary pressure**, decreases. When intrapulmonary pressure is lower than the **atmospheric pressure**, inspiration occurs, and air rushes into the lungs. Because the intrapulmonary pressure must be *lower* than the atmospheric pressure for inspiration to occur, it sometimes is referred to as *negative pressure*.

Expiration is achieved primarily by the elastic recoil of the lungs. As the inspiratory muscles relax, the lungs' elastic tissue causes them to recoil to their original, smaller size. This decreases the volume of the lungs and increases the intra-pulmonary pressure. Once the intrapulmonary pressure is higher than the atmospheric pressure, air exits the lungs, and expiration occurs. In the event of forced expiration, several accessory muscles of expiration, including the internal intercostals, will decrease the height and diameter of the thoracic cavity.

## **Procedure** Model Ventilation with the Bell-Jar Model

In this procedure we will use a bell-jar model of the lungs, shown in **Figure 22.3**, to view the effects of pressure and volume on ventilation. The bell-jar model has two balloons, each representing one lung, and a flexible membrane on the bottom that represents the diaphragm. After you have completed this activity, answer Check Your Understanding questions 1 and 2 (p. 547).

**1** Apply upward pressure to the diaphragm. This represents how the diaphragm looks when it is relaxed. What has happened to the pressure of the system (has it increased or decreased)? What happened to the volume of the lungs?

_____

_____

_____

_____

_____

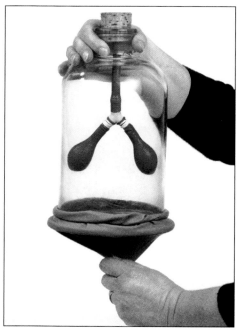

FIGURE **22.3** Bell-jar model of the lungs, simulated inspiration

**2** Now slowly release the diaphragm. This represents the diaphragm flattening out as it contracts. What is happening to the pressure as you release the diaphragm? What happened to the volume of the lungs?

_____

_____

_____

**3** If your bell-jar model has a rubber stopper in the top, you can use it to demonstrate the effects of a **pneumothorax** on lung tissue. A pneumothorax generally is caused by a tear in the pleural membranes that allows air to enter the pleural cavity. With the diaphragm flat and the lungs (balloons) inflated, loosen the rubber stopper. What happens to the lungs? Why?

_____

_____

_____

# Exercise 22-2

## Measuring Pulmonary Volumes and Capacities

### MATERIALS
❑ Wet or computerized spirometer
❑ Disposable mouthpiece

**Respiratory volumes** are the volumes of air exchanged with each breath. These volumes of air are measured with an instrument called a **spirometer** and include the following (**Figure 22.4**):

1. **Tidal volume (TV).** The tidal volume, or TV, is the amount of air exchanged with each breath during normal, quiet breathing. It measures about 500 ml in a healthy adult.

2. **Expiratory reserve volume (ERV).** The expiratory reserve volume, or the ERV, is the volume of air that may be expired after a tidal expiration. It averages between 700 and 1,200 ml of air.

3. **Inspiratory reserve volume (IRV).** The inspiratory reserve volume, or the IRV, is the amount of air that may be inspired after a tidal inspiration. It averages between 1,900 and 3,100 ml of air.

Note that there is also a fourth volume, called the **residual volume (RV)**, that cannot be measured with a spirometer. It is defined as the amount of air that remains in the lungs after maximal expiration and is generally equal to about 1,100–1,200 ml of air. This amount of air accounts for the difference between the IRV and the ERV.

Two or more respiratory volumes may be combined to give **respiratory capacities**. As you can see in **Figure 22.4**, there are four respiratory capacities:

1. **Inspiratory capacity (IC).** The inspiratory capacity, or IC, is equal to the TV plus the IRV and is the amount of air a person can maximally inspire after a tidal expiration. It averages between 2,400 and 3,600 ml of air.

2. **Functional residual capacity (FRC).** The functional residual capacity, or the FRC, is the amount of air normally left in the lungs after a tidal expiration. It is the sum of the ERV and the RV and averages between 1,800 and 2,400 ml of air. It is not measurable with general spirometry.

3. **Vital capacity (VC).** The vital capacity, or VC, represents the total amount of exchangeable air that moves in and out of the lungs. It is equal to the sum of the TV, IRV, and the ERV and averages between 3,100 and 4,800 ml of air.

4. **Total lung capacity (TLC).** The total lung capacity, or TLC, represents the total amount of exchangeable and non-exchangeable air in the lungs. It is the total of all four respiratory volumes and is not measurable with general spirometry. It averages between 4,200 and 6,000 ml of air.

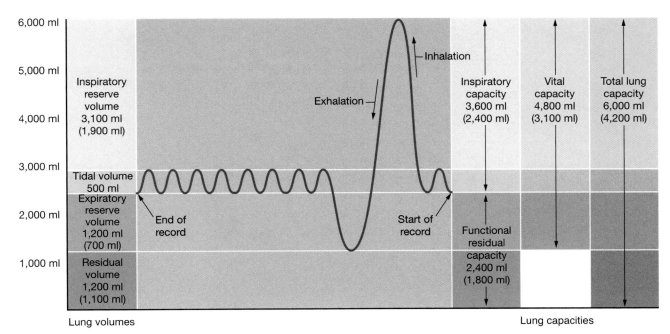

FIGURE **22.4** Respiratory volumes and capacities (note that the top value in the figure is the average value for males, and the value in parentheses is the average value for females)

**22**

Spirometry is a useful tool with which to assess pulmonary function. The respiratory volumes and capacities are especially helpful in differentiating the two primary types of respiratory disorders—**restrictive diseases** and **obstructive diseases**. Restrictive diseases, such as pulmonary fibrosis, are characterized by a loss of elasticity (decreased compliance) of the lung tissue. As a result, patients' ability to inspire is affected adversely. This decreases the IRV, IC, VC, and TLC.

Obstructive diseases, such as chronic obstructive pulmonary disease (COPD) and asthma, are characterized by increased airway resistance caused by narrowing of the bronchioles, increased mucus secretion, and/or an obstructing body such as a tumor. It may seem counterintuitive, but obstructive diseases make expiration difficult. This is because the increased intrapulmonary pressure during expiration naturally tends to shrink the diameter of the bronchioles. When the bronchioles are already narrowed, as in an obstructive disease, the increased intrapulmonary pressure can actually collapse the bronchioles and trap oxygen-poor air in the distal respiratory passages. Therefore, patients with obstructive diseases often exhale slowly and through pursed lips to minimize the pressure changes and maximize the amount of air exhaled. This decreases the ERV and VC and increases the RV and FRC.

## **Procedure** Measure Respiratory Volumes with a Wet Spirometer

Three types of spirometers are commonly found in anatomy and physiology labs:

1. a bell, or "wet" spirometer,

2. a computerized spirometer, and

3. a handheld spirometer (shown in **Figure 22.5**).

FIGURE **22.5** A handheld spirometer

Many types of wet and handheld spirometers allow you to assess only expiratory volumes, whereas computerized spirometers typically allow you to assess both inspiratory and expiratory volumes. All types of devices allow measurement of one or more respiratory capacities.

The following procedure is intended for a wet or handheld spirometer. If your lab has a computerized spirometer, however, you can either follow the procedure below or follow the prompts by the computer program. When you have completed the activity, answer Check Your Understanding questions 3 and 4 (pp. 547–548).

> **Safety Note**
> You should not be the subject of this experiment if you have cardiovascular or pulmonary disease or are prone to dizziness or fainting.

**1** Obtain a disposable mouthpiece, and attach it to the end of the tube.

**2** Before you begin, practice exhaling through the tube several times. Note that you are supposed to only *exhale* into the tube, because it likely has no filter.

**3** *Measure the tidal volume:* Sit in a chair with your back straight and your eyes closed. Inhale a normal tidal inspiration, and exhale this breath through the tube. Getting a true representation of the tidal volume is often difficult, because people have a tendency to force the expiration. To get the most accurate tidal volume, take several measurements and average the numbers.

Measurement 1: _____

Measurement 2: _____

Measurement 3: _____

Average Tidal Volume: _____

> Note that some handheld spirometers are not terribly sensitive, so they give a value for the tidal volume that is too low. If you find that this is the case, you may need to take extra readings to get an approximately accurate measurement.

**4** *Measure the expiratory reserve volume*: Before taking this measurement, inhale and exhale a series of tidal volumes. Then inspire a normal tidal inspiration, breathe out a normal tidal expiration, put the mouthpiece to your mouth, and exhale as forcibly as possible. Don't cheat by taking in a large breath first! As before, perform several measurements and average the numbers.

Measurement 1: _____

Measurement 2: _____

Measurement 3: _____

Average Expiratory Reserve Volume: _____

**5** *Measure the vital capacity*: As before, inhale and exhale a series of tidal volumes. Then bend over, and exhale maximally. Once you have exhaled as much air as you can, raise yourself upright, and inhale as much air as you possibly can (until you feel like you are about to "pop"). Quickly place the mouthpiece to your mouth, and exhale as forcibly and as long as possible. Take several measurements (you may want to give yourself a minute to rest between measurements), and record the data below.

Measurement 1: _____

Measurement 2: _____

Measurement 3: _____

Average Vital Capacity: _____

**6** *Calculate the inspiratory reserve volume*: Even though the bell spirometer cannot measure inspiratory volumes, you can calculate this volume now that you have the vital capacity (VC), tidal volume (TV), and expiratory reserve volume (ERV). Recall that VC = TV + IRV + ERV. Rearrange the equation: IRV = VC − (TV + ERV).

Average IRV: _____

**7** How do your values compare with the average values? What factors, if any, do you think may have affected your results?

_____

_____

_____

_____

**8** Pool the results for your class, and divide the results into four categories: female cigarette smokers, female nonsmokers, male cigarette smokers, and male nonsmokers. Calculate the average TV, ERV, and VC for each group, and record these data in **Table 22.2**.

TABLE **22.2** Respiratory Volumes and Capacities for Smokers and Nonsmokers

| Group | Tidal Volume | Expiratory Reserve Volume | Inspiratory Reserve Volume | Vital Capacity |
|---|---|---|---|---|
| Female smokers | | | | |
| Female nonsmokers | | | | |
| Male smokers | | | | |
| Male nonsmokers | | | | |

**9** Interpret your results:

**a** How did the average values differ for males and females?

**b** How did the average values differ for smokers and nonsmokers?

**c** What disease pattern would you expect to see with the smokers? Did your results follow this expectation?

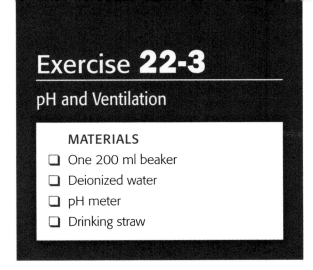

# Exercise 22-3

## pH and Ventilation

**MATERIALS**

☐ One 200 ml beaker
☐ Deionized water
☐ pH meter
☐ Drinking straw

Pulmonary ventilation is required not only to bring oxygen into the body but also to *release carbon dioxide*. Carbon dioxide is a nonpolar gas that does not fully dissolve in the water of plasma. For this reason, carbon dioxide is transported in another form that is water soluble: the weak base called the **bicarbonate ion** ($HCO_3^-$). Carbon dioxide reacts with water to form **carbonic acid** ($H_2CO_3$), which dissociates to form hydrogen and bicarbonate ions. This forms a system known as the **carbonic acid–bicarbonate buffer system**. This reversible reaction is catalyzed by an enzyme called **carbonic anhydrase** and proceeds as follows:

$$H_2O + CO_2 \rightleftharpoons H_2CO_3 \rightleftharpoons HCO_3^- + H^+$$

Because carbon dioxide reacts to form an acid, the concentration of carbon dioxide in the blood is a major determinant of blood pH. The concentration of carbon dioxide in the blood is determined directly by the rate and depth of ventilation, so anything that affects ventilation can also affect blood pH. When the rate and depth of ventilation increase during **hyperventilation,** more carbon dioxide is expelled from the body, and the carbon dioxide concentration of the blood decreases. This in turn decreases the amount of carbonic acid formed, which raises the pH of the blood. Conversely, when the rate and depth of ventilation decrease during **hypoventilation,** carbon dioxide is retained, and its concentration in the blood increases. This increases the amount of carbonic acid formed and decreases the blood pH.

When the pH of the blood increases or decreases because of carbon dioxide or other factors unrelated to ventilation, the body attempts to return the blood pH to normal by adjusting the rate and depth of ventilation. If the blood pH decreases, the brainstem triggers an increase in respiratory rate to remove carbon dioxide from the blood and raise the blood pH. This is something you witness every time you engage in even minor exercise—when you exercise, your metabolic rate and carbon dioxide production increase, and your brainstem triggers an increase in respiratory rate to remove the excess carbon dioxide from the blood. If the pH of the blood increases, the brainstem will slow the rate of ventilation to retain carbon dioxide and lower the blood pH.

## Procedure Measure the Effect of Carbon Dioxide on the pH of a Solution

We can't measure the effects of ventilatory rate on blood pH directly in the lab because it requires an analysis of arterial blood. However, we can observe and measure the effects of carbon dioxide on the pH of water and correlate this with the changes you would see in the blood, which is what you will do in the following experiment. When you have completed the activity, answer Check Your Understanding questions 5–7 (p. 548).

**1** Measure your partner's respiratory rate while he or she is seated and resting. Measuring the respiratory rate can be tricky because as soon as you tell your partner to breathe normally, he or she automatically becomes conscious of breathing and alters the rate. A trick you may want to try is to place two fingers on your partner's radial artery and take his or her pulse. This will cause your partner to focus more on his or her pulse rate, and he or she will breathe more naturally. Record this value in **Table 22.3.**

**2** Obtain a 200 ml beaker about half-full of deionized water. Measure the pH of the water with a pH meter. Record this value below:

pH of deionized water: _____

**3** Place a drinking straw into the water, and have your partner exhale through the straw into the water (the water should bubble) for 20 seconds. Repeat the pH measurement of the water with the pH meter, and record the value in **Table 22.3.**

**4** Empty your beaker, and refill it with new water. Have your partner do vigorous exercise for at least 5 minutes (running up and down stairs usually does the job). Immediately upon finishing the exercise, have your partner blow through the straw into the beaker of water for 20 seconds. Measure the pH of the water with the pH meter, and record the value in **Table 22.3.**

**5** After you record the pH, measure your partner's respiratory rate (you may want to use the pulse trick again), and record the value in **Table 22.3**.

**6** Empty your beaker again, and refill it with new water. Allow your partner to have a (well-deserved) rest for 5 minutes. Again, have your partner blow through the straw into the beaker of water for 20 seconds, and measure the pH of the water with the pH meter. Record the value in **Table 22.3**.

**7** After you record the pH, measure your partner's respiratory rate, and record the value in **Table 22.3**.

TABLE **22.3** pH of Water and Ventilatory Rate

| Setting | Respiratory Rate | pH |
|---|---|---|
| At rest | | |
| Immediately after exercise | | |
| After 5 minutes of rest | | |

**8** Interpret your results:

   **a** Start with the basics: What happens to the metabolism during exercise? What is generated as a by-product of glucose metabolism?

   **b** What does this by-product do to the pH of the blood? Why?

   **c** What sort of ventilatory response does this change in pH cause (hyperventilation or hypoventilation)? Why?

   **d** Now apply this to your results:

      (1) What happened to the respiratory rate during and after exercise? Why? (*Hint:* See item c. above.)

      (2) What happened to the pH of the water after exercise? Why?

      (3) What happened to the respiratory rate and the pH of the water after resting? Why?

22

# Check Your Recall

**1** Which of the following is *not* a basic physiological process of respiration?

a. Pulmonary gas exchange

b. Glycolytic catabolism

c. Pulmonary ventilation

d. Gas transport

e. Tissue gas exchange

**2** *Fill in the blanks:* According to Boyle's law, as the volume of a container increases, the pressure _____.

Conversely, as the volume of a container decreases, the pressure _____.

**3** Air moves into the lungs when

a. intrapulmonary pressure is less than atmospheric pressure.

b. intrapulmonary pressure is more than atmospheric pressure.

c. blood pressure is more than intrapulmonary pressure.

d. blood pressure is less than intrapulmonary pressure.

**4** Passive expiration is achieved primarily by the

a. contraction of the diaphragm and external intercostals.

b. decrease in intrapulmonary pressure.

c. increase in atmospheric pressure.

d. elastic recoil of the lungs.

**5** *Matching:* Match the following pulmonary volumes and capacities with their correct definition or description.

_____ Total lung capacity

_____ Residual volume

_____ Functional residual capacity

_____ Vital capacity

_____ Inspiratory reserve volume

_____ Tidal volume

_____ Inspiratory capacity

_____ Expiratory reserve volume

A. Volume of air in the lungs after a tidal expiration

B. Volume of air forcibly expired after a tidal expiration

C. Volume of air forcibly inspired after a tidal inspiration

D. Maximum amount of exchangeable and nonexchangeable air in the lungs

E. Maximum amount of air that can be inhaled with a forced inspiration

F. Volume of air remaining in the lungs after forceful expiration

G. Maximum amount of exchangeable air that can be ventilated with forced inspiration and expiration

H. Volume of air exchanged with normal, quiet breathing

**6** How do restrictive and obstructive diseases differ?

_____

_____

_____

_____

_____

**7** Calculate the inspiratory capacity, vital capacity, and total lung capacity for a patient with the following respiratory volumes:

- Tidal Volume = 450 ml
- Expiratory Reserve Volume = 1,100 ml
- Inspiratory Reserve Volume = 3,000 ml
- Residual Volume = 1,200 ml

Inspiratory Capacity = _____

Vital Capacity = _____

Total Lung Capacity = _____

**8** Carbon dioxide is transported through the blood primarily as
a. dissolved gas.
b. bicarbonate.
c. carbonic anhydrase.
d. carbaminohemoglobin.

**9** _Fill in the blanks:_ Hyperventilation _____ the amount of carbon dioxide exhaled and therefore _____ the blood pH. Hypoventilation _____ the amount of carbon dioxide exhaled and therefore _____ the blood pH.

**10** The enzyme that converts carbon dioxide and water into carbonic acid is called
a. bicarbonate dehydrogenase.
b. adenosine deaminase.
c. carbonic anhydrase.
d. hemoglobin carboxylase.

# Check Your Understanding

**Critical Thinking and Application Questions**

**1** The condition *emphysema* results in loss of elastic recoil of the lung tissue. Would this make inspiration or expiration difficult? Explain.

_____

_____

_____

_____

_____

_____

**2** Certain chest deformities prevent normal expansion of the chest. Explain why this makes ventilation difficult in terms of pressure and volume.

_____

_____

_____

_____

_____

_____

**3** A male patient presents with the following respiratory volumes and capacities: TV = 500 ml, ERV = 1,200 ml, IRV = 1,100 ml. What is this patient's VC? Are these values normal? If not, are they more consistent with an obstructive or restrictive disease pattern? Explain.

22

_____

_____

_____

_____

_____

_____

**4** The disease *asthma* is a disease characterized by bronchoconstriction (narrowing of the airways), airway inflammation, and excess secretion of thick mucus. Is asthma likely a restrictive disease or an obstructive disease? Predict the values you would expect to see on spirometry for a patient with asthma.

**5** Why should a person breathe into a paper bag when he or she is hyperventilating abnormally?

**6** A patient presents in a state of *ketoacidosis* (a type of metabolic acidosis in which the pH of the blood drops) caused by poorly controlled *diabetes mellitus*. Will the patient be hyperventilating or hypoventilating? Why?

**7** Why does the ventilatory rate increase during exercise?

# Digestive System

## OBJECTIVES

*Once you have completed this unit, you should be able to:*

- Identify structures of the digestive system.

- Identify histological structures of the digestive system.

- Demonstrate and describe the action of emulsifying agents on lipids.

- Demonstrate and describe the functions of digestive enzymes.

- Trace the pathway of physical and chemical digestion of carbohydrates, proteins, and lipids.

Name _____ Section _____ Date _____

# PRE-LAB EXERCISES

Complete the following exercises prior to coming to lab, using your textbook and lab manual for reference.

## Pre-Lab Exercise 23-1

### ✎ Key Terms

You should be familiar with the following terms before coming to lab. Please note that this list is not all-inclusive, because the terminology for the digestive system is extensive. Digestive enzymes are covered separately in Pre-Lab Exercise 23-3 (p. 556).

| Term | Definition |
| --- | --- |

**Digestive System Structures**

Alimentary canal _____

_____

Accessory organ _____

_____

Peritoneal cavity _____

_____

Gastroesophageal sphincter _____

_____

Pyloric sphincter _____

_____

Duodenum _____

_____

Jejunum _____

_____

Ileum _____

_____

Colon _____

_____

Salivary glands _____

_____

Pancreas _____

_____

23

Liver _____

_____

Gallbladder _____

_____

## Digestive Histology

Mucosa _____

_____

Submucosa _____

_____

Muscularis externa _____

_____

Serosa _____

_____

Acinar cells _____

_____

Pancreatic islet _____

_____

Liver lobule _____

_____

## Digestive Physiology

Digestive enzyme _____

_____

Chemical digestion _____

_____

**23**

Emulsification _____

_____

Bile _____

_____

## Pre-Lab Exercise **23-2**

Anatomy of the Digestive System

Label and color the structures of the digestive system depicted in **Figures 23.1–23.4** with the terms from Exercise 23-1 (p. 557). Use your text and Exercise 23-1 in this unit for reference.

FIGURE **23.1** The organs of the digestive system

23

FIGURE **23.2** The stomach

FIGURE **23.3** Gross and microscopic anatomy of the small intestine

23

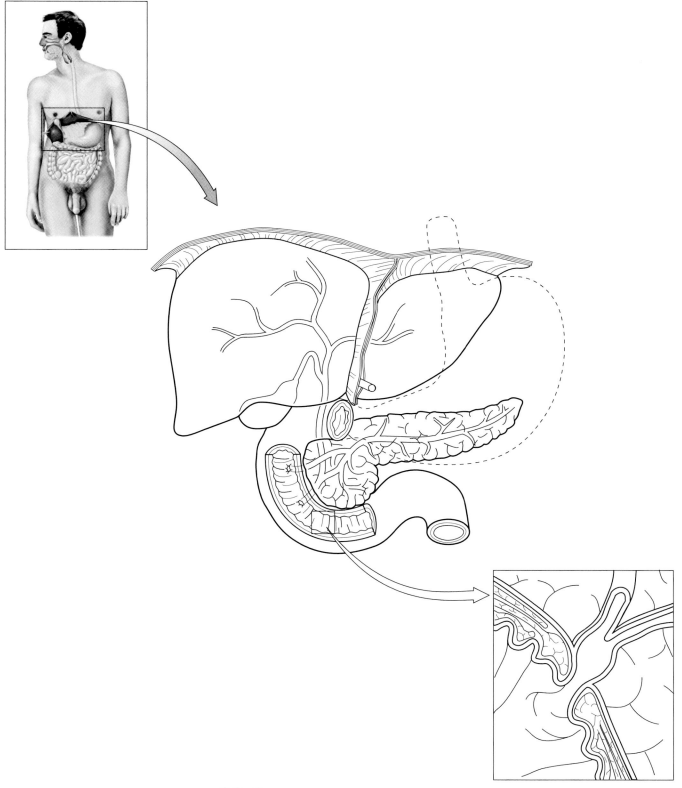

FIGURE **23.4** The liver, gallbladder, pancreas, and duodenum

23

## Digestive Enzymes

Chemical digestion is one of the main functions of the digestive system and is carried out with the help of numerous digestive enzymes. Record the organ that produces each enzyme and the function of each enzyme in **Table 23.1**.

TABLE **23.1** Digestive Enzymes

| Enzyme | Source | Function |
|---|---|---|
| Salivary amylase | | |
| Pepsin | | |
| Trypsin | | |
| Pancreatic lipase | | |
| Brush border enzymes | | |

23

# EXERCISES

The food we eat contains nutrients our cells use to build and repair body tissues and to make ATP. Food macromolecules, however, are typically too large for the body to absorb and utilize, so the body must break them down into smaller molecules. This process of breaking down foods into smaller substances that can enter body cells is called **digestion** and is carried out by the **digestive system**. In general, the functions of the digestive system include taking in food, breaking down food mechanically and chemically into nutrients, absorbing these nutrients into the bloodstream, and eliminating indigestible substances.

We begin this unit with an introduction to the anatomy and histology of the organs of the digestive system. Next we examine the physiological processes of chemical digestion and emulsification. We conclude with a "big picture" view of digestion through a tracing exercise.

## Exercise 23-1

### Digestive System Anatomy

> **MATERIALS**
> ❏ Digestive system models
> ❏ Head and neck models
> ❏ Human torso models
> ❏ Human skulls with teeth
> ❏ Intestinal villus model
> ❏ Models of the tissue layers of the alimentary canal
> ❏ Digestive organ models (stomach, pancreas, liver, and duodenum)

The digestive system is composed of two types of organs: (1) the organs of the **alimentary canal**, also known as the **gastrointestinal** or **GI tract,** through which food travels, and (2) the **accessory organs**, which assist in mechanical or chemical digestion. The alimentary canal consists of the following organs (**Figure 23.5**):

1. **Mouth.** The alimentary canal begins with the mouth. In and around the mouth we find numerous accessory organs, including the **teeth**, the **tongue**, and the **salivary glands**. These accessory organs help the mouth carry out mechanical digestion and chemical digestion.

2. **Pharynx.** The food next enters the pharynx, also known as the throat. The muscles surrounding the pharynx propel swallowed food into the next portion of the alimentary canal.

3. **Esophagus.** The esophagus is a narrow tube posterior to the heart and the trachea in the thoracic cavity. The esophagus contains both smooth and skeletal muscle fibers that contract via rhythmic contractions called **peristalsis,** which propels food into the stomach. A sphincter at the inferior end of the esophagus called the **gastroesophageal sphincter** (also known as the *lower esophageal sphincter*) prevents the contents of the stomach from regurgitating into the esophagus.

4. **Stomach.** The stomach, shown in **Figure 23.6,** has five regions:
   a. the dome-shaped **fundus,**
   b. the **cardia** near the gastroesophageal sphincter,
   c. the middle **body,**
   d. the inferior **pyloric antrum,** and
   e. the **pylorus.** A sphincter called the **pyloric sphincter** separates the stomach from the initial portion of the small intestine.

   Note in **Figure 23.6** that the stomach has interior folds called **rugae** that allow it to expand considerably when it is filled with food. In addition, note that the wall of the stomach has three layers of smooth muscle (inner oblique, middle circular, and outer longitudinal layers) that work together to pummel ingested food into a liquid material called **chyme.**

5. **Small intestine.** The small intestine is the portion of the alimentary canal where most chemical digestion and absorption take place. It has three portions:
   a. the initial **duodenum,**

**23**

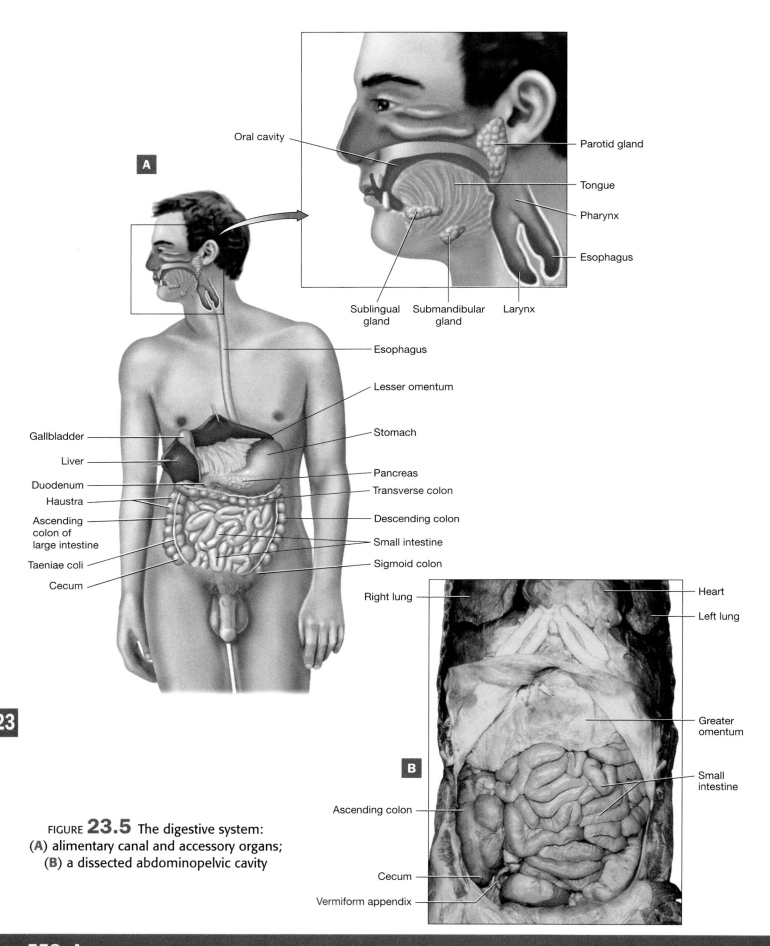

Oral cavity

Parotid gland

Tongue

Pharynx

Esophagus

Sublingual gland

Submandibular gland

Larynx

**A**

Esophagus

Lesser omentum

Gallbladder

Stomach

Liver

Pancreas

Duodenum

Transverse colon

Haustra

Ascending colon of large intestine

Descending colon

Small intestine

Taeniae coli

Sigmoid colon

Cecum

Right lung

Heart

Left lung

Greater omentum

**B**

Small intestine

Ascending colon

Cecum

Vermiform appendix

FIGURE **23.5** The digestive system:
(**A**) alimentary canal and accessory organs;
(**B**) a dissected abdominopelvic cavity

**23**

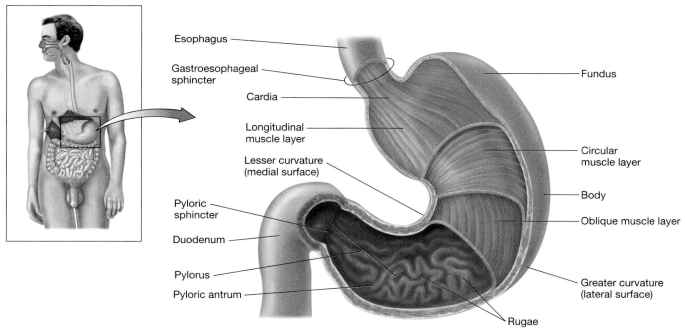

FIGURE **23.6** The stomach

b. the middle **jejunum**, and

c. the terminal **ileum**.

Of the three divisions, the duodenum is the shortest, measuring only about 10 inches. The jejunum and the ileum measure about 8 feet and 12 feet long, respectively.

6. **Large intestine.** The large intestine is named for its large diameter rather than its length, which only measures about 5.5 feet. Much of the large intestine features bands of longitudinal smooth muscle called **taeniae coli** that have a drawstring effect on the large intestine, pulling it into pouches called **haustra**. Extending from the external surface of the large intestine are fat-filled pouches called **epiploic appendages**. The large intestine may be divided into five regions:

a. the **cecum** (Figure 23.7), the pouch that receives contents from the ileum, from which it is separated by the **ileocecal valve**,

b. the **vermiform appendix**, also visible in **Figure 23.7**, a blind-ended sac that extends from the cecum and contains lymphatic nodules,

c. the **colon**, which itself has four divisions (the *ascending*, *transverse*, *descending*, and *sigmoid colon*),

d. the **rectum**, the straight part of the large intestine that runs along the sacrum, and

e. the **anal canal**, the terminal portion of the large intestine with two sphincters: the involuntary **internal anal sphincter** and the voluntary **external anal sphincter**.

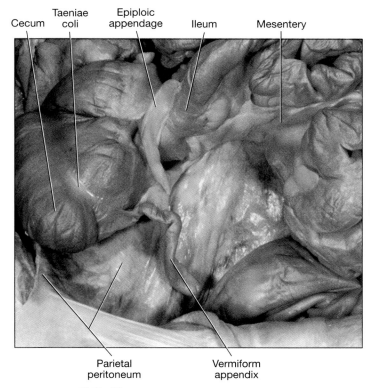

FIGURE **23.7** The cecum and vermiform appendix

The accessory digestive organs generally do not come into direct contact with the ingested food (the teeth and tongue are exceptions). Most of them instead secrete substances such as bile salts and enzymes that travel through a duct to the alimentary canal. The accessory digestive organs are the following:

1. **Teeth and tongue.** The teeth and the tongue are accessory organs located in the mouth that assist in mechanical digestion of ingested food. The tongue contains keratinized papillae called **filiform papillae** that provide a rough surface, which helps to break down food physically. Note that filiform papillae do not contain taste buds, which are located only on circumvallate, foliate, and fungiform papillae.

2. **Salivary glands.** The three types of salivary glands are located around the mouth, and their secretions pass through a duct into the mouth.

    a. The largest salivary glands are the paired **parotid glands** superficial to the masseter muscle.

    b. The smaller **submandibular glands** are located just medial to the mandible.

    c. The **sublingual gland** is located under the tongue, as its name implies.

    All three types of glands secrete saliva, which contains water, mucus, an enzyme called *salivary amylase*, and antimicrobial molecules such as *lysozyme*.

3. **Liver and gallbladder.** The liver and gallbladder are organs located on the right side of the abdominal cavity (**Figure 23.8**).

    a. The liver consists of four lobes—the large *right* and *left lobes* and the small, posterior *caudate* and *quadrate lobes*. The liver has a multitude of functions in the body, most of which are metabolic in nature. For example, recall from Unit 17 that all blood from the digestive organs and the spleen travels via the hepatic portal vein to the hepatic portal system of the liver. There, the absorbed nutrients and chemicals are processed before they enter the general circulation. One of the liver's main digestive functions is to produce a chemical called *bile*, required for the digestion and absorption of fats.

    b. Bile leaves the liver via a duct called the **common hepatic duct,** after which much of it enters the gallbladder on the posterior side of the liver for storage. When stimulated by certain hormones, the gallbladder contracts,

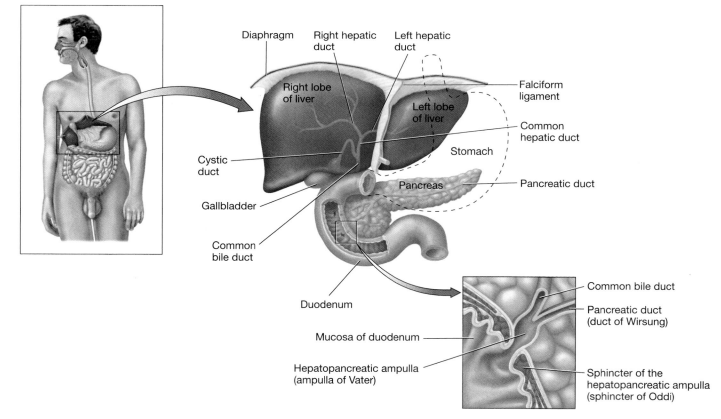

FIGURE **23.8** The liver, gallbladder, pancreas, and duodenum

and the bile contained within it is ejected through the **cystic duct**. Note in **Figure 23.8** that the cystic duct joins with the common hepatic duct to form the **common bile duct**, which empties into the duodenum at the **hepatopancreatic ampulla**.

4. **Pancreas**. The pancreas is an exocrine and endocrine gland that sits posterior and inferior to the stomach. Its exocrine functions are digestive, whereas its endocrine functions are metabolic. The exocrine portion of the pancreas produces a fluid called **pancreatic juice** that contains water, bicarbonate to neutralize the acid produced by the stomach, and a plethora of digestive enzymes. Pancreatic juice is released through the **pancreatic duct** and enters the duodenum at the hepatopancreatic ampulla.

Much of the alimentary canal and many of the accessory organs reside inside a cavity known as the **peritoneal cavity** (**Figure 23.9**). Like the pleural and pericardial cavities, the peritoneal cavity is found between a double-layered serous membrane, which secretes **serous fluid** that allows the organs to slide over one another without friction. The two peritoneal layers include the:

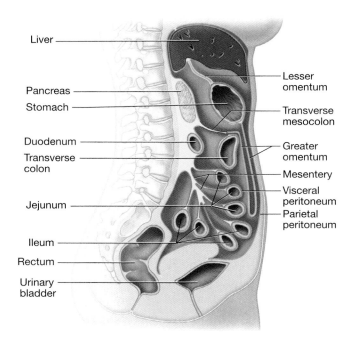

FIGURE **23.9** The peritoneal membranes and cavity

1. **Parietal peritoneum**. The outer parietal peritoneum is a thin membrane functionally fused to the abdominal wall and certain organs.

2. **Visceral peritoneum**. The inner visceral peritoneum adheres to the surface of many digestive organs. The visceral peritoneum around the intestines folds over on itself to form a thick membrane known as the **mesentery** (**Figure 23.10**). The mesentery houses blood vessels, nerves, and lymphatic vessels and anchors these structures and the intestines in place. Some mesenteries have specific names, including the **greater omentum**, which covers the abdominal organs like an apron. The smaller lesser omentum runs from the liver to the lesser curvature of the stomach.

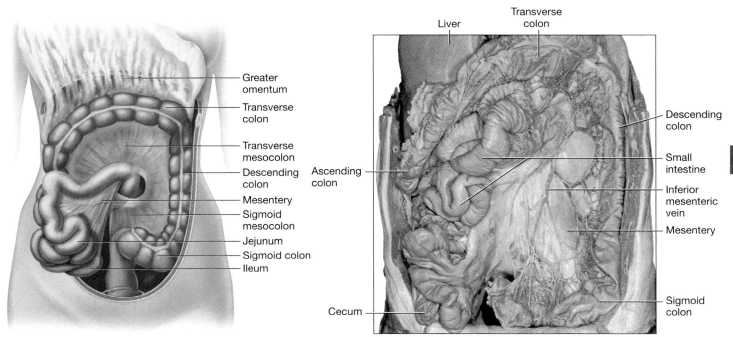

FIGURE **23.10** The mesentery and greater omentum

Identify the following structures of the digestive system on models and diagrams, using your textbook and this unit for reference. As you examine the anatomical models and diagrams, record the name of the model and the structures you were able to identify on the model inventory in **Table 23.2.** After you have completed the activity, answer Check Your Understanding question 1 (p. 583).

## Alimentary Canal

1. Mouth:
   a. Lips
   b. Cheeks
   c. Vestibule
   d. Oral cavity proper
   e. Hard palate
   f. Soft palate
   g. Uvula
2. Pharynx
   a. Oropharynx
   b. Laryngopharynx
3. Esophagus
   a. Gastroesophageal sphincter (or lower esophageal sphincter)
4. Stomach
   a. Cardia
   b. Fundus
   c. Body
   d. Pyloric antrum
   e. Pylorus
   f. Pyloric sphincter
   g. Rugae

5. Small intestine
   a. Duodenum
      (1) Hepatopancreatic ampulla
   b. Jejunum
   c. Ileum
   d. Ileocecal valve
6. Large intestine
   a. Haustra
   b. Taeniae coli
   c. Epiploic appendages
   d. Cecum
   e. Vermiform appendix
   f. Ascending colon
   g. Hepatic flexure
   h. Transverse colon
   i. Splenic flexure
   j. Descending colon
   k. Sigmoid colon
   l. Rectum
   m. Anal canal

## Accessory Organs

1. Teeth
2. Tongue
   a. Filiform papillae
   b. Fungiform papillae
   c. Circumvallate papillae
   d. Foliate papillae
3. Salivary glands
   a. Parotid gland with parotid duct
   b. Submandibular gland
   c. Sublingual gland
4. Pancreas
   a. Pancreatic duct

5. Liver
   a. Right lobe, left lobe, caudate lobe, and quadrate lobe
   b. Falciform ligament
   c. Common hepatic duct
   d. Hepatic portal vein
   e. Hepatic arteries
   f. Hepatic veins
6. Gallbladder
   a. Cystic duct
   b. Common bile duct

## Other Structures

1. Peritoneal cavity
2. Peritoneum
   a. Visceral peritoneum
   b. Parietal peritoneum
   c. Mesentery
3. Greater omentum
4. Blood vessels:
   a. Celiac trunk

   b. Left gastric artery
   c. Common hepatic artery
   d. Superior mesenteric artery
   e. Inferior mesenteric artery
   f. Superior mesenteric vein
   g. Inferior mesenteric vein
   h. Gastric veins
   i. Hepatic portal vein

**23**

Note: Your instructor may wish to omit certain structures or add structures not included in these lists. List any additional structures below:

TABLE **23.2** Model Inventory for the Digestive System

| Model | Digestive Structures Identified |
|---|---|
|  |  |
|  |  |
|  |  |
|  |  |
|  |  |
|  |  |
|  |  |
|  |  |

# Exercise 23-2

## Digestive System Histology

### MATERIALS

- ☐ Model of the layers of the alimentary canal
- ☐ Esophagus slide
- ☐ Stomach slide
- ☐ Duodenum slide
- ☐ Pancreas slide
- ☐ Liver slide
- ☐ Light microscope
- ☐ Colored pencils

The organs of the alimentary canal follow the same basic pattern we have seen for other hollow organs: an inner layer of epithelial tissue that rests on connective tissue, a middle layer of smooth muscle, and an outer layer of supportive connective tissue. The tissue layers of the alimentary canal, shown in **Figure 23.11**, are named as follows:

1. **Mucosa.** The inner epithelial tissue lining of the alimentary canal is called the mucosa. It consists of simple columnar epithelium overlying a lamina propria and a thin layer of smooth muscle called the **muscularis mucosa.** The mucosa contains a collection of lymphoid nodules called **mucosal-associated lymphoid tissue (MALT)**, mucus-secreting goblet cells, and other glands that secrete products such as hydrochloric acid and enzymes. The mucosa is generally covered with a layer of mucus that helps protect the underlying epithelium from the effects of the acid and digestive enzymes.

2. **Submucosa.** Deep to the mucosa we find the submucosa, a layer of connective tissue that houses blood vessels, nerves, lymphatics, and elastic fibers.

3. **Muscularis externa.** The muscularis externa actually contains two layers of smooth muscle—an inner circular layer and an outer longitudinal layer. These two layers contract alternately to produce the rhythmic contractions of **peristalsis.** Recall from Exercise 23-1 that the muscularis externa of the stomach has three layers of smooth muscle, with an additional oblique layer.

4. **Serosa.** The outer connective tissue layer is called the serosa, which is partially composed of the visceral peritoneum, throughout much of the alimentary canal. It is also known as the **adventitia.**

Although this pattern is followed throughout most of the alimentary canal, there are some notable differences, particularly in the esophagus (**Figure 23.12**). The mucosa of the esophagus is composed of stratified squamous epithelium with no goblet cells or thick mucus layer. The esophagus's muscularis externa changes as it progresses toward the stomach. The upper one-third of the esophagus is *skeletal muscle*, the middle one-third is about *half skeletal muscle and half smooth muscle*, and the lower one-third is *smooth muscle*. These differences in the muscularis externa allow you to determine the location of the section you are examining.

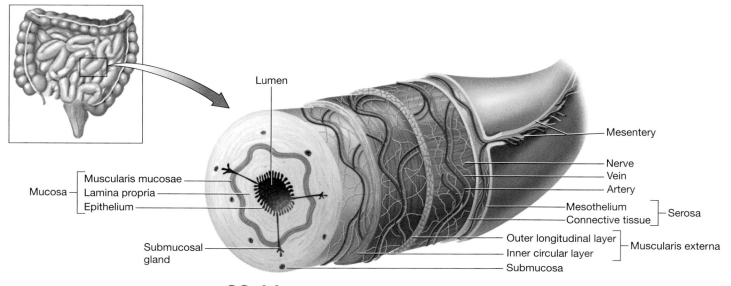

FIGURE **23.11** The tissue layers of the alimentary canal

Stratified squamous epithelium

Submucosa

Smooth muscle

Seromucous glands

FIGURE **23.12** The esophagus, photomicrograph

The mucosae of both the stomach and the small intestine are folded into ridges. The stomach mucosa is heavily indented, which reflects the presence of numerous **gastric pits** that house **gastric glands** (Figure 23.13). These glands secrete products for digestion known collectively as **gastric juice**. The stomach mucosa between the gastric pits contains a large number of goblet cells.

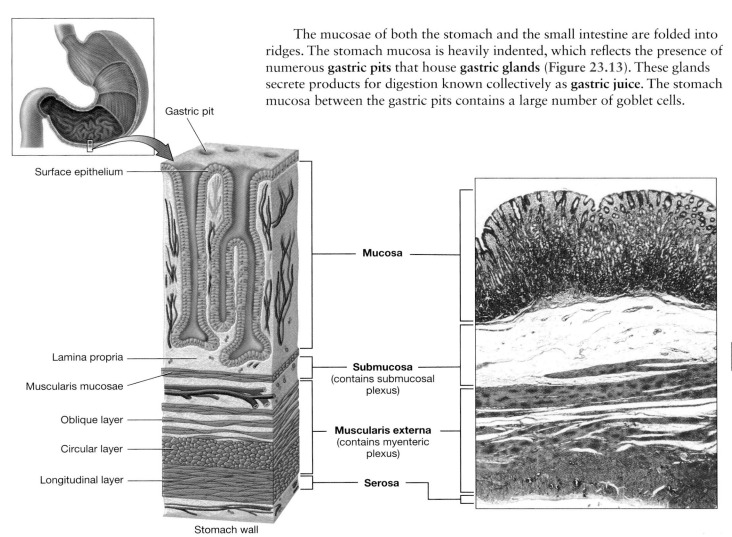

Gastric pit

Surface epithelium

Lamina propria

Muscularis mucosae

Oblique layer

Circular layer

Longitudinal layer

Stomach wall

**Mucosa**

**Submucosa**
(contains submucosal plexus)

**Muscularis externa**
(contains myenteric plexus)

**Serosa**

FIGURE **23.13** The stomach and gastric glands

The small intestinal mucosae are also highly folded, but in a different manner and with a different purpose: They all increase the surface area available for absorption. The three sets of folds in the small intestine, shown in **Figure 23.14**, include the:

1. **Plicae circulares.** The plicae circulares, also known as *circular folds*, are folds of the mucosa and submucosa.

FIGURE **23.14** The small intestine:
(**A**) folds of the small intestine;
(**B**) duodenum, photomicrograph

23

2. **Villi.** The fingerlike villi are the "fringe" on the plicae and are composed of folds of mucosa. Note in **Figure 23.14** that between the villi are glands called **intestinal crypts**. Each villus is lined with intestinal cells called **enterocytes** that surround a core containing blood vessels and a lymphatic vessel called a **lacteal**.

3. **Microvilli.** Microvilli are folds of the enterocytes' plasma membranes.

The two accessory digestive organs we will examine in this exercise—the pancreas and the liver—are not hollow organs, so they do not follow the same histological pattern as the alimentary canal. The pancreas is a gland that consists of clusters of cells that produce both exocrine and endocrine secretions (**Figure 23.15**). The exocrine cells of the pancreas, called **acinar cells**, produce and secrete pancreatic juice (digestive enzymes, bicarbonate, and water) into ducts. The endocrine cells of the pancreas, called **pancreatic islets** or **islets of Langerhans**, secrete hormones such as insulin into the bloodstream. They are visible as small, circular groups of cells that stain less darkly than the surrounding acinar cells. We will reexamine these groups of cells in Unit 26.

The liver consists of hepatocytes organized into hexagonal plates of cells called **liver lobules** (**Figure 23.16**). At the center of each lobule is a central vein that will eventually drain into the hepatic veins. At each of the six corners of a liver lobule we find three small vessels collectively called **portal triads**. The three vessels are: (1) a **bile duct**, which carries bile made by hepatocytes and drains into the hepatic duct, (2) a **portal vein**, a tiny branch off the hepatic portal vein that delivers nutrient-rich blood to the liver for processing and detoxification, and (3) a **hepatic artery** (arteriole), which delivers oxygen-rich blood to the hepatocytes.

FIGURE **23.15** The pancreas, photomicrograph

23

Sinusoids

Interlobular veins
(to hepatic vein)

Plates of
hepatocytes

**A**

Hepatic
macrophages
in sinusoid walls

Blood flow
from portal vein

Bile canaliculi

Bile duct
(receives bile from
bile canaliculi)

Portal triad:

Bile duct

Portal venule

Hepatic arteriole

Central vein

Liver lobule

Central vein

FIGURE **23.16** Microscopic structure of the liver: (**A**) liver lobule; (**B**) liver, photomicrograph

## Procedure Model Inventory for the Histologic Structures of the Digestive System

Identify on models and diagrams the following histologic structures of the digestive system, using your textbook and this unit for reference. As you examine the anatomical models and diagrams, record the name of the model and the structures you were able to identify on the model inventory in **Table 23.2** (p. 563—note that this is the model inventory in Exercise 23-1).

1. Mucosa
   a. Epithelial tissue (simple columnar epithelium)
   b. Glands
   c. MALT
   d. Muscularis mucosae

2. Submucosa
   a. Blood vessels
   b. Nerves
   c. Lymphatic vessels
3. Muscularis externa
   a. Inner circular layer
   b. Outer longitudinal layer

4. Serosa
5. Intestinal villi
   a. Lacteal
   b. Enterocytes
   c. Microvilli

## Procedure Microscopy of Digestive Organs

Obtain prepared slides of the esophagus, the fundus of the stomach, the duodenum, the pancreas, and the liver. Examine the slides initially with the low-power objective, and advance to higher-powered objectives to see more details. Use your colored pencils to draw what you see, and label your drawing with the following terms. When you have completed the activity, answer Check Your Understanding questions 2–4 (pp. 583–584).

### Esophagus

1. Stratified squamous epithelium
2. Submucosa with blood vessels
3. Muscularis externa
   a. Smooth muscle
   b. Skeletal muscle
4. Serosa

### Fundus of the Stomach

1. Mucosa with goblet cells and gastric glands
2. Submucosa with blood vessels
3. Muscularis externa with smooth muscle layers
4. Serosa

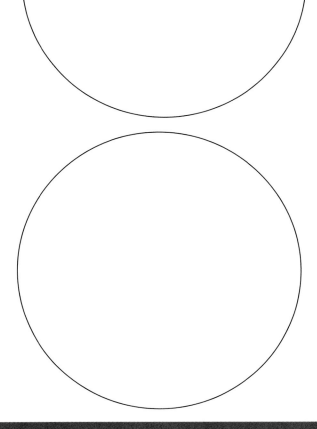

23

## Duodenum

1. Villi
2. Mucosal-associated lymphoid tissue
3. Microvilli (visible as small "hairs" projecting from the cells); may only be visible on oil immersion
4. Intestinal glands

## Pancreas

1. Acinar cells
2. Pancreatic ducts
3. Pancreatic islets (islets of Langerhans)

## Liver

1. Liver lobule
2. Portal triad
3. Central vein

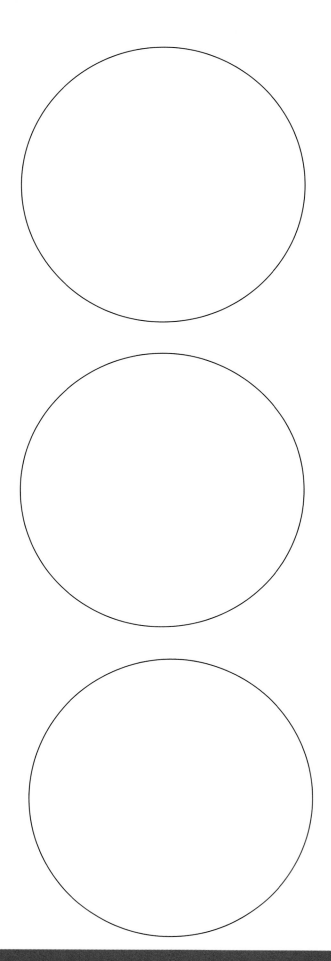

23

# Exercise 23-3

## Digestion

### MATERIALS

- ❏ Graduated cylinders
- ❏ Starch solution
- ❏ Amylase enzyme
- ❏ Distilled water
- ❏ Water bath set to 37°C
- ❏ Glass stirring rod
- ❏ Lugol's iodine solution
- ❏ Benedict's reagent
- ❏ Boiling water bath
- ❏ Small glass test tubes
- ❏ Egg white
- ❏ Pepsinogen in KCI buffer
- ❏ 0.5M hydrochloric acid
- ❏ Biuret reagent
- ❏ 15 glass test tubes
- ❏ Test tube rack
- ❏ Vegetable oil
- ❏ Rubber stopper
- ❏ Sudan red stain
- ❏ Liquid detergent
- ❏ Phenol red
- ❏ 0.1M sodium hydroxide
- ❏ Lipase
- ❏ Bile

We now turn our attention to the physiology of food breakdown and absorption. Three major types of nutrients are broken down and absorbed in the alimentary canal: *carbohydrates*, *proteins*, and *lipids*. Carbohydrate digestion begins in the mouth with the enzyme **salivary amylase**, which catalyzes the reactions that digest polysaccharides into smaller oligosaccharides. The remaining polysaccharides and oligosaccharides are digested in the small intestine with the help of many enzymes, including pancreatic amylase and enzymes associated with the enterocytes of the small intestine called **brush border enzymes** (including lactase, maltase, and sucrase).

Protein digestion begins in the stomach with the enzyme **pepsin**, secreted from chief cells in the gastric glands as the inactive pre-enzyme **pepsinogen**. Pepsinogen becomes the active enzyme pepsin when it encounters an acidic environment. Pepsin begins digesting proteins into polypeptides and some free amino acids. The remainder of protein digestion occurs in the small intestine with pancreatic enzymes such as **trypsin** and **chymotrypsin** and also brush border enzymes such as **carboxypeptidase** and **aminopeptidase**.

Lipid digestion does not begin until the lipids reach the small intestine. It is more complicated than protein or carbohydrate digestion because lipids are nonpolar molecules that do not dissolve in the water-based environment of the small intestine. This causes the lipids to clump together and form large fat "globules" in the small intestine. In this form, it is nearly impossible for the enzyme **pancreatic lipase** to catalyze the breakdown of lipids into monoglycerides and free fatty acids because it has only a small surface area on which to work. Therefore, the first step in lipid digestion is to break up this large fat globule into smaller pieces, a process called **emulsification** (**Figure 23.17**).

Emulsification is accomplished by chemicals called **bile salts**, produced by the liver and stored in the gallbladder. Bile salts are amphipathic molecules that have both nonpolar and polar parts. Notice in **Figure 23.17** what happens to the fat globule when bile salts come into contact with it: The bile salts' nonpolar parts interact with the lipids, but their polar parts repel other lipids. The overall effect is that the fat globule is physically broken down into smaller fat pieces. This gives pancreatic lipase much more surface area on which to work, and the fats are digested into monoglycerides and free fatty acids.

Once chemical digestion is complete, all three nutrients are absorbed through the plasma membranes of the enterocytes lining the intestinal villi. Both monosaccharides and amino acids are absorbed into the enterocytes via facilitated diffusion or secondary active transport mechanisms. These nutrients then exit the other side of the enterocytes and enter the capillaries in the core of the villus. Remember from Exercise 23-1 that all blood from the small intestine and other abdominal organs is delivered to the hepatic portal system before being delivered to the general circulation. Therefore, both monosaccharides and amino acids travel first to the liver via the hepatic portal vein, after which they enter the blood in the hepatic veins and the inferior vena cava.

Lipid absorption, like lipid digestion, is more complicated because lipids are nonpolar and cannot be transported as free fatty acids through the water-filled plasma of the blood. In the small intestinal lumen, digested lipids remain associated with bile salts to form clusters of lipids and other nonpolar chemicals called **micelles**. Micelles escort the lipids to the enterocytes' plasma membranes; they are able to do so because of their polar outer surface, which allows them to move freely between the enterocytes' microvilli. The lipids then leave the micelles and enter the enterocytes by simple diffusion. Note that in the absence of bile salts, nearly all of the lipids will simply pass through the small intestine unabsorbed and be excreted in the feces.

23

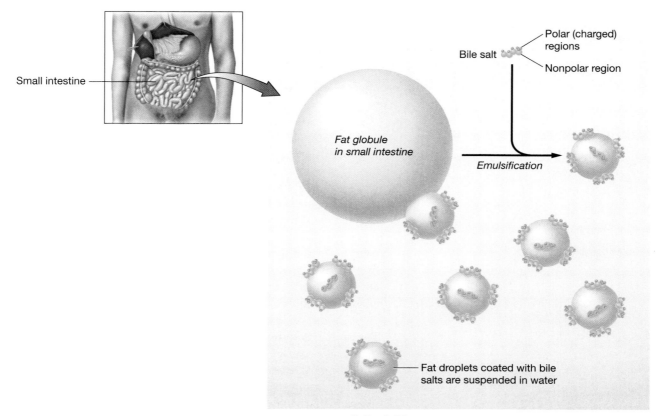

Small intestine

Polar (charged) regions

Bile salt

Nonpolar region

Fat globule in small intestine

Emulsification

Fat droplets coated with bile salts are suspended in water

FIGURE **23.17** Emulsification of fats by bile salts

In the enterocytes, lipids associate with other nonpolar substances and proteins to form protein-coated droplets called **chylomicrons**. Chylomicrons exit the enterocytes into the core of the villus by exocytosis, but they are too large to enter the blood capillaries. Instead, they enter the lacteal and join the lymph in the lymphatic system. For this reason, lipids are not delivered to the hepatic portal system like carbohydrates and amino acids. The chylomicrons travel within the lymphatic vessels until they enter the blood at the junction of the left subclavian vein and the left internal jugular vein.

## Procedure Test Carbohydrate Digestion

In the following procedure you will test for carbohydrate digestion using a starch solution and the enzyme amylase. You will check for digestion using two reagents: Lugol's iodine, which turns dark blue in the presence of starch, and Benedict's reagent, which forms a precipitate (solid) in the presence of simple sugars. You may interpret the results of your tests as follows:

■ If the solution turns dark blue with Lugol's iodine, *no* or limited carbohydrate digestion occurred because starch is still present.

■ If the solution develops a greenish-brownish red precipitate with Benedict's reagent, monosaccharides are present, and carbohydrate digestion *did* occur.

**Safety Note**

**Safety glasses and gloves are required!**

**1** Obtain six glass test tubes, and number them 1 through 6.

**2** Place 3 ml of starch solution into tube 1 and tube 2.

**3** Add 3 ml of amylase to tube 1.

**4** Add 3 ml of distilled water to tube 2.

**5** Set tubes 1 and 2 into a 37°C water bath, and leave them in there for 30 minutes.

**6** Remove the tubes from the water bath, and mix the contents of each tube thoroughly with stirring rods (be certain to use two separate rods so you don't contaminate the contents of the tubes).

**7** Divide the mixture in tube 1 equally into tubes 3 and 4 (each tube should contain about 3 ml of the mixture).

**8** Divide the mixture in tube 2 equally into tubes 5 and 6 (each tube should contain about 3 ml of the mixture).

**9** Add two drops of Lugol's iodine solution to tube 3 and tube 5. Record the results in **Table 23.3**.

**10** Add 10 drops of Benedict's reagent to tube 4 and tube 6. Swirl the tubes gently to mix the contents.

**11** Place tubes 4 and 6 into a boiling water bath for 3 minutes. *Use caution* to prevent the mixture from splattering and burning you or your classmates.

**12** Remove the tubes from the boiling water, and record the results of the test in **Table 23.3**.

TABLE **23.3** Results of Carbohydrate Digestion Experiment

| Tube | Reaction with Lugol's Iodine |
|------|------------------------------|
| Tube 3 (starch + amylase) | |
| Tube 5 (starch + water) | |

| Tube | Reaction with Benedict's Reagent |
|------|----------------------------------|
| Tube 4 (starch + amylase) | |
| Tube 6 (starch + water) | |

**13** Interpret your results:

   **a** In which tube(s) did carbohydrate digestion occur? How do you know?

_____

_____

_____

   **b** In which tube(s) did no carbohydrate digestion occur? How do you know?

_____

_____

_____

   **c** What conclusions can we draw?

_____

_____

_____

## Procedure Test Protein Digestion

Now you will examine protein digestion with the enzyme pepsin. Recall that pepsin is produced by the chief cells as the inactive enzyme *pepsinogen*, and pepsinogen requires certain conditions to become active. You will assess for protein digestion with the chemical Biuret reagent, which turns purple-blue in the presence of undigested proteins and lavender-pink in the presence of amino acids. When you have completed the activity, answer Check Your Understanding question 5 (p. 584).

**1** Obtain four test tubes, and number them 1 through 4.

**2** Add about 3 ml of egg white (or albumin) to each test tube.

**3** Add 3 ml of pepsinogen in 0.1M NaOH (pH about 8) to tube 1.

**4** Add 3 ml of pepsinogen in 0.5M HCl (pH about 2–3) to tube 2.

**5** Add 3 ml of 0.1M NaOH to tube 3.

**6** Add 3 ml of 0.5M HCl to tube 4.

**7** Place each tube in a 37°C water bath for one hour.

**8** Remove the tubes and add five drops of Biuret reagent to each tube. Record the color change you observe in each tube in Table 23.4.

**Safety Note**
⚠ Safety glasses and gloves are required!

TABLE **23.4** Results of Protein Digestion Experiment

| Tube | Color of Mixture after Adding Biuret Reagent |
|---|---|
| Tube 1 (egg white + pepsinogen + NaOH) | |
| Tube 2 (egg white + pepsinogen + HCl) | |
| Tube 3 (egg white + NaOH) | |
| Tube 4 (egg white + HCl) | |

**9** Interpret your results:

**a** In which tube(s) did protein digestion occur? How do you know?

_____

_____

_____

**b** In which tube(s) did no protein digestion occur? How do you know?

_____

_____

_____

**c** What effect does pH have on protein digestion with pepsin? Why?

_____

_____

_____

**d** What other effects does acid have in the stomach?

_____

_____

_____

## Procedure Demonstrate Lipid Emulsification

Before we examine the chemical digestion of lipids, let's examine what happens during the process of emulsification. In this exercise you will use four compounds to observe emulsification in action:

■ *Lipids*: The source of lipids for this exercise is vegetable oil.

■ *Emulsifying agent*: Detergents are considered emulsifiers because they have both polar parts and nonpolar parts, similar to bile salts. The emulsifying agent in this exercise, therefore, will be a liquid detergent.

■ *Distilled water.*

■ *Sudan red stain*: This is a stain that binds only to lipids. The sole purpose of this stain is to make the lipids more visible during the procedure.

When you have completed the activity, answer Check Your Understanding questions 6 and 7 (p. 584).

**Safety Note**

⚠ Safety glasses and gloves are required!

**1** Obtain a glass test tube, and add approximately 2 ml of distilled water to it.

**2** Add about 2 ml of vegetable oil to the water.

**3** Place a rubber stopper into the tube, and shake it vigorously for 15 seconds. Allow it to stand for 2 minutes. What happens to the oil and water?

_____

_____

**4** Add three or four drops of Sudan red stain, and shake the tube again for 15 seconds. What color is the oil? What color is the water?

_____

_____

**5** Add about 1 ml of liquid detergent to the mixture, and shake the tube vigorously for 15 seconds. Allow the tube to stand for 2 minutes. What has happened to the solution? Is it still two distinct colors? Explain your results.

_____

_____

_____

_____

_____

_____

_____

23

## Procedure Test Lipid Digestion

In the following procedure you will be comparing the ability of four solutions to digest vegetable oil: lipase and bile, lipase alone, bile alone, and water. You will check for the presence of digestion using an indicator called **phenol red**. Phenol red appears pink at an alkaline (basic) pH, changes to an orange-red color at a neutral pH, and changes to a yellow color when the pH becomes acidic. If fat digestion has occurred, fatty acids will be released that will decrease the pH of the contents of your tube and turn them orange-red or yellow. You may interpret your results in the following way:

Pink color = the pH is basic, and no (or limited) fat digestion occurred

Red-orange color = the pH is neutral, and some fat digestion occurred

Yellow color = the pH is acidic, and significant fat digestion occurred

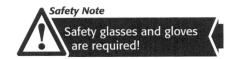
**Safety Note**
Safety glasses and gloves are required!

**1** Obtain four glass test tubes, and number them 1 through 4.

**2** Add 3 ml of vegetable oil to each test tube.

**3** Add 8–10 drops of the pH indicator phenol red to each test tube. The oil now should appear pink. If it does not appear pink, add drops of 0.1M NaOH to each tube until the indicator turns pink.

**4** Add 1 ml of lipase and 1 ml of bile to tube 1.

**5** Add 1 ml of lipase to tube 2.

**6** Add 1 ml of bile to tube 3.

**7** Add 1 ml of distilled water to tube 4.

**8** Place each tube in a 37°C water bath for 30 minutes.

**9** Remove the tubes from the water bath, and record the color of each tube in **Table 23.5**.

TABLE **23.5** Results of Lipid Digestion Experiment

| Tube | Color | pH (acidic, neutral, or alkaline) |
|---|---|---|
| Tube 1 (oil + lipase + bile) | | |
| Tube 2 (oil + lipase) | | |
| Tube 3 (oil + bile) | | |
| Tube 4 (oil + water) | | |

**10** Interpret your results:

**a** In which tube(s) did lipid digestion occur? How do you know?

23

**b** In which tube(s) did limited or no lipid digestion occur? How do you know?

_____

_____

_____

**c** What effect does bile have on lipid digestion? Why?

_____

_____

_____

23

# Exercise 23-4

## Time to Trace!

### MATERIALS
☐ Laminated outline of the human body
☐ Water-soluble marking pens

Now it's time to put all of the digestive anatomy and physiology together to get a "big picture" view of the digestive system. In this exercise you will trace the pathways three different nutrients take from their ingestion at the mouth to their arrival at the heart. You will trace a cookie (primarily carbohydrates), an egg (primarily protein), and greasy fried food (primarily lipids).

Along the way, detail the following for each:

1. The *anatomical pathway* each takes from ingestion through its passage through the alimentary canal, to its absorption into the blood, and finally to its passage through the blood until it reaches the heart.

2. The *physical* and *chemical processes* that break down each substance, including enzymatic breakdown, churning, chewing, and emulsification.

*Some hints*:

▌ Don't forget that carbohydrates and proteins travel through the hepatic portal system before they enter the general circulation.

▌ Remember that digestion and absorption are quite different for lipids. For example, fats are not absorbed into the same structures as proteins and carbohydrates.

▌ Use the text in Exercise 23-3 (p. 570) and your list of enzymes that you completed in Pre-Lab Exercise 23-3 (p. 556) for reference.

▌ Refer to the tracing exercises from Unit 17 (p. 411) and Unit 20 (p. 487) to review the pathway of blood and lymph flow through the body.

## Tracing Steps

**1** Cookie: **Start:** mouth → _____

_____

_____

_____

_____

_____

_____

_____

_____

_____ heart **End**

**2** Egg: **Start:** mouth →  _____

_____

_____

_____

_____

_____

_____

_____

_____ heart **End**


**3** Greasy fried food: **Start:** mouth →  _____

_____

_____

_____

_____

_____

_____

_____

_____ heart **End**

23

UNIT
**23**
REVIEW

## Check Your Recall

**1** Label the following structures on **Figure 23.18**.

- ☐ Cecum
- ☐ Esophagus
- ☐ Gallbladder
- ☐ Liver
- ☐ Parotid gland
- ☐ Sigmoid colon
- ☐ Sublingual gland
- ☐ Submandibular gland
- ☐ Transverse colon

FIGURE **23.18** The digestive system

23

**2** Label the following structures on **Figure 23.19.**

- ☐ Body
- ☐ Cardia
- ☐ Circular muscle layer
- ☐ Duodenum
- ☐ Esophagus
- ☐ Fundus
- ☐ Gastroesophageal sphincter
- ☐ Pyloric sphincter
- ☐ Pylorus
- ☐ Rugae

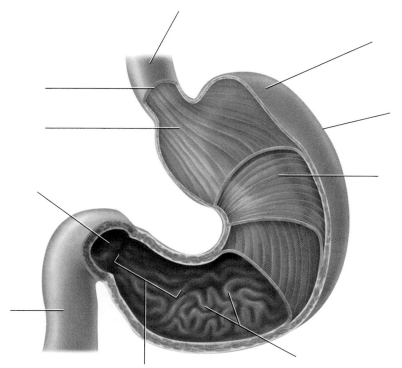

FIGURE **23.19** The stomach

**3** Label the following structures on **Figure 23.20.**

- ☐ Common bile duct
- ☐ Common hepatic duct
- ☐ Cystic duct
- ☐ Duodenum
- ☐ Gallbladder
- ☐ Hepatopancreatic ampulla
- ☐ Liver
- ☐ Pancreas
- ☐ Pancreatic duct

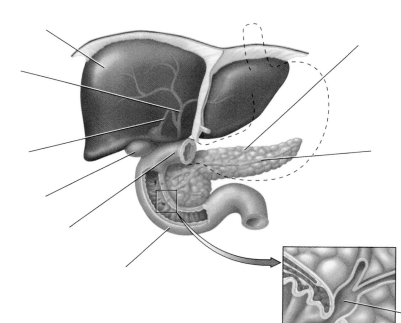

FIGURE **23.20** The liver, gallbladder, pancreas, and duodenum

23

**4** Which of the following organs is *not* part of the alimentary canal?

a. Esophagus

b. Gallbladder

c. Cecum

d. Ileum

**5** *Mark the following statements as true (T) or false (F).* If the statement is false, correct it so it becomes a true statement.

_____ a. The peritoneal cavity is located between the visceral peritoneum and the mesentery.

_____ b. The longest segment of the small intestine is the duodenum.

_____ c. The stomach has three layers of smooth muscle that contract to churn food into chyme.

_____ d. The gallbladder produces and stores bile.

_____ e. The small intestine features three sets of progressively smaller folds that increase surface area for absorption.

_____ f. The liver consists of plates of hexagonal liver lobules.

**6** Label the following structures on **Figure 23.21**.

☐ Circular layer of smooth muscle ☐ Mucosa ☐ Serosa
☐ Longitudinal layer of smooth muscle ☐ Muscularis mucosae ☐ Submucosa
☐ Lumen

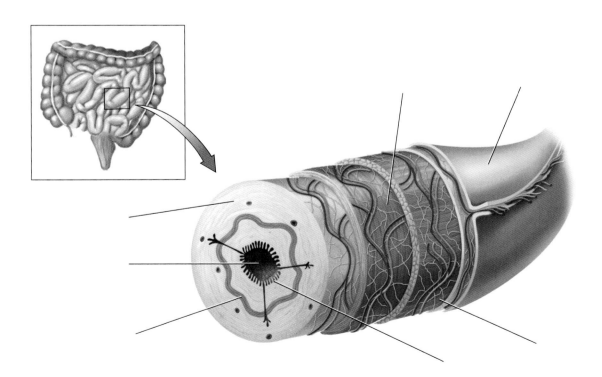

FIGURE **23.21** The tissue layers of the alimentary canal

23

**7** *Fill in the blanks:* The exocrine cells of the pancreas are called _____ and secrete _____.

The endocrine cells of the pancreas are called _____ and secrete _____.

**8** *Matching:* Match the following with the correct definition.

_____ Salivary amylase

_____ Bile salts

_____ Micelles

_____ Lacteal

_____ Brush border enzymes

_____ Pancreatic lipase

_____ Hydrochloric acid

_____ Trypsin

_____ Chylomicrons

_____ Pepsin

A. Protein-coated lipid droplets that are absorbed
B. Digestive enzyme(s) associated with the enterocytes of the small intestine
C. Emulsifies/emulsify fats
D. Protein-digesting enzyme(s) produced by the pancreas
E. Required to activate pepsinogen
F. Begin(s) carbohydrate digestion in the mouth
G. The structure(s) into which lipids are absorbed
H. Protein-digesting enzyme(s) produced by the stomach
I. Clusters of bile salts and digested lipids
J. Enzyme(s) that digest(s) lipids into free fatty acids and monoglycerides

**9** Why is the digestion and absorption of lipids more complex than the digestion and absorption of carbohydrates and proteins?

_____

_____

_____

_____

_____

_____

_____

**10** How does the absorption of lipids differ from the absorption of carbohydrates and proteins?

_____

**23**

_____

_____

_____

_____

_____

_____

# Check Your Understanding

### Critical Thinking and Application Questions

**1** *Peritonitis* is an infection of the fluid in the peritoneal cavity. Why do you think this condition might be difficult to treat? Why would this condition have wide-ranging effects on the abdominal organs?

_____

_____

_____

_____

_____

_____

**2** Explain how the structures of the epithelium and muscularis of the small intestine, stomach, and esophagus follow their functions.

_____

_____

_____

_____

_____

_____

**3** The condition known as heartburn is most often caused by acid regurgitating from the stomach into the esophagus. Why do you think the acid tends to burn the esophagus and produce pain but does not similarly burn the stomach?

23

_____

_____

_____

_____

_____

**4** If you go into your local grocery store, you are bound to find a host of gluten-free products made with synthetic proteins instead of the natural wheat-based protein gluten. About 1–3% of the population is sensitive to gluten. The most severe form of gluten intolerance is the autoimmune disease *celiac disease* (note that it is not caused by gluten, and you cannot get it from eating gluten). In individuals with celiac disease, the immune response to gluten antigens causes the flattening of the villi in the small intestine. How would this affect the ability of the small intestine to function? Predict the symptoms you would expect to see from celiac disease.

_____

_____

_____

_____

_____

**5** Many dietary supplements contain digestive enzymes the manufacturers claim are necessary to digest food properly. What will happen to these enzymes in the stomach? (*Hint*: Enzymes are proteins.) Will the enzymes continue to function once they have reached the small intestine? Why or why not?

_____

_____

_____

_____

_____

**6** One of the common consequences of gallstones is blockage of the common bile duct, which prevents bile from being emptied into the duodenum. Predict the possible consequences of this condition.

_____

_____

_____

_____

_____

**7** Explain why detergents, which have an amphipathic structure like bile salts, are able to remove both water-based and oil-based stains.

_____

_____

_____

_____

_____

23

# Urinary System Anatomy

# 24

## OBJECTIVES

*Once you have completed this unit, you should be able to:*

- Identify gross structures of the urinary system.

- Identify structures of the nephron.

- Identify histological structures of the urinary system.

Name _____ Section _____ Date _____

# PRE-LAB EXERCISES

Complete the following exercises prior to coming to lab, using your textbook and lab manual for reference.

## Pre-Lab Exercise 24-1

✎ **Key Terms**

You should be familiar with the following terms before coming to lab.

| Term | Definition |
|------|------------|

### Gross Structures of the Kidney

Renal cortex _____

_____

Renal medulla _____

_____

Medullary pyramid _____

_____

Renal column _____

_____

Renal pelvis _____

_____

Major and minor calyces _____

_____

### Other Structures of the Urinary System

Ureter _____

_____

Urinary bladder _____

_____

Urethra _____

_____

### Blood Vessels and Tubules of the Nephron

Nephron _____

_____

Glomerulus _____

_____

Peritubular capillaries _____

_____

Proximal tubule _____

_____

Nephron loop (loop of Henle) _____

_____

Distal tubule _____

_____

Collecting duct _____

_____

# Pre-Lab Exercise **24-2**

## Structures of the Urinary System

Label and color the structures of the urinary system depicted in **Figures 24.1** and **24.2** with the terms from Exercise 24-1 (p. 593). Use your text and Exercise 24-1 in this unit for reference.

FIGURE **24.1** Right kidney, frontal section

FIGURE **24.2** Organs of the urinary system: **(A)** anterior view; **(B)** posterior view

**24**

Label and color the structures of the nephron depicted in **Figure 24.3** with the terms from Exercise 24-1 (p. 593). Use your text and Exercise 24-1 in this unit for reference.

FIGURE **24.3** The nephron

# EXERCISES

The urinary system performs many functions critical to homeostasis, including removing waste products from the blood; regulating the body's fluid, electrolyte, and acid-base balance; and producing the hormone erythropoietin, which regulates blood cell formation. In addition to these roles, the urinary system helps the liver detoxify certain compounds and makes glucose during times of starvation. These critical functions are carried out primarily by the main organs of the urinary system—the kidneys. The other organs in the urinary system include the ureters, the urinary bladder, and the urethra. In this unit you will become acquainted with the anatomy and the histology of the urinary system. In Unit 25 you will use this knowledge to study urinary system physiology.

## Exercise **24-1**

### Urinary System Anatomy

#### MATERIALS
- ❏ Urinary system models
- ❏ Kidney models
- ❏ Nephron models
- ❏ Preserved kidney
- ❏ Dissection equipment
- ❏ Dissecting tray

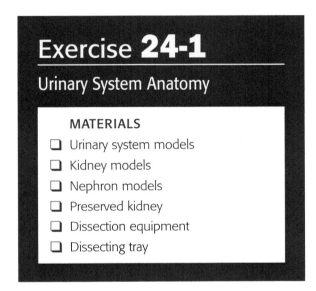

Renal column

Papilla of pyramid

Major calyx

Renal medulla

Minor calyx

Fibrous capsule

Renal artery

Renal vein

Renal pelvis

Ureter

Renal pyramid of medulla

Renal cortex

FIGURE **24.4** Right kidney, frontal section

The paired **kidneys** are situated against the posterior body wall posterior to the peritoneal membranes (i.e., they are *retroperitoneal*) that surround many of the digestive organs. They are encased within three layers of connective tissue, the thickest of which is a middle layer of adipose tissue. The innermost layer of connective tissue encases each kidney like plastic wrap and is called the **renal capsule**.

Internally, each kidney has three distinct regions that can be seen in a frontal section (**Figure 24.4**):

1. **Renal cortex.** The most superficial region is known as the renal cortex. It is dark brown because it consists of many blood vessels that serve the tiny blood-filtering structures of the kidney, the **nephrons**.

2. **Renal medulla.** The kidney's middle region is known as the medulla and consists of triangular structures known as **renal** or **medullary pyramids**. The medullary pyramids are separated from one another by inward extensions of the renal cortex called **renal columns**. Like the renal cortex, the renal columns contain many blood vessels. Each pyramid contains looping tubules of the nephron as well as structures that drain fluid from the nephron. These tubes give the pyramids a striped (or *striated*) appearance.

3. **Renal pelvis.** The tubes that drain the fluid from the nephron drain into larger tubes called **minor calyces** that in turn drain into even larger **major calyces**. The major calyces drain into the kidney's innermost region, called the renal pelvis, that serves as a basin for collecting urine.

The blood flow through the kidney follows a unique pattern that allows it to carry out its function of maintaining the homeostasis of the blood (**Figure 24.5**). The large **renal arteries** deliver about 1,200 milliliters of blood per minute to the kidney to be filtered. The renal arteries branch into progressively smaller arteries as they pass through the medulla to the cortex, including the **segmental arteries** in the renal pelvis, the **interlobar arteries** between the medullary pyramids, the **arcuate arteries** that curve around the top of the pyramids, and finally the small **interlobular arteries** in the renal cortex. The interlobular arteries branch into tiny **afferent arterioles**, each of which supplies a ball of capillaries known as the **glomerulus**. This is where the blood is filtered. Note that the glomerulus is not the primary site for gas and nutrient exchange for the tissues of the kidneys.

You learned in Unit 17 that a capillary bed generally drains into a venule, but note in **Figure 24.5** that the capillaries of the glomerulus drain into a second *arteriole* called the **efferent arteriole**. The efferent arteriole then branches to form a second capillary bed known as the **peritubular capillaries**. These capillaries surround the tubules of the nephron, where they provide them with oxygen and nutrients and also take substances reabsorbed by the tubules back into the blood. In some nephrons, a second set of capillaries called the **vasa recta** is found surrounding the nephron loop. This second capillary bed then drains out through the small **interlobular veins**, which drain into **arcuate veins**, then into **interlobar veins**, and finally into the large **renal vein**. This pattern of blood flow allows the kidneys both to filter blood and to reclaim most of the fluid and solutes filtered.

Microscopically, each kidney is composed of more than a million tiny units called **nephrons** (**Figure 24.6**). Blood first enters the high-pressure circuit of the glomerulus, where fluid and small solutes are forced out of the capillaries and into a space called the **capsular space**. The capsular space is found within a structure known as the **glomerular capsule** or **Bowman's capsule**. Note in **Figure 24.6A** that the glomerular capsule has two layers: (a) the outer **parietal layer**, which is simple squamous epithelium, and (b) the inner **visceral layer**, which consists of cells called **podocytes** that surround the capillaries of the glomerulus. The glomerulus and its surrounding capsule together are known as the **renal corpuscle**.

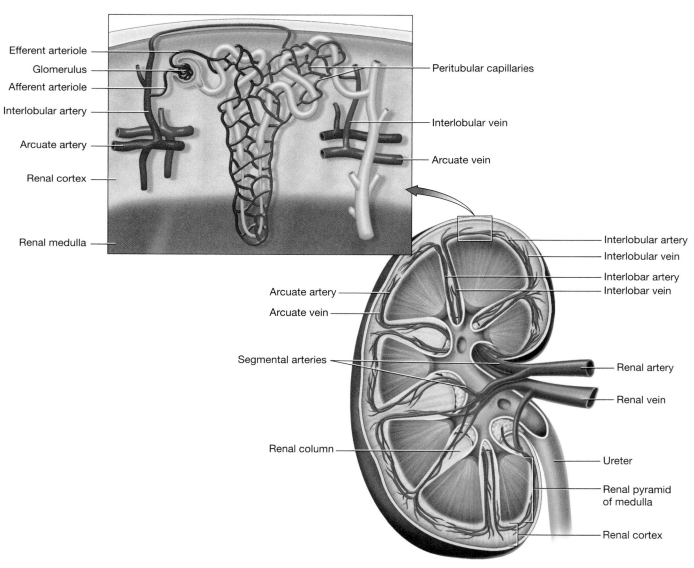

FIGURE **24.5** Pathway of blood flow and microanatomy of the kidney

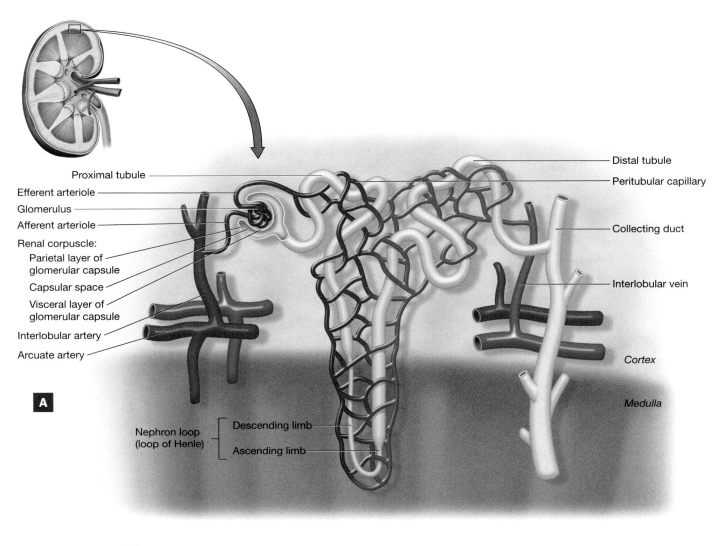

Proximal tubule

Efferent arteriole

Glomerulus

Afferent arteriole

Renal corpuscle:

  Parietal layer of glomerular capsule

  Capsular space

  Visceral layer of glomerular capsule

Interlobular artery

Arcuate artery

Distal tubule

Peritubular capillary

Collecting duct

Interlobular vein

*Cortex*

*Medulla*

Nephron loop (loop of Henle) { Descending limb

Ascending limb

**A**

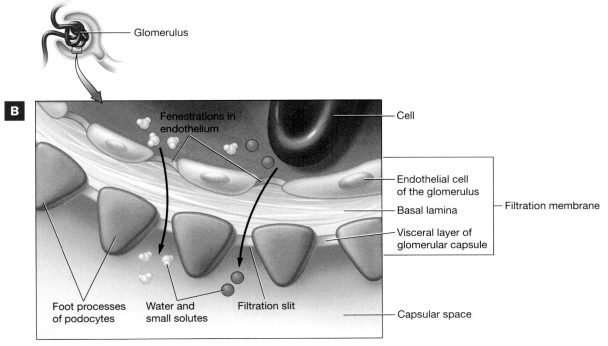

Glomerulus

**B**

Fenestrations in endothelium

Cell

Endothelial cell of the glomerulus

Basal lamina

Filtration membrane

Visceral layer of glomerular capsule

Foot processes of podocytes

Water and small solutes

Filtration slit

Capsular space

24

FIGURE **24.6** Microanatomy of the kidney: (**A**) the nephron; (**B**) the filtration membrane

The podocytes of the visceral layer have extensions called **foot processes** that interlock to form narrow **filtration slits**. As you can see in **Figures 24.6B** and **24.7**, the glomerular endothelial cells, the podocytes, and their shared basal lamina together are called the **filtration membrane**. The filtration membrane prevents large substances in the blood, such as blood cells and most proteins, from exiting the glomerular capillaries.

The fluid in the capsular space, known as **filtrate**, next enters a series of small tubes or *tubules* that can be likened to the "plumbing" of the kidney. These tubes are called collectively the **renal tubule**, which consists of three parts:

1. the **proximal tubule** (sometimes called the *proximal convoluted tubule*),

2. the ascending and descending limbs of the **nephron loop** (also called the **loop of Henle**), and

3. the **distal tubule** (sometimes called the *distal convoluted tubule*).

Several distal tubules drain into one **collecting duct**. Note that the majority of the renal tubule is confined to the renal cortex; only the nephron loops of certain nephrons dip down into the renal medulla.

At the junction between the ascending limb of the nephron loop and the distal tubule we find a group of tall, closely packed cells known as the **macula densa**. The macula densa comes into contact with a portion of the afferent arteriole that contains specialized cells called **juxtaglomerular (JG) cells**. The JG cells and macula densa together are called the **juxtaglomerular apparatus (JGA; Figure 24.7)**. The JGA plays a role in controlling the flow of filtrate through the nephron and the blood pressure within the glomerulus.

Filtrate in the renal tubule and collecting duct is modified, and most of the water and solutes are reclaimed. From the collecting ducts, filtrate drains into larger tubules called **papillary ducts**, where more water is reclaimed. After the fluid leaves the papillary ducts, it is known as **urine** and drains into the minor calyces, major calyces, and finally the renal pelvis. From the renal pelvis, urine enters the next organs of the urinary system—the tubes called the **ureters (Figure 24.8)**. Ureters are lined by a type of epithelium called **transitional epithelium**, and their walls contain smooth muscle that massages the urine inferiorly via peristalsis. The ureter exits the kidney through a slit in its medial surface known as the **hilum**.

The ureters drain urine into the posteroinferior wall of the organ known as the **urinary bladder (Figure 24.9)**. Like the ureters, the urinary bladder is lined with transitional epithelium and contains smooth muscle (sometimes called the **detrusor muscle**) in its wall. The majority of the urinary bladder contains folds called **rugae** that allow it to expand when it is filled with urine. The smooth inferior portion of the urinary bladder wall, a triangular-shaped area known as the **trigone**, contains the opening called the **internal urethral orifice**. This opens into the final organ of the urinary system—the **urethra**. The urethra contains two rings of smooth muscle—the involuntary **internal urethral sphincter** and the voluntary **external urethral sphincter**. When both of these sphincters relax, urine is expelled from the body via a process called **micturition**.

FIGURE **24.7** Juxtaglomerular apparatus

Anterior view

Posterior view

FIGURE **24.8** Organs of the female urinary system: **(A)** anterior view; **(B)** posterior view

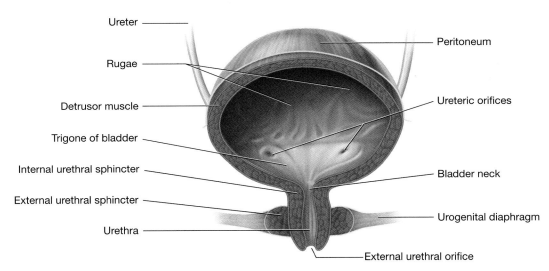

FIGURE **24.9** Female urinary bladder, frontal section

24

Identify the following structures of the urinary system on models and diagrams, using your textbook and this unit for reference. As you examine the anatomical models and diagrams, record the name of the model and the structures you were able to identify on the model inventory in **Table 24.1**. When you have completed the activity, answer Check Your Understanding questions 1–3 (p. 609).

## Kidney Anatomy

1. Surrounding connective tissue:
   a. Renal fascia
   b. Adipose capsule
   c. Renal capsule
2. Hilum
3. Regions:
   a. Renal cortex
   b. Renal medulla
      (1) Renal columns
      (2) Renal pyramids
      (3) Renal papilla
   c. Minor calyces
   d. Major calyces
   e. Renal pelvis

4. Blood supply:
   a. Renal artery
      (1) Segmental artery
      (2) Lobar artery
      (3) Interlobar artery
      (4) Arcuate artery
      (5) Interlobular artery
   b. Renal vein
      (1) Interlobular vein
      (2) Arcuate vein
      (3) Interlobar vein

## Structures of the Nephron

1. Renal corpuscle
   a. Glomerulus
      (1) Afferent arteriole
      (2) Efferent arteriole
      (3) Peritubular capillaries
      (4) Vasa recta
   b. Glomerular (Bowman's) capsule
      (1) Parietal layer
      (2) Visceral layer
      (3) Podocytes
      (4) Capsular space

2. Renal tubule
   a. Proximal tubule
   b. Nephron loop (loop of Henle)
      (1) Descending limb
      (2) Ascending limb
   c. Distal tubule
3. Juxtaglomerular apparatus
   a. JG cells
   b. Macula densa
4. Collecting duct
5. Papillary duct

## Other Urinary Structures

1. Ureter
2. Urinary bladder
   a. Detrusor muscle
   b. Ureteral orifices
   c. Internal urethral orifice
   d. Trigone

3. Urethra
   a. Internal urethral sphincter
   b. External urethral sphincter
   c. External urethral orifice

TABLE **24.1** Model Inventory for Urinary Anatomy

| Model/Diagram | Structures Identified |
|---|---|
|  |  |
|  |  |
|  |  |
|  |  |
|  |  |

## Procedure Kidney Dissection

In this exercise you will identify several of the structures you just identified on models and diagrams by dissecting a preserved kidney.

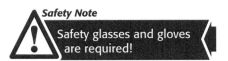

*Safety Note*
Safety glasses and gloves are required!

**1** Obtain a fresh or preserved kidney specimen and dissection supplies.

**2** If the thick surrounding connective tissue coverings are intact, note their thickness and amount of adipose tissue.

**3** Use scissors to cut through the connective tissue coverings, and remove the kidney.

**4** List surface structures you are able to identify (see **Figure 24.10A** for reference):

_____

_____

24

**5** Distinguishing between the ureter, the renal artery, and the renal vein is often difficult. Following are some hints to aid you:

a. The renal artery typically has the thickest and most muscular wall, and it branches into several segmental arteries prior to entering the kidney.

b. The renal vein is thinner, flimsier, and often larger in diameter than the renal artery.

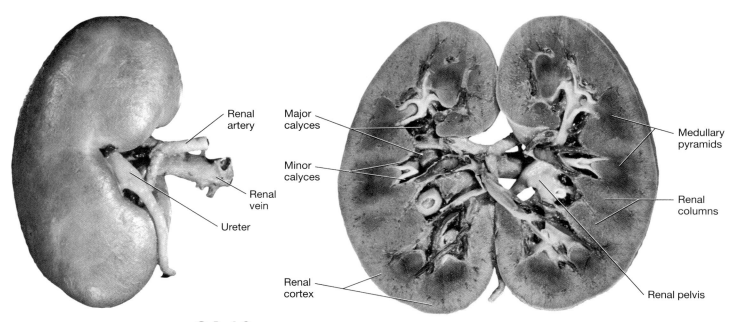

FIGURE **24.10** Preserved kidney: (**A**) anterior view; (**B**) frontal section

Labels: Renal artery, Renal vein, Ureter, Major calyces, Minor calyces, Renal cortex, Medullary pyramids, Renal columns, Renal pelvis

c. The ureter has a thick, muscular wall, too, but it does not branch after it leaves the kidney. Also, its diameter is usually smaller than either the renal artery or the renal vein.

Keeping these points in mind, determine the location of the renal artery, the renal vein, and the ureter on your specimen. Sketch the arrangement of the three structures below:

**24** **6** Use a scalpel to make a frontal section of the kidney. List the internal structures you are able to identify (**Figure 24.10B**):

# Exercise 24-2

## Urinary Organ Histology

### MATERIALS

- ❑ Urinary bladder slide
- ❑ Renal cortex slide
- ❑ Renal medulla slide
- ❑ Light microscope
- ❑ Colored pencils
- ❑ Ureter slide

Like all hollow organs, the ureters and urinary bladder consist of several tissue layers (**Figures 24.11** and **24.12**). The innermost layer is the mucosa, composed of transitional epithelium and a thin layer of loose connective tissue. Recall from Unit 5 that transitional epithelium is stratified epithelium with cells that differ in appearance on the apical and basal sides. The cells at the apical edge are dome-shaped (or sometimes squamous) in appearance, and those nearer the basal edge are typically cuboidal. The apical cells can change shape to accommodate stretching of the urinary bladder.

The middle layer of the ureters and urinary bladder is the submucosa, composed of loose connective tissue and numerous glands that secrete a watery mucus. This mucus plays a critical protective role, because it prevents the chemical in the urine from damaging and ulcerating the epithelium. The third layer is the muscularis, composed of smooth muscle that contracts to propel urine through the urinary tract. The outermost layer is the adventitia, composed of dense irregular connective tissue.

In this exercise you will also examine the tissues in two different regions of the kidney—the renal cortex and the renal medulla. Both regions are composed primarily of blood vessels and tubules made of simple epithelia. The simple epithelia of the renal tubules provide an example of form following function: A main function of these tubules is to reabsorb the water and solutes in the filtrate. Were the tubules made of stratified epithelium, reabsorption would not take place efficiently, and excessive water and solutes would be lost to the urine. Another structural feature that aids in reabsorption is the presence of microvilli on the proximal tubule cells (and to a lesser extent, the distal tubule). Microvilli dramatically increase the surface area available for reabsorption, which ensures that solutes and large volumes of water are reabsorbed rapidly from the filtrate.

Note in **Figure 24.13A** the abundance of small, ball-shaped glomeruli in the renal cortex. The glomeruli are surrounded by a space (the capsular space) lined by the ring of simple squamous epithelial cells that make up the parietal layer of the glomerular capsule. In between the glomeruli, note the presence of nephron tubules (the proximal and distal tubules and sections of nephron loops).

Note in **Figure 24.13C** that the renal medulla is composed mostly of collecting ducts and the nephron loops. You will not see any glomeruli in the medulla, because they are confined to the cortex. This makes the renal cortex and renal medulla easily distinguishable.

FIGURE **24.11** Tissues of the ureter

FIGURE **24.12** Urinary bladder: (**A**) bladder wall; (**B**) close-up of transitional epithelium

FIGURE **24.13** Tissues of the kidney: (**A**) renal cortex; (**B**) glomerulus; (**C**) renal medulla

## **Procedure** Microscopy

Now you will view prepared slides of sections of the ureter, the urinary bladder, the renal cortex, and the renal medulla. Begin your examination of the slides on low power, and advance to medium and high power to observe details. Use colored pencils to draw what you see under the microscope, and label your drawing with the terms indicated. When you have completed the activity, answer Check Your Understanding questions 4–7 (p. 610).

### **Ureter**

Label the following on your drawing:

1. Transitional epithelium
2. Submucosa
3. Muscularis (smooth muscle)
4. Adventitia

### **Transitional Epithelium (Urinary Bladder)**

Label the following on your drawing:

1. Transitional epithelium
2. Submucosa
3. Muscularis (detrusor muscle)

## Renal Cortex

Label the following on your drawing:

1. Glomerulus
2. Glomerular capsule with simple squamous epithelium
3. Capsular space
4. Kidney tubules

## Renal Medulla

Label the following on your drawing:

1. Nephron loop
2. Collecting duct

## Check Your Recall

**1** Label the following parts of the kidney on **Figure 24.14**.

- ☐ Major calyx
- ☐ Minor calyx
- ☐ Renal artery
- ☐ Renal capsule
- ☐ Renal column
- ☐ Renal cortex
- ☐ Renal medulla
- ☐ Renal pelvis
- ☐ Renal pyramid
- ☐ Renal vein
- ☐ Ureter

FIGURE **24.14** Right kidney, frontal section

24

**2** Label the following parts of the nephron on **Figure 24.15**.

☐ Afferent arteriole
☐ Collecting duct
☐ Distal tubule
☐ Efferent arteriole
☐ Glomerular capsule
☐ Glomerulus
☐ Nephron loop
☐ Peritubular capillaries
☐ Proximal tubule

FIGURE **24.15** The nephron

24

**3** Which of the following is *not* one of the urinary system's functions?

    a. Regulating fluid, electrolyte, and acid-base balance

    b. Regulating blood cell formation

    c. Regulating production of insulin and glucagon

    d. Removing waste products from the blood

**4** The blood flow through the kidney is special because

    a. its first capillary beds drain into arterioles.

    b. its second capillary beds drain into arterioles.

    c. it is supplied by three renal arteries.

    d. it contains no capillary beds.

**5** Number the following from the point the filtrate is first formed (with a number 1) to the point it drains into the renal pelvis (with a number 7).

    _____ Minor calyx

    _____ Proximal tubule

    _____ Collecting duct

    _____ Capsular space

    _____ Nephron loop

    _____ Papillary duct

    _____ Distal tubule

**6** *Fill in the blanks:* At the junction between the ascending limb of the nephron loop and the distal tubule, we find a group of densely packed cells called the _____. These cells come into contact with a portion of the afferent arteriole that contains cells called the _____. Together, the two are called the

_____.

**7** Urine drains from the kidneys via the

    a. urinary bladder.

    b. urethra.

    c. ureters.

    d. papillary calyces.

**8** Urine is expelled from the body by a process called

a. micturition.

b. parturition.

c. defecation.

d. procrastination.

**9** The urinary bladder and ureters are lined by

a. simple squamous epithelium.

b. transitional epithelium.

c. pseudostratified columnar epithelium.

d. stratified cuboidal epithelium.

**10** How can one easily discern the renal cortex from the renal medulla on a microscope slide?

_____

_____

UNIT
24
REVIEW

# Check Your Understanding

**Critical Thinking and Application Questions**

**1** Predict the effects renal failure might have on the body's overall homeostasis.

_____

_____

_____

_____

_____

_____

**2** Why do the medullary pyramids appear different from the renal cortex and renal columns? Why do the renal columns and renal cortex appear similar in color and texture?

_____

_____

_____

_____

_____

_____

**3** How is the pattern of blood flow unique in the kidney? How does this pattern of blood flow allow the kidney's form to follow its function?

_____

_____

_____

_____

_____

_____

**4** Why do you think the epithelium of the urinary bladder is stratified? What might happen if the epithelial tissue were simple rather than stratified? (*Hint*: Remember that form follows function.)

_____

_____

_____

_____

_____

_____

**5** The condition *interstitial cystitis* is characterized by insufficient mucus production by the mucosa and submucosa of the urinary bladder. What signs and symptoms would you expect to see with interstitial cystitis? Why?

_____

_____

_____

_____

_____

_____

**6** How does the form of the simple epithelial tissues that make up the renal tubules follow its function?

_____

_____

_____

_____

_____

_____

**7** Predict how the functioning of the kidney might be affected if the microvilli in the proximal and distal tubules were destroyed. Explain your reasoning.

_____

_____

_____

_____

# Urinary
# System
# Physiology

## OBJECTIVES

*Once you have completed this unit, you should be able to:*

■ Model the physiology of the kidney, and test for chemicals in the filtrate.

■ Perform and interpret urinalysis on simulated urine specimens.

■ Research the possible causes of abnormalities detected on urinalysis.

■ Trace an erythrocyte, a glucose molecule, and a urea molecule through the gross and microscopic anatomy of the kidney.

# PRE-LAB EXERCISES

Complete the following exercises prior to coming to lab, using your textbook and lab manual for reference.

## Pre-Lab Exercise 25-1

### ✎ Key Terms

You should be familiar with the following terms before coming to lab.

| Term | Definition |
|------|------------|
| Glomerular filtration | |
| Glomerular filtration rate (GFR) | |
| Tubular reabsorption | |
| Filtration membrane | |
| Filtrate | |
| Tubular secretion | |
| Urine | |
| Urinalysis | |
| Specific gravity | |
| Urobilinogen | |
| Ketones | |

**25**

## Pre-Lab Exercise **25-2**

Nephron Structure and Function

Review the structure of the nephron by labeling and color-coding the diagram in **Figure 25.1**. Then use your textbook to determine which substances are absorbed or secreted from each region of the nephron, and enter this information in the blank boxes to the side of the tubules. Place substances secreted into the urine into the boxes whose arrows are pointing toward the tubules. Place substances reabsorbed into the bloodstream into boxes whose arrows are pointing away from the tubules.

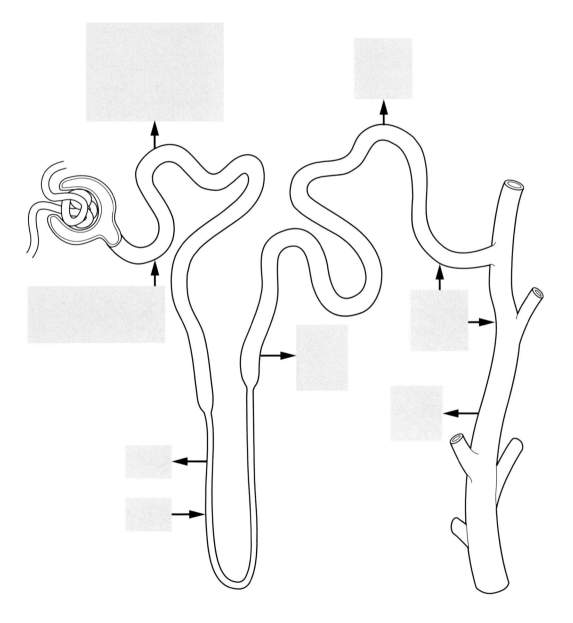

FIGURE **25.1** The nephron

# Pre-Lab Exercise 25-3

## Glomerular Filtration and Tubular Reabsorption

Table 25.1 lists chemicals and cells found in blood. Do the following for each chemical, using your textbook for reference:

**1** Determine whether each listed substance is filtered at the glomerulus (i.e., if it is able to leave the blood and enter the nephron to become part of the filtrate).

**2** If the substance is found in the filtrate, determine if the substance is reabsorbed into the blood or if it is found in the urine.

**3** If the substance is reabsorbed into the blood, determine where this substance is reabsorbed in the nephron.

TABLE **25.1** Substances Found in the Blood and Their Filtration and Reabsorption by the Nephron

| Substance | Filtered in the Glomerulus? (yes/no) | Reabsorbed? (yes/no) | Location Where Reabsorption Takes Place |
|---|---|---|---|
| Erythrocyte | | | |
| Leukocyte | | | |
| Water | | | |
| Glucose | | | |
| Proteins | | | |
| Amino acids | | | |
| Urea | | | |
| Creatinine | | | |
| Electrolytes (sodium, potassium, chloride) | | | |
| Uric acid | | | |
| Ketones | | | |

# EXERCISES

The glomeruli filter the blood delivered to them. Fluid is forced from the glomerular capillaries into the capsular space by a pressure called the net filtration pressure (NFP). The NFP is the sum of three forces:

1. **Glomerular hydrostatic pressure (GHP).** The GHP is equal to the blood pressure in the glomerulus and averages about 50 mmHg. This pressure forces fluid out of the glomerulus and into the capsular space.

2. **Colloid osmotic pressure (COP).** The COP is the pressure created by the presence of proteins and other solutes in the blood and averages about 30 mmHg; this pressure tends to oppose filtration and draw water *into* the glomerular capillaries by osmosis.

3. **Capsular hydrostatic pressure (CHP).** The final pressure, the CHP, is the force of the fluid in the capsular space and averages about 10 mmHg; like the COP, it tends to oppose filtration.

To get the NFP, simply subtract the two forces that oppose filtration from the one force that favors filtration:

$$\text{GHP} - (\text{COP} + \text{CHP}) = \text{NFP}$$
$$50 - (30 + 10) = 10 \text{ mmHg}$$

So, in the glomerulus, there is a net pressure of 10 mmHg that drives fluid out of the glomerulus into the capsular space.

The fluid that results from this filtration process enters the proximal tubule and is called **filtrate**. Filtrate is similar in composition to blood, but it lacks most of the proteins and cells we find in blood. This selectivity is possible because of the **filtration membrane**, which consists of the glomerular endothelial cells, the basal lamina, and specialized cells called **podocytes**. The filtration membrane acts in a similar manner to the filter in your coffeemaker: It prevents large substances such as proteins and cells from leaving the blood while allowing small substances such as water, glucose, amino acids, electrolytes, and waste products to leave the blood and enter the filtrate.

As you can see, the filtrate contains a number of substances our bodies need to reclaim, which occurs by a process known as **tubular reabsorption**. About 99% of the water in the filtrate and most of the solutes are reabsorbed through the epithelium of the nephron tubules and returned to the blood. The importance of this function cannot be overstated. If the water were not reabsorbed in the renal tubule and the collecting duct, we would lose our entire plasma volume in less than 30 minutes! In addition to tubular reabsorption, the nephron tubules remove substances from the blood not filtered at the glomerulus and put them into the filtrate, a process known as **tubular secretion**.

The rate of filtrate production averages about 120 ml/min. and is called the **glomerular filtration rate (GFR)**. The GFR is determined largely by the three pressures that factor into the NFP. Basically, if the NFP increases, then the GFR generally also increases, and the opposite is also true. Note that this is only true up to a point because there are many compensatory mechanisms that work to maintain a stable GFR. It is an important determinant of kidney function; if the GFR is too low, the kidneys will not be able to carry out their homeostatic functions adequately. Conversely, if the GFR is too high, the nephron tubules will not have time to reabsorb all of the substances in the filtrate.

In the following exercises you will examine the processes of filtration and reabsorption by constructing a model kidney and testing samples of simulated urine. In the final exercise you will trace the pathway of substances through the general circulation and the microanatomy of the kidney to contrast the filtration and reabsorption of three substances: erythrocytes, urea, and glucose.

25

# Exercise 25-1

## The Model Kidney

### MATERIALS

- ❏ 4-inch piece of dialysis tubing
- ❏ 2 pieces of string
- ❏ Animal blood or simulated blood
- ❏ 200 ml beaker
- ❏ Deionized water
- ❏ URS-10 and bottle with key

In this exercise you will examine the process of glomerular filtration by constructing a model kidney and testing for substances that appear in the filtrate. A kidney is easily modeled with either animal blood or simulated blood and simple dialysis tubing, which has a permeability similar to that of the filtration membrane. Glomerular filtration is mimicked by immersing the blood-filled tubing in water. Substances to which the "filtration membrane" is permeable will enter the surrounding water (the "filtrate"), whereas those to which the membrane is not permeable will remain in the tubing. Recall from Unit 24 (p. 587) that the filtration membrane consists of the glomerular endothelial cells, the basal lamina, and specialized cells called **podocytes** (**Figure 25.2**).

The results of this test are analyzed using **urinalysis reagent test strips** (URS-10; **Figure 25.3**). Each strip consists of 10 small, colored pads that change color in the presence of certain chemicals. The strip is interpreted by watching the pads for color changes and comparing the color changes with a color-coded key on the side of the bottle. The color closest to the color on the strip is recorded as your result. Please note that for this exercise you will read only 4 of the 10 boxes: blood, leukocytes, protein, and glucose.

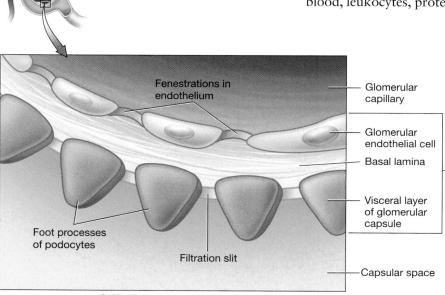

— Glomerular capsule

— Glomerulus

Fenestrations in endothelium

Glomerular capillary

Glomerular endothelial cell

Basal lamina

Filtration membrane

Visceral layer of glomerular capsule

Foot processes of podocytes

Filtration slit

Capsular space

FIGURE **25.2** The filtration membrane

FIGURE **25.3** URS-10 vial and strip

25

## **Procedure** Making a Model Kidney and Testing Glomerular Filtration

Now let's construct our model kidneys. The model kidney must sit in water for 25 minutes while you wait for your results. While you wait, answer Check Your Understanding questions 1–4 (pp. 627–628).

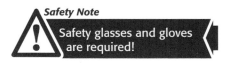

**Safety Note**

Safety glasses and gloves are required!

**1** Cut a 4-inch piece of dialysis tubing.

**2** Securely tie off one end of the tubing with string. Open the other end of the tubing by wetting the end of the tube and rubbing it between your fingers.

**3** Fill the dialysis tubing about half-full with either animal blood or simulated blood, and securely tie off the open end of the tube with string.

**4** Place the tied-off tube in a 200 ml beaker containing about 150 ml of deionized water.

**5** Leave the tubing in the water for approximately 25 minutes.

**6** After 25 minutes have passed, remove the tubing from the water.

**7** Dip a URS-10 in the water, and remove it quickly. Turn the bottle on its side, and compare the colors of the pads for glucose, blood, leukocytes, and protein. You will notice that on the side of the bottle, time frames are listed for each substance tested. Wait this amount of time to watch for a reaction; otherwise, you could obtain a false negative result. If you wait too long to read the results, though, the colors will tend to darken and may blend with adjacent colors.

**8** Record your results below.

   **a** Glucose: _____

   **b** Erythrocytes (blood): _____

   **c** Leukocytes: _____

   **d** Protein: _____

**9** Which substances were filtered out? Which substances stayed in the tubing?

_____

_____

_____

_____

# Exercise 25-2

## Urinalysis

### MATERIALS
- ☐ Samples of simulated urine
- ☐ Graduated cylinder or test tube
- ☐ URS-10 and bottle with key

For hundreds of years, health-care providers have recognized the utility of urinalysis as a diagnostic tool. Historically, the urine was evaluated for color, translucency, odor, and taste (yes, taste!). Today, although these characteristics are still examined (except, thankfully, taste), we also utilize urinalysis test strips, as we did in Exercise 25-1, to test for the presence of various chemicals in the urine. Both normal and abnormal readings give a health-care provider a wealth of information about a patient's renal function and overall health.

In this exercise you will analyze different urine samples using a URS-10, as in Exercise 25-1. For each urine sample, one or more results will be read as abnormal. In the second part of the exercise, you will research potential causes of the abnormalities you detected.

## Procedure Urinalysis Part One

Test a minimum of five samples of simulated urine using your urinalysis test strips. Use a new strip for each sample, and take care to clean your graduated cylinder thoroughly between each sample you test. Please note that this is *simulated* urine rather than real urine, although it contains the same chemicals as real urine and should be handled with equal caution.

**1** Randomly choose one sample of simulated urine, and pour approximately 3 ml of the sample into a test tube or graduated cylinder.

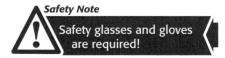

Safety Note

Safety glasses and gloves are required!

**2** Submerge one URS-10 in the urine, and quickly remove it.

**3** Compare the resulting colors and patterns on the test strip with those on the key on the bottle. Be sure to wait the appropriate amount of time to read the results.

**4** Record your results in **Table 25.2**, and note any anomalous results.

**5** Repeat this procedure for the remaining samples.

25

TABLE **25.2** Urinalysis Results

| Reading | Sample 1 | Sample 2 | Sample 3 | Sample 4 | Sample 5 |
|---|---|---|---|---|---|
| pH | | | | | |
| Leukocytes | | | | | |
| Nitrite | | | | | |
| Urobilinogen | | | | | |
| Protein | | | | | |
| Blood | | | | | |
| Specific gravity | | | | | |
| Ketones | | | | | |
| Bilirubin | | | | | |
| Glucose | | | | | |

## Procedure Urinalysis Part Two

Each sample you tested should have had at least one abnormality. Use your textbook and/or the Internet as a resource to determine which disease state(s) could lead to the abnormality you detected in each sample. Record your results in **Table 25.3**. After you have completed the activity, answer Check Your Understanding questions 5 and 6 (p. 628).

TABLE **25.3** Urinalysis Abnormalities and Disease States

| Sample | Primary Abnormality | Potential Causes |
|--------|---------------------|------------------|
| 1 | | |
| 2 | | |
| 3 | | |
| 4 | | |
| 5 | | |

25

# Exercise 25-3

## Time to Trace!

### MATERIALS

❏ Anatomical models of the kidney and nephron

It's time to trace again! In this exercise you will trace the pathway of different molecules through the general circulation, the vasculature of the kidney, and the microanatomy of the kidney. You will examine three different substances:

1. An erythrocyte,
2. A molecule of glucose, and
3. A molecule of urea.

Following are some hints to use as you trace:

▌ Don't forget the basic rules of blood flow: You must pass through a capillary bed in order to get from an artery to a vein.

▌ Refer to **Table 25.1** in Pre-Lab Exercise 25-3 (p. 614) before you trace each substance. This will help you determine if the substance gets filtered at the glomerulus or if it remains in the blood, if a filtered substance gets reabsorbed, and where the reabsorption occurs.

▌ Refer to anatomical models of the kidney and nephron and **Figure 25.4** for help with the pathway of blood flow through the kidney and its microanatomy.

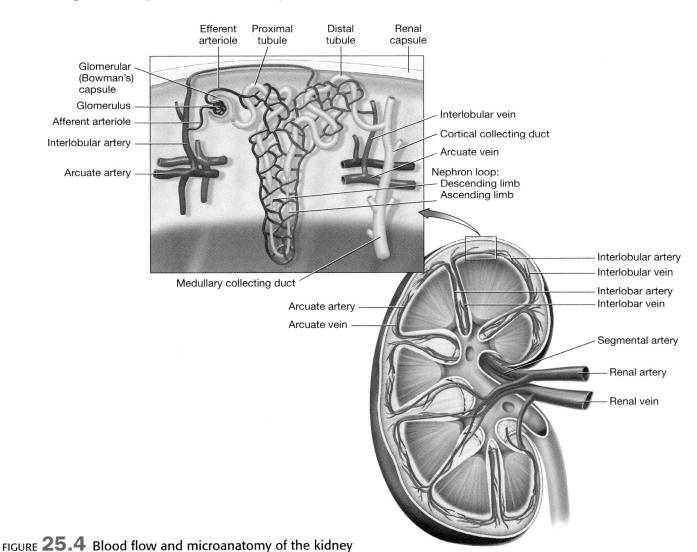

FIGURE **25.4** Blood flow and microanatomy of the kidney

Note that urea is sometimes recycled, being reabsorbed in the medullary collecting duct and secreted back into the filtrate at the proximal tubule. Your instructor may want you to take this into account in your tracing of the urea molecule. When you complete the tracing, answer Check Your Understanding question 7 (p. 628).

## Part 1: Erythrocyte

Trace an erythrocyte from the renal *artery* to the renal *vein*.

_____ → _____ →

_____ → _____ →

_____ → _____ →

_____ → _____ →

_____ → _____ →

_____ → _____ →

_____ .

## Part 2: Glucose

Trace a molecule of glucose from the renal artery to the renal vein.

_____ → _____ →

_____ → _____ →

_____ → _____ →

_____ → _____ →

_____ → _____ →

_____ → _____ →

_____ → _____ .

**25**

## Part 3: Urea

Trace a molecule of urea from the renal artery to its final destination outside the body.

_____ → _____ → _____

_____ → _____ → _____

_____ → _____ → _____

_____ → _____ → _____

_____ → _____ → _____

_____ → _____ → _____

_____ → _____ → _____

_____ → _____ → _____

_____ → _____ → _____

_____ → _____ → _____

_____ → _____ .

## Check Your Recall

**1** What is filtrate, and how is it formed?

_____

_____

_____

_____

_____

_____

_____

**2** *Fill in the blanks:* The rate of filtrate formation is known as the _____

and averages about _____ .

**3** What is the purpose of tubular reabsorption?
   a. To remove substances in the blood and place them into the filtrate
   b. To reclaim 99% of the water from the filtrate
   c. To reclaim electrolytes, glucose, and amino acids from the filtrate
   d. Both b and c are correct.
   e. Both a and b are correct.

**4** How do tubular reabsorption and tubular secretion differ?

_____

_____

_____

_____

_____

_____

25

**5** Which of the following is *not* a component of the filtration membrane?

a. Glomerular endothelial cells

b. Podocytes

c. Smooth muscle

d. Basal lamina

**6** How does the filtration membrane prevent certain substances from entering the filtrate?

_____

_____

_____

_____

_____

_____

_____

**7** Which of the following is *true*?

a. The fluid and solutes in the filtrate have been removed from the blood and are located in the renal tubules.

b. The filtration membrane allows blood cells and other large substances to pass but prevents small substances from moving through it.

c. Tubular secretion involves the reclamation of substances in the filtrate and their return to the blood.

d. Substances such as glucose, proteins, and erythrocytes are secreted into the filtrate.

**8** Which of the following substances would you *not* expect to find in the filtrate in significant amounts?

a. Erythrocytes

b. Glucose

c. Electrolytes

d. Amino acids

**9** Urinalysis can tell you

a. if there is an infection in the kidney or urinary tract.

b. if blood sugar is elevated.

c. if a person is dehydrated.

d. if a person's kidneys function normally.

e. All of the above.

**10** Which of the following substances would you expect to find in normal urine?

a. Glucose

b. Proteins

c. Urea

d. Ketones

e. Nitrites

Name _____

Section _____ Date _____

# Check Your Understanding

### Critical Thinking and Application Questions

**1** Why didn't all of the substances exit your model kidney to enter the "filtrate"?

_____

_____

_____

_____

_____

_____

_____

**2** Should glucose, erythrocytes, leukocytes, and protein appear in the urine in large amounts? Why or why not? What happens to each of these substances in the kidney?

_____

_____

_____

_____

_____

_____

_____

_____

**3** What other substances would you expect to find in the "filtrate" surrounding your model kidney? Explain.

_____

_____

_____

_____

_____

_____

25

**4** Calculate the net filtration pressure if the glomerular hydrostatic pressure measures 56 mmHg, the colloid osmotic pressure 29 mmHg, and the capsular hydrostatic pressure 8 mmHg. Does this differ from the normal value? If so, how? What effect, if any, would this have on the GFR?

**5** The presence of blood and protein in the urine can indicate inflammation of the glomerulus. Why would glomerular inflammation result in these substances passing into the filtrate and the urine? (*Hint*: What happens to capillaries during an inflammatory response?)

**6** Poorly controlled diabetes mellitus, which results in high blood glucose, often causes glucose to be found in the urine. Explain this finding.

**7** In Exercise 25-3 (p. 621), you traced a molecule of urea from the renal artery out of the body. How would the steps of this tracing differ if the urea were recycled? Write out the steps of this alternative pathway.

25

# Endocrine System

# 26

## OBJECTIVES

*Once you have completed this unit, you should be able to:*

▮ Identify endocrine organs and structures.

▮ Identify microscopic structures of endocrine organs.

▮ Trace the functions, stimulus for secretion, and target tissues of various hormones.

▮ Apply principles of the endocrine system to clinical cases.

# PRE-LAB EXERCISES

Complete the following exercises prior to coming to lab, using your textbook and lab manual for reference.

## Pre-Lab Exercise **26-1**

 Key Terms

You should be familiar with the following terms before coming to lab. Please note that key hormones are covered in Pre-Lab Exercise 26-3 (p. 633).

| Term | Definition |
|------|------------|
| **General Terms** | |
| Endocrine organ (gland) | |
| Hormone | |
| Target tissue | |
| Negative feedback | |
| **Endocrine Organs** | |
| Hypothalamus | |
| Anterior pituitary | |
| Posterior pituitary | |
| Thyroid gland | |
| Parathyroid glands | |
| Pineal gland | |
| Thymus | |

**26**

Pancreas _____

_____

Adrenal cortex _____

_____

Adrenal medulla _____

_____

Ovaries _____

_____

Testes _____

_____

**26**

Label and color the structures of the endocrine system depicted in **Figure 26.1** with the terms from Exercise 26-1 (p. 635). Use your text and Exercise 26-1 in this unit for reference.

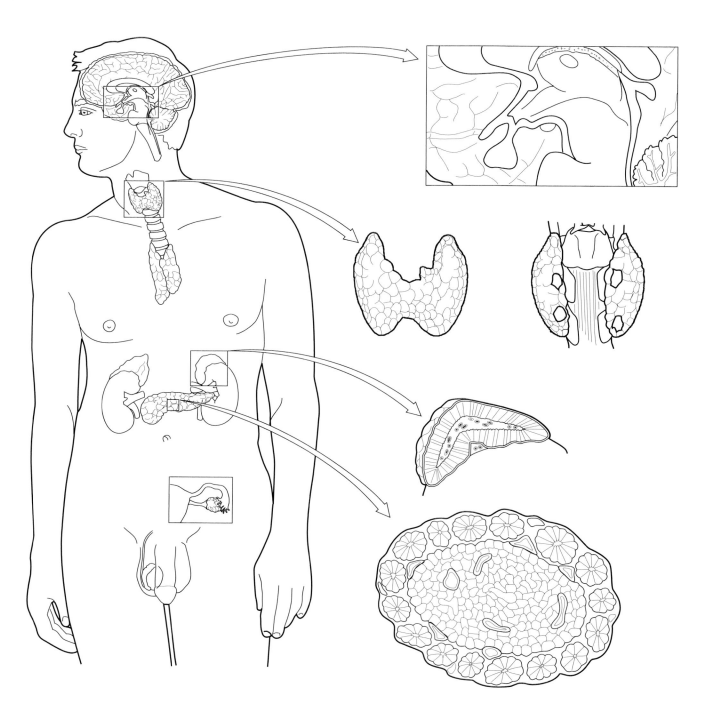

FIGURE **26.1** The endocrine system

26

# Pre-Lab Exercise 26-3

## Hormones: Target Tissues and Effects

Fill in **Table 26.1** with the organ that secretes each hormone, the stimulus for that hormone's secretion, and the hormone's target tissue.

TABLE **26.1** Properties of Hormones

| Hormone | Organ That Secretes the Hormone | Stimulus for Secretion | Target Tissue |
|---|---|---|---|
| Antidiuretic hormone | | | |
| Oxytocin | | | |
| Thyroid-stimulating hormone | | | |
| Adrenocorticotropic hormone | | | |
| Growth hormone | | | |
| Prolactin | | | |
| Melatonin | | | |
| Thyroxine and triiodothyronine (T4 and T3) | | | |
| Calcitonin | | | |

*(continues)*

TABLE **26.1** Properties of Hormones *(cont.)*

| Hormone | Organ That Secretes the Hormone | Stimulus for Secretion | Target Tissue |
|---|---|---|---|
| Thymosin and thymopoietin | | | |
| Parathyroid hormone | | | |
| Cortisol | | | |
| Aldosterone | | | |
| Epinephrine and norepinephrine | | | |
| Insulin | | | |
| Glucagon | | | |

26

# EXERCISES

The **endocrine system** is a diverse group of ductless glands that plays a major role in maintaining the body's homeostasis. It works closely with the other system that maintains homeostasis—the nervous system. Although these two systems both work toward the same goal, you will notice that the methods by which they do so differ. The nervous system works via action potentials (nerve impulses) and releases **neurotransmitters** that directly affect target cells. The effects are nearly immediate, but they are temporary. In contrast, the endocrine system works via secretion of **hormones**—chemicals secreted into the bloodstream that typically act on distant targets. The effects of hormones are not immediate, but they are longer lasting than those of nerve impulses.

In general, hormones function to regulate the processes of other cells, including inducing the production of enzymes or other hormones, changing the metabolic rate of the cell, and altering permeability of the plasma membranes. You might think of hormones as the "middle managers" of the body, because they communicate the messages from their "bosses" (the endocrine glands) and tell other cells what to do. Some endocrine glands (e.g., the thyroid and anterior pituitary glands) secrete hormones as their primary function. Others, however, secrete hormones as a secondary function, examples of which are the heart (atrial natriuretic peptide), adipose tissue (leptin), kidneys (erythropoietin), and stomach (gastrin).

This unit introduces you to the anatomy, histology, and physiology of the endocrine organs and hormones. To close out this unit, you will play "endocrine detective" and try to solve three "endocrine mysteries" involving our favorite cast of characters—Ms. Magenta and company.

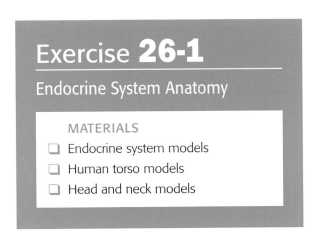

## Exercise **26-1**

### Endocrine System Anatomy

> **MATERIALS**
> ☐ Endocrine system models
> ☐ Human torso models
> ☐ Head and neck models

The 10 organs in the body that have hormone secretion as a primary function are the hypothalamus, the pituitary gland, the pineal gland, the thyroid gland, the parathyroid glands, the thymus, the adrenal gland, the pancreas, and the ovaries or testes (**Figure 26.2**). The testes and ovaries are discussed further in Units 27 and 28.

1. **Hypothalamus.** The hypothalamus, the inferior part of the diencephalon, can be likened to the endocrine system's chief executive officer (CEO). It releases hormones that inhibit and stimulate secretion—called **inhibiting** and **releasing hormones,** respectively—from the anterior pituitary gland. The anterior pituitary in turn releases hormones that stimulate other endocrine and exocrine glands in the body. In addition to inhibiting and releasing hormones, the hypothalamus makes the hormone **oxytocin,** which triggers uterine contraction and milk ejection from the mammary gland, and **antidiuretic hormone (ADH),** which causes water retention from the collecting ducts of the kidneys.

2. **Pituitary gland.** Note in **Figure 26.2** that the pituitary gland is actually two separate structures:
   a. The **anterior pituitary gland,** or **adenohypophysis,** is composed of glandular epithelium and secretes a variety of hormones that affect other tissues in the body. Examples include **thyroid-stimulating hormone (TSH),** which stimulates growth of and secretion from the thyroid, **growth hormone (GH),** which increases the rate of cell division and protein synthesis in all tissues, **prolactin,** which stimulates milk production from mammary glands, **adrenocorticotropic hormone (ACTH),** which stimulates secretion from the adrenal cortex, and two reproductive hormones, **luteinizing hormone** and **follicle-stimulating hormone,** which affect primarily the testes and ovaries.
   b. The other component of the pituitary gland is the **posterior pituitary** or **neurohypophysis,** which is actually composed of nervous tissue rather than glandular tissue. In fact, the posterior pituitary doesn't produce any hormones at all and functions merely as a place for storing the oxytocin and ADH produced by the hypothalamus.

3. **Pineal gland.** Recall from Unit 13 (p. 308) that the tiny pineal gland is located in the posterior and superior diencephalon. It secretes the hormone **melatonin** in response to decreased light levels and acts on the reticular formation of the brainstem to trigger sleep.

26

4. **Thyroid gland.** The thyroid gland, located in the anterior and inferior neck superficial to the larynx, is composed of hollow spheres called **thyroid follicles**. The cells that line the thyroid follicles are simple cuboidal cells called **follicle cells**, and they surround a gelatinous, iodine-rich substance called **colloid**. The follicle cells respond to TSH from the anterior pituitary by secreting a chemical into the colloid that reacts with iodine to produce two different hormones: **thyroxine** or **T4**, which has four iodine molecules, and **triiodothyronine** or **T3**, which has three iodine molecules. T3 is the most active of the two hormones and acts on essentially all cells in the body to increase the metabolic rate, increase protein synthesis, and regulate the heart rate and blood pressure, among other things. About 10 times as much T4 is produced as T3, and the body converts T4 to T3 when T3 levels in the blood drop.

   In between the follicles we find another cell type called the **C cells** or **parafollical cells**. These cells produce the hormone **calcitonin**, one of the body's main hormones that regulate calcium homeostasis. Calcitonin is secreted when calcium ion levels in the blood rise, and it triggers osteoblast activity and bone deposition.

5. **Parathyroid glands.** As you can see in **Figure 26.2**, the small parathyroid glands are located on the posterior thyroid gland. They secrete the hormone **parathyroid hormone** (**PTH**), which maintains calcium ion homeostasis. PTH is secreted in response to decreased levels of calcium ions in the blood, and it triggers osteoclast activity and resorption of bone tissue, increased calcium ion absorption from the gut, and increased calcium ion reabsorption from the kidneys. Hormones such as PTH and calcitonin that have opposite actions are called **antagonists**.

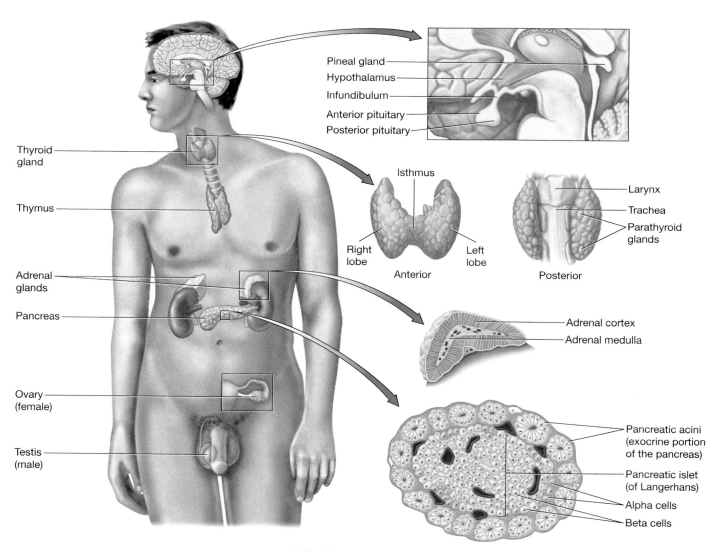

FIGURE **26.2** The endocrine system

6. **Thymus**. The thymus gland sits in the superior mediastinum. It is largest and most active in infancy and early childhood, during which time it secretes the hormones **thymosin** and **thympoietin**. Both of these hormones stimulate the development of T lymphocytes within the thymus. In adults most of the thymic tissue is gradually replaced by fat and other connective tissue.

7. **Adrenal gland**. As the name implies, the adrenal glands sit atop the superior pole of each kidney. Like the pituitary gland, the adrenal gland is actually two separate glands.

   a. The superficial region consists of glandular tissue called the **adrenal cortex**, and it secretes **steroid hormones** in response to stimulation by ACTH and other factors. The outermost zone of the adrenal cortex secretes steroids called **mineralocorticoids,** such as **aldosterone**, that regulate fluid and electrolyte homeostasis. The middle zone of the adrenal cortex secretes steroids called **glucocorticoids,** such as **cortisol**, that regulate the stress response, blood glucose, fluid homeostasis, and inflammation. The innermost zone secretes glucocorticoids and steroids called **gonadocorticoids** that affect the gonads and other tissues.

   b. The deep region of the adrenal gland, called the **adrenal medulla**, consists of modified postsynaptic sympathetic neurons that secrete **epinephrine** and **norepinephrine** in response to sympathetic stimulation.

8. **Pancreas**. In Unit 23 you learned about the exocrine functions of the pancreas, and now we turn to its endocrine functions. These functions are carried out by cells in small, round "islands" called **pancreatic islets,** or **islets of Langerhans**. The cells within the pancreatic islets secrete the hormones **insulin** and **glucagon**, which play a major role in regulating blood sugar levels. Insulin triggers the uptake of glucose by cells, which decreases blood glucose, and glucagon triggers the release of stored glucose from the liver, which increases blood glucose. Note that glucagon and insulin are antagonists.

9. **Testes**. The testes are the male reproductive organs that produce sperm cells, the male gametes. Cells within the testes are called **interstitial cells** (or **Leydig cells**) produce a steroid hormone called **testosterone**. This hormone promotes the production of sperm cells and the development of male secondary sex characteristics such as a deeper voice, greater bone and muscle mass, and facial hair.

10. **Ovaries**. The ovaries are the female reproductive organs that produce ooctyes, the female gametes. The ovaries produce steroid hormones called **estrogens** and **progesterone**. Estrogens play a role in the development of oocytes, female secondary sex characteristics such as breast development and the development of subcutaneous fat stores, and a variety of other processes. Progesterone has a number of effects that prepare the body for pregnancy.

Let's now examine these structures on models and charts. Note that human torso models are typically a good place to start when studying the endocrine system, because most of the organs are easy to find. The one exception is the thymus; many torsos and models choose not to show this structure because it is fairly inactive in adults.

Identify the following structures of the endocrine system on models and diagrams, using your textbook and this unit for reference. As you examine the anatomical models and diagrams, record on the model inventory in Table 26.2 the name of the model and the structures you were able to identify. After you have completed this activity, answer Check Your Understanding questions 1 and 2 (p. 651).

## Endocrine Glands

1. Hypothalamus
   a. Infundibulum
2. Pituitary gland
   a. Anterior pituitary
   b. Posterior pituitary
3. Pineal gland
4. Thyroid gland
   a. Isthmus
   b. Right and left lobes
5. Parathyroid glands
6. Thymus gland
7. Adrenal glands
   a. Adrenal cortex
   b. Adrenal medulla
8. Pancreas
9. Ovaries
10. Testes

TABLE **26.2** Model Inventory for the Endocrine System

| Model/Diagram | Structures Identified |
|---|---|
|  |  |
|  |  |
|  |  |
|  |  |
|  |  |

26

# Exercise 26-2

## Endocrine Organ Histology

### MATERIALS

- ❏ Thyroid gland slide
- ❏ Adrenal gland slide
- ❏ Pancreas slide
- ❏ Light microscope
- ❏ Colored pencils

In this exercise you will examine the histology of three endocrine organs: the thyroid gland, the adrenal gland, and the pancreas. (We will examine the ovary and testis in the next unit.) Following are keys to identification of each tissue:

1. **Thyroid gland.** Thyroid follicles are perhaps one of the easiest structures to identify (**Figure 26.3**). Look for a ring of simple cuboidal epithelium surrounding the reddish-brown, acellular colloid. Note that in between the follicles, you can see large C cells.

2. **Adrenal gland.** The adrenal cortex may be subdivided into three zones: the outer *zona glomerulosa*, a thin layer where the cells are arranged in clusters; the middle *zona fasciculata*, where the cells are stacked on top of one another and resemble columns; and the inner *zona reticularis*, where the cells stain more darkly and are tightly packed (**Figure 26.4**).

The innermost adrenal medulla is distinguished from the zones of the cortex by its numerous, loosely arranged blood vessels. The regions of the adrenal gland are best viewed on low or medium power. Switch to high power to get more detail of the individual cells.

3. **Pancreas.** In Unit 23 you examined the exocrine portion of the pancreas, the enzyme-secreting *acinar cells*. Most of the pancreas is composed of acinar cells; however, note the small "islands" of tissue in **Figure 26.5**. These islands are the insulin- and glucagon-secreting pancreatic islets. Typically, the islets are lighter in color than the surrounding acinar cells. On high power, the differences between the two tissues are easily seen: The acinar cells are slightly cuboidal and are arranged around a duct, whereas the islet cells have no distinctive arrangement (there also may be larger, stain-free spaces between the islet cells).

**FIGURE 26.3** The thyroid gland, photomicrograph

26

Capsule

Zona glomerulosa

Zone fasciculata

Zona reticularis

Adrenal medulla

FIGURE **26.4** The zones of the adrenal gland, photomicrograph

Pancreatic ducts

Pancreatic islet

Acinar cells

FIGURE **26.5** The pancreas, photomicrograph

## Procedure Microscopy

Obtain prepared slides of the thyroid gland, the adrenal gland, and the pancreas. Place each slide on the stage of the microscope, and scan it on low power. Advance to higher power to see the cells and associated structures in greater detail. Use your colored pencils to draw what you see in the field of view, and label your drawing with the terms indicated.

### Thyroid

1. Follicle cells
2. Colloid
3. C (parafollicular) cells

### Adrenal Gland

1. Zones of the adrenal cortex:
   a. Zona glomerulosa
   b. Zona fasciculata
   c. Zona reticularis
2. Adrenal medulla

### Pancreas

1. Acinar cells (exocrine portion)
2. Pancreatic islets (endocrine portion)

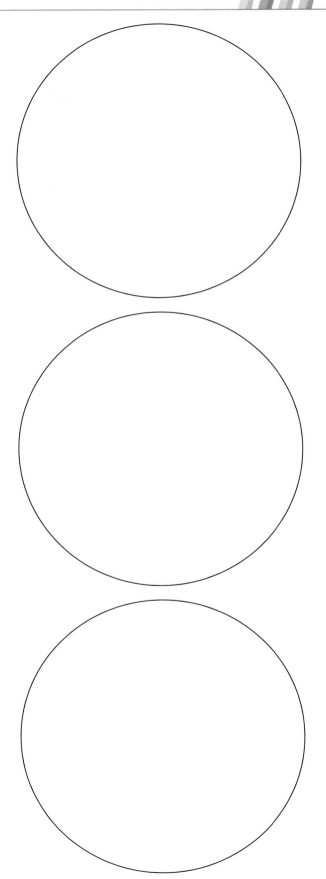

26

# Exercise 26-3

## Time to Trace: Negative Feedback Loops

Earlier in this unit we pointed out that each hormone has its own stimulus for secretion. The stimulus for secretion is generally a disturbance of homeostasis such as a change in body temperature, a change in the concentration of blood glucose or electrolytes, or a stressor. The hormone's response is to act on distant target cells to cause changes that restore homeostasis. When homeostasis is restored, the concentration of the hormone and activity of the glands declines. This type of response is called a **negative feedback loop**.

In this exercise you will be tracing a hormone's negative feedback loop from the initial homeostatic disturbance through the hormone's effects on its target cells to restore homeostasis. Following is an example:

**Start:** Blood glucose falls → the pancreas releases glucagon → glucagon levels in the blood rise → glucagon causes glycogenolysis by the liver → blood glucose levels rise → glucagon levels in the blood decline → **End**

Some of the negative feedback loops are slightly more complex because they involve multiple organs, such as the following:

**Start:** body temperature and/or metabolic rate drops → thyrotropin-releasing hormone is released from the hypothalamus → thyroid-stimulating hormone is released from the anterior pituitary → T3 and T4 are produced and released from the thyroid gland → T4 and T3 concentrations rise in the blood → T3 increases the metabolic rate of cells and increases heat production → body temperature and metabolic rate return to normal → levels of thyrotropin-releasing hormone in the blood decrease → levels of thyroid-stimulating hormone in the blood decrease → levels of T3 and T4 in the blood decrease → **End**

Now it's your turn! Note that sometimes the effects of the hormones cross over, so I've given you a hint at the start of each negative feedback loop as well as the general number of steps in each loop. After you have completed the activity, answer Check Your Understanding questions 3–7 (pp. 651–652).

**1  Start:** the concentration of the blood increases (i.e., there is *inadequate water* in the blood) → the hypothalamus

releases _____ and stores it in the _____ →

_____ level in the blood _____ →

_____ increases _____ →

the concentration of the blood _____ → levels of _____ in

the blood _____ **End**

**2  Start:** blood glucose increases → the pancreas releases _____ →

_____ level in the blood _____ →

_____ causes _____ → blood glucose

levels _____ → levels of _____

in the blood _____ **End**

26

**3 Start:** blood calcium ion concentration decreases → the parathyroid glands release

_____ → _____ level in the blood

→ _____

causes _____ → the concentration of calcium ions in the blood

→ levels of _____

in the blood _____ **End**


**4 Start:** blood potassium and hydrogen ion concentrations increase → the adrenal cortex produces

_____ → _____ level in the blood

_____ → _____ causes

_____ → the concentration of potassium and hydrogen ions in the blood

→ levels of _____

in the blood _____ **End**


**5 Start:** the body is exposed to a stressor → the hypothalamus releases _corticotropin-releasing hormone_ →

CRH stimulates _____ → _____ stimulates

_____ release → _____

level in the blood _____ → _____

triggers _____ → the body adapts to the stressor → levels of

_____ in the blood _____ **End**

## Exercise 26-4

### Endocrine "Mystery Cases"

In this exercise you will be playing the role of "endocrine detective" to solve endocrine disease mysteries with our favorite cast of characters. In each case you will have a victim who has suddenly fallen ill with a mysterious malady. You will be presented a set of "witnesses," each of whom will give you a clue as to the nature of the illness. Other clues will come from samples you send off to the lab for analysis. You will solve the mystery by providing the victim a diagnosis. You may wish to use your textbook for assistance with these cases.

### Case 1: The Cold Colonel

You are called upon to visit the ailing Col. Lemon. Before you see him, you speak with three witnesses who were with him when he fell ill.

*Witness statements:*

- *Ms. Magenta:* "Col. Lemon has been hot-blooded for as long as I've known him. But I noticed that he couldn't seem to keep warm. He kept complaining about being cold. . . ."
- *Mr. Olive:* "Just between you and me, I've noticed that the old chap has put on quite a bit of weight lately."
- *Professor Purple:* "The colonel and I used to go on major expeditions together. Now he just doesn't seem to have the energy to do much of anything."

What are your initial thoughts about the witnesses' statements? Does one hormone come to mind that may be the cause?

You see the colonel and collect some blood to send off to the lab. The analysis comes back as follows:

T3 (triiodothyronine): 0.03 ng/dl (normal: 0.2–0.5 ng/dl)

T4 (thyroxine): 1.1 µg/dl (normal: 4–7 µg/dl)

TSH (thyroid-stimulating hormone): 86 mU/l (normal: 0.3–4.0 mU/l)

Analyze the results. Why do you think the T3 and T4 are low and the TSH is elevated?

Based upon the witness statements and the laboratory analysis, what is your final diagnosis?

26

## Case 2: The Bloated Mrs. Blanc

Your next call is to the home of Mrs. Blanc. As before, you have three witnesses to interview.

*Witness statements:*

▌ *Ms. Magenta*: "Between you and me, she's really let herself go. She has fat deposits in all of these strange places, like around her face and trunk, and a weird hump on her back. If you ask me, some exercise may do her some good!"

▌ *Col. Lemon*: "Not that I want to talk about such things, but she's seemed bloated, a bit swollen, lately. She even told me that her blood pressure has gone up!"

▌ *Ms. Feather*: "The poor dear. She's been sick so much lately with all kinds of infections. Personally, I think her immune system needs boosting. But my herbal teas don't seem to help."

What are your initial thoughts, based upon the witnesses' statements? Does one hormone come to mind that could produce these effects?

Your next step is to speak with Mrs. Blanc. During the interview you notice a bottle of pills on her nightstand. You log the pills as evidence and send them off to the lab for analysis. The lab report shows the pills to be the medication *prednisone*, which you know to be a glucocorticoid similar to the hormone cortisol. Is this finding significant? Why or why not?

Based upon the witness statements and the laboratory analysis, what is your final diagnosis?

26

## Case 3: The Parched Professor

Your last call is to the aid of Professor Purple. Three witnesses are present from whom to take statements.

*Witness statements:*

- *Mr. Olive*: "I swear that I saw him drink a full glass of water every half an hour today. He kept saying how thirsty he was!"

- *Mrs. Blanc*: "He must be going to the . . . well, you know, the little boys' room, two or three times every hour!"

- *Ms. Feather*: "He's been saying lately that his mouth is dry and that he feels weak. Personally, I think he's just not following a healthy diet! He should be drinking some of my herbal teas!"

Based upon the witnesses' statements, what are your initial thoughts? Does one hormone come to mind that could produce these effects?

You interview Professor Purple and collect blood and urine specimens to be sent off to the lab for analysis. The lab reports that the urine osmolality is 150 mOsm/kg, which means the urine is overly dilute (too *much* water in the urine). The blood osmolality is 300 mOsm/kg, meaning the blood is overly concentrated (too *little* water in the blood). The lab also reports that his blood glucose is completely *normal*. What is the significance of these clues?

Based upon the witness statements and the laboratory analysis, what is your final diagnosis? (*Hint:* Think of the hormone that is supposed to trigger water retention from the kidneys. Is there a disease where this hormone is deficient?)

Name _____

Section _____ Date _____

## Check Your Recall

**1** Label **Figure 26.6** with the terms below.

- ☐ Adrenal cortex
- ☐ Adrenal medulla
- ☐ Anterior pituitary
- ☐ Hypothalamus

- ☐ Ovary
- ☐ Pancreas
- ☐ Pancreatic islet
- ☐ Parathyroid glands

- ☐ Posterior pituitary
- ☐ Testis
- ☐ Thyroid gland

FIGURE **26.6** The endocrine system

26

**2** *Fill in the blanks:* The nervous system works through secretion of _____ , whereas the endocrine system works via secretion of _____ .

**3** Which of the following is not a function of the hypothalamus?
   a. Produces antidiuretic hormone and oxytocin
   b. Stimulates production and release of hormones from the anterior pituitary
   c. Stimulates production and release of hormones from the posterior pituitary
   d. Inhibits the production and release of hormones from the anterior pituitary

**4** Which of the following endocrine organs are part of the diencephalon?
   a. Hypothalamus
   b. Pineal gland
   c. Thymus
   d. Both a and b are correct.
   e. Both b and c are correct.

**5** Which of the following sets of hormones are antagonists?
   a. T3 and T4
   b. Calcitonin and parathyroid hormone
   c. Epinephrine and cortisol
   d. Growth hormone and thyroxine

**6** *Matching:* Match the following endocrine organs with the hormone(s) each secretes.

| | |
|---|---|
| _____ Pineal gland | A. Parathyroid hormone |
| _____ Thyroid gland | B. Insulin and glucagon |
| _____ Pancreas | C. Steroid hormones |
| _____ Thymus | D. Epinephrine and norepinephrine |
| _____ Parathyroid glands | E. Thyroxine, triiodothyronine, and calcitonin |
| _____ Adrenal cortex | F. Thyroid-stimulating hormone, growth hormone |
| _____ Anterior pituitary | G. Melatonin |
| _____ Adrenal medulla | H. Thymosin and thymopoietin |

**7** Which of the following is *not* true regarding endocrine organ histology?
   a. The thyroid gland consists of rings of simple cuboidal follicle cells surrounding colloid.
   b. The pancreas has an exocrine portion consisting of pancreatic islets and an endocrine portion consisting of acinar cells.
   c. The adrenal cortex has three zones of cells that secrete three different types of hormones.
   d. The adrenal medulla is modified nervous tissue of the sympathetic nervous system.

26

**8** What is a negative feedback loop? Cite an example of a negative feedback loop in the endocrine system.

**9** List the stimulus for secretion for each of the following hormones (note that some hormones are stimulated by other hormones):

a Glucagon _____

b Calcitonin _____

c Parathyroid hormone _____

d Thyroxine and triiodothyronine (T4 and T3) _____

e Cortisol _____

f Antidiuretic hormone _____

**10** List the target tissue and effects of each of the following hormones:

a Insulin: _____

b Adrenocorticotropic hormone (ACTH): _____

c Growth hormone: _____

26

d Prolactin: _____

e    Oxytocin: _____

_____

_____


f    Thymosin: _____

_____

_____

# Check Your Understanding

**Critical Thinking and Application Questions**

**1** Endocrine glands are often called "ductless glands" because their products are secreted directly into the bloodstream. How is this an example of structure following function?

**2** How are the endocrine system and nervous system similar? How do they differ?

**3** Explain why an increase in thyroid hormone generally leads to a decrease in thyroid-stimulating hormone (TSH).

**4** Tumors of the parathyroid gland often result in secretion of excess parathyroid hormone. Considering the function of this hormone, predict the effects of such a tumor.

**5** The hormone calcitonin is prescribed to treat the disease *osteoporosis*. Explain why this hormone would help to reduce bone loss in patients affected with this disease.

**6** Drugs known as potassium-sparing diuretics work by blocking the effects of aldosterone on the kidney. What effect would these drugs have on the amount of urine produced, on the blood pressure, and on potassium ion concentration in the blood? Could these drugs have a negative impact on the acid-base balance of the body? Explain.

**7** The disease *diabetes mellitus*, type I, is characterized by destruction of the cells that produce insulin in the pancreatic islets. Predict the symptoms and effects of this disease.

26

# Reproductive System

## OBJECTIVES

*Once you have completed this unit, you should be able to:*

- Identify structures of the male and female reproductive system.

- Identify the stages of meiosis and describe how meiosis differs from mitosis.

- Describe the histology of the ovary, seminiferous tubules, and spermatozoa.

- Trace the pathway a developing gamete takes from its development to the time of fertilization.

Name _____ Section _____ Date _____

Complete the following exercises prior to coming to lab, using your textbook and lab manual for reference.

## Pre-Lab Exercise **27-1**

### ✎ Key Terms

You should be familiar with the following terms before coming to lab.

| Term | Definition |
|------|------------|

**Structures of the Male Reproductive System**

Testes _____

_____

Seminiferous tubules _____

_____

Epididymis _____

_____

Vas (ductus) deferens _____

_____

Spermatic cord _____

_____

Seminal vesicle _____

_____

Prostate gland _____

_____

Corpus spongiosum _____

_____

Corpora cavernosa _____

_____


**Structures of the Female Reproductive System**

Ovaries _____

_____

Ovarian follicles _____

_____

27

Uterine tube _____

Uterus _____

Cervix _____

Vagina _____

Vulva _____

Mammary glands _____

**Gametogenesis Terms**

Meiosis I _____

Meiosis II _____

Haploid _____

Spermatogenesis _____

Oogenesis _____

27

Label and color the structures of the male reproductive system depicted in **Figure 27.1** with the terms from Exercise 27-1 (p. 659). Use your text and Exercise 27-1 in this unit for reference.

FIGURE **27.1** Male reproductive organs: (**A**) midsagittal section of the male pelvis; (**B**) midsagittal section through the testis

## Pre-Lab Exercise **27-3**

Female Reproductive Anatomy

Label and color the structures of the female reproductive system depicted in **Figure 27.2** with the terms from Exercise 27-2 (p. 662). Use your text and Exercise 27-2 in this unit for reference.

FIGURE **27.2** Female reproductive tract: (**A**) midsagittal section of the female pelvis; (**B**) posterior view of the female reproductive organs

# Pre-Lab Exercise 27-4

## Stages of Mitosis

In this unit we will discuss the process of meiosis. Meiosis proceeds in a fashion similar to mitosis, so it's a good idea to review the discussion of mitosis before you begin the lab. Fill in **Table 27.1** with the events occurring in the cell in each of the stages of mitosis.

TABLE **27.1** Stages of Mitosis

| Stage of Mitosis | Events Occurring in the Cell |
|---|---|
| Prophase | |
| Metaphase | |
| Anaphase | |
| Telophase | |

# EXERCISES

The other organ systems in the human body we have discussed all function in some manner to help maintain homeostasis of the body. The reproductive system, however, plays little role in maintaining homeostasis and instead functions to perpetuate the species. The main organs of the reproductive system are the **gonads**—the testes and the ovaries—which produce **gametes**, or sex cells, for reproduction.

We begin this unit with the anatomy of the male (Exercise 27-1) and female (Exercise 27-2) reproductive systems. Then we turn to the main functions of these organs: **gametogenesis**, the formation of new gametes.

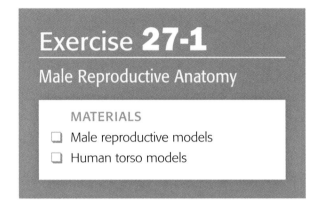

## Exercise **27-1**

### Male Reproductive Anatomy

MATERIALS
- ☐ Male reproductive models
- ☐ Human torso models

The **testes**, the gamete-producing organs of the male, are situated outside the body in a sac of skin and connective tissue called the **scrotum** (**Figure 27.3**). They are located externally because sperm production requires a temperature lower than body temperature of about 34°C (about 94°F). Smooth muscle tissue called the **cremaster muscle** surrounds each testis in the scrotum, helping control their height and therefore their temperature.

The testes are surrounded by an outer connective tissue sheath called the **tunica vaginalis** and an inner sheath called the **tunica albuginea**, which divides the interior of the testes into **lobules**. Each lobule contains a tightly coiled **seminiferous tubule** where **spermatogenesis** takes place.

The seminiferous tubules converge near the superior part of the testis to form a structure called the **rete testis**. The rete testis exits the testis to join the first segment of the duct system of the male reproductive tract, the **epididymis** (**Figures 27.3** and **27.4**). Immature sperm produced by the seminiferous tubules migrate to the epididymis to finish their maturation, after which they exit via a long tube called the **vas** (or **ductus**) **deferens**. The vas deferens travels superiorly through the **spermatic cord**, a structure that also carries the testicular artery, testicular veins (the pampiniform plexus), and nerves.

Note in **Figure 27.4** that once the vas deferens enters the pelvic cavity, it crosses superiorly and posteriorly over the urinary bladder to join a gland called the **seminal vesicle**. At this point, the vas deferens merges with the duct from the seminal vesicle to form the **ejaculatory duct**. This duct passes through the **prostate gland**, where it joins with the **prostatic urethra**. The prostatic urethra becomes the **membranous urethra** as it exits the prostate, and then becomes the **spongy urethra** as it enters the corpus spongiosum of the penis.

The male reproductive tract consists of three exocrine glands: the prostate gland, seminal vesicles, and **bulbourethral glands**. The seminal vesicles and the prostate gland together produce about 90% of the volume of **semen**, a fluid that contains chemicals to nourish and activate sperm. The smaller bulbourethral glands produce an alkaline secretion released prior to the release of sperm during ejaculation.

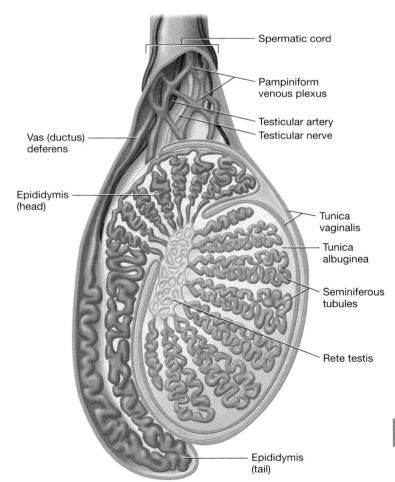

FIGURE **27.3** Midsagittal section through the testis

27

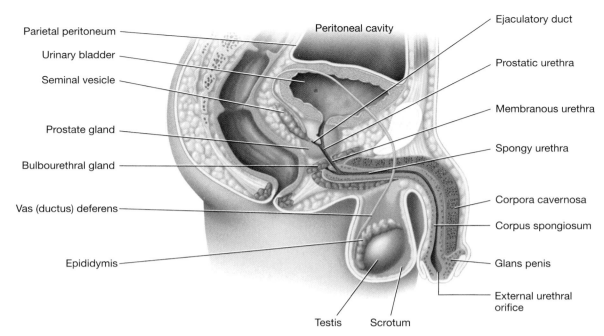

FIGURE **27.4** Midsagittal section through the male pelvis

The **penis** is composed of three erectile bodies: the single **corpus spongiosum** and the paired, dorsal **corpora cavernosa**. The corpus spongiosum, which surrounds the spongy urethra, enlarges distally to form the **glans penis**. All three bodies consist of vascular spaces that fill with blood during an erection.

## Procedure  Model Inventory of the Male Reproductive System

Identify the following structures of the male reproductive system on models and diagrams using your textbook and this unit for reference. As you examine the anatomical models and diagrams, record on the model inventory in **Table 27.2** the name of the model and the structures you were able to identify. When you have completed the activity, answer Check Your Understanding questions 1–3 (p. 677).

1. Scrotum
2. Testes
   a. Tunica albuginea
   b. Tunica vaginalis
   c. Seminiferous tubules
   d. Rete testis
3. Epididymis
4. Spermatic cord
   a. Testicular arteries
   b. Pampiniform plexus
5. Cremaster muscle
6. Vas (ductus) deferens
7. Ejaculatory duct

8. Urethra
   a. Prostatic urethra
   b. Membranous urethra
   c. Spongy urethra
   d. External urethral orifice
9. Glands:
   a. Seminal vesicle
   b. Prostate gland
   c. Bulbourethral gland
10. Penis
   a. Corpora cavernosa
   b. Corpus spongiosum
      (1) Glans penis

TABLE **27.2** Model Inventory for the Male Reproductive System

| Model/Diagram | Structures Identified |
|---|---|
|  |  |
|  |  |
|  |  |
|  |  |
|  |  |

# Exercise 27-2

## Female Reproductive Anatomy

### MATERIALS

❑ Female reproductive models
❑ Human torso models

The female reproductive organs lie in the pelvic cavity, with the exception of the almond-shaped **ovaries**, located in the peritoneal cavity (**Figures 27.5** and **27.6**). The ovaries, the female gonads, produce gametes (**oocytes**, or eggs) that travel through the reproductive tract to be fertilized. The ovaries are held in place by several ligaments, including the **ovarian ligament**, a part of the **broad ligament** called the **mesovarium**, and the **suspensory ligament**. The suspensory ligaments extend from the lateral body wall and carry with them the ovaries' blood supply.

Developing oocytes within the ovaries are encased in structures called **follicles** (**Figure 27.7**). Follicles are present in various stages, ranging from the immature **primordial follicles** to the mature **vesicular** (or **Graafian**) **follicle**, from which an oocyte is released during ovulation. Notice in **Figure 27.5** that, unlike the male reproductive tract, the tubule system of the female reproductive tract is not continuous. Therefore, the oocyte actually is released into the pelvic cavity, and the fingerlike extensions of the **uterine (fallopian) tube**, called **fimbriae**, must "catch" the oocyte and bring it into the uterine tube.

The uterine tubes join the superolateral portion of the **uterus**. The uterus is situated between the urinary bladder and the rectum. It is held in place by the broad ligament and an anterior ligament called the **round ligament**. The uterus consists of three portions: the dome-shaped **fundus**, the central **body**, and the narrow **cervix**. The wall of the uterus has three layers: the inner epithelial and connective tissue lining called the **endometrium**, in which a fertilized ovum implants; the middle, muscular **myometrium**; and the outermost connective tissue lining, the **perimetrium** (which is actually just the visceral peritoneum).

The inferiormost portion of the cervix is the **cervical (external) os**. The **vagina** extends inferiorly from the cervical os and terminates at the vaginal orifice. Flanking the vaginal orifice are the **greater vestibular** (Bartholin's) **glands**, which secrete mucus to lubricate the vaginal canal during coitus.

The external anatomy of the female is collectively called the **vulva**, and consists of the **labia majora** and **labia minora** (singular: labium major and minus, respectively) the **clitoris**, and the urethral and vaginal orifices. The labia are paired skin folds analogous to the scrotum in the male. They enclose a region called the **vestibule** that houses the clitoris and the urethral and vaginal orifices.

The **mammary glands** are not true reproductive organs but do have an associated reproductive function in milk production (**Figure 27.8**). Mammary glands are present in both males and females (males can produce milk too), but their anatomy is most appropriately discussed with female anatomy. Internally, mammary glands consist of 15–25 **lobes**,

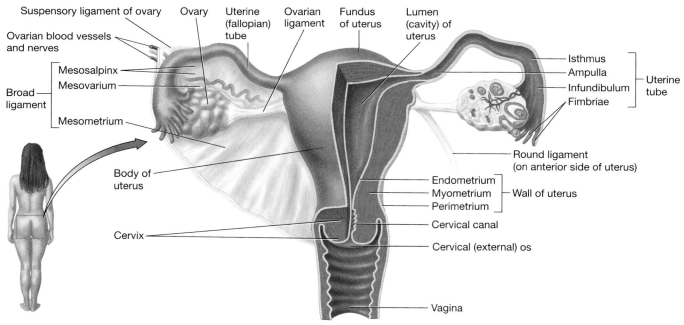

FIGURE **27.5** Posterior view of the female reproductive organs

27

each of which has smaller **lobules** that contain milk-producing **alveoli**. Milk leaves the alveoli through **alveolar ducts**, which join to form storage areas called **lactiferous sinuses**. Milk leaves through the **nipple**, which is surrounded by a darkly pigmented area called the **areola**.

FIGURE **27.6** Midsagittal section of the female pelvis

FIGURE **27.7** The ovary

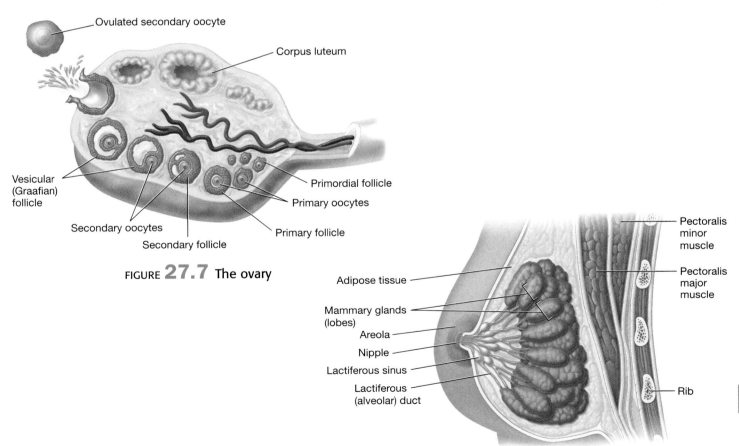

FIGURE **27.8** The mammary gland

## Procedure Model Inventory of the Female Reproductive System

Identify the following structures of the female reproductive system on models and diagrams, using your textbook and this unit for reference. As you examine the anatomical models and diagrams, record on the model inventory in **Table 27.3** the name of the model and the structures you were able to identify. After you have completed the activity, answer Check Your Understanding questions 4 and 5 (p. 678).

1. Ovary
2. Ovarian follicles
   a. Primordial follicle
   b. Primary follicle
   c. Secondary follicle
   d. Graafian (vesicular) follicle
3. Uterine (fallopian) tube
   a. Fimbriae
4. Uterus
   a. Fundus
   b. Body
   c. Cervix
   d. Layers
      (1) Endometrium
      (2) Myometrium
      (3) Perimetrium

5. Vagina
6. Vulva
   a. Vestibule
   b. Labia majora
   c. Labia minora
   d. Clitoris
   e. Urethral orifice
7. Ligaments
   a. Broad ligament
      (1) Mesometrium
      (2) Mesovarium
      (3) Mesosalpinx
   b. Round ligament
   c. Suspensory ligament of the ovary
   d. Ovarian ligament

8. Mammary glands
   a. Areola
   b. Nipple
   c. Lobe
   d. Alveolar duct
   e. Lactiferous sinus
   f. Adipose tissue

TABLE **27.3** Model Inventory for the Female Reproductive System

| Model/Diagram | Structures Identified |
|---|---|
|  |  |
|  |  |
|  |  |
|  |  |
|  |  |
|  |  |

27

# Exercise 27-3

## Meiosis

### MATERIALS

☐ Meiosis models

☐ Mitosis models

☐ 4 colors of pop beads

As you may recall from Unit 4, somatic cells divide by a process called mitosis. During mitosis, a cell replicates its 23 pairs of homologous chromosomes and divides its DNA and organelles into two identical daughter cells (see Pre-Lab Exercise 27-4 (p. 658) for a review of the stages). Each new **diploid** cell is identical to the original cell. **Gametogenesis** (oogenesis and spermatogenesis), however, must proceed in a different way for two reasons:

1. If each gamete were to have the same genetic material, we all would be genetically identical to our siblings.

2. If each gamete were to have two sets of chromosomes, our offspring would have four sets of chromosomes, and we would be tetraploid.

For this reason, gametes undergo a process known as **meiosis** rather than mitosis.

In meiosis, gametes proceed through two rounds of cell division, and each gamete ends up with only one set of chromosomes (a **haploid** cell). As you can see in **Figure 27.9**, meiosis begins in a manner similar to mitosis: The homologous chromosomes have replicated during the S phase of the cell cycle to yield two pairs of sister chromatids. The cell then goes through **meiosis I**, during which replicated chromosomes line up next to one another so closely that they overlap in several places—a phenomenon called **synapsis**.

Each synapse has two sets of sister chromatids, and therefore the entire structure is called a **tetrad**. As the tetrads align themselves on the equator of the cell and prepare to attach to spindle fibers during metaphase I, the areas of the chromosomes that overlap form points called **chiasmata**, or **crossover**. As anaphase I begins, the chromosomes exchange pieces of genetic material at the points of crossover.

Once telophase I and cytokinesis are complete, the gametes have 23 sets of sister chromatids, and the amount of genetic material is halved because of the separation of homologous chromosomes. They then go through the second round of division, **meiosis II**, during which the sister chromatids separate. Meiosis II is similar to meiosis I with one major exception: The DNA does *not* replicate prior to the start of prophase II. At the end of telophase II, each gamete has only 23 chromosomes and is a haploid cell.

## Procedure Model Meiosis

In the following procedure you will model meiosis with either a set of cell meiosis models or with pop-bead chromosomes. When you have completed the activity, answer Check Your Understanding question 6 (p. 678).

**1** Obtain a set of meiosis and mitosis models.

**2** Arrange the mitosis models in the correct order.

**3** Now arrange the meiosis models in the correct order.

**4** Compare the models of the two processes. List all of the differences you can see between meiosis and mitosis below.

27

**Meiosis I (first division)**

**Meiosis II (second division)**

**Early prophase I**

Diploid cell with four sets of sister chromatids

Sister chromatids

Nucleus

Centrioles

**Prophase II**

**Mid- to late prophase I**

Tetrad

Crossing over

Spindle fibers

**Metaphase II**

**Metaphase I**

Centromere

**Anaphase II**

**Anaphase I**

**Telophase II**

Cleavage furrow

**Telophase I**

Haploid cells with two sets of sister chromatids

Haploid cells with two chromosomes

FIGURE **27.9** Meiosis

# Exercise 27-4

## Spermatogenesis and Oogenesis

### MATERIALS

- ❑ Testis slide
- ❑ Sperm slide
- ❑ Ovary slide
- ❑ Light microscope
- ❑ Colored pencils

**Gametogenesis** is really two separate processes: **spermatogenesis**, which takes place in the seminiferous tubules of the testes, and **oogenesis**, which takes place in the ovary and uterine tube. Spermatogenesis, shown in **Figure 27.10**, begins with stem cells located at the outer edge of the seminiferous tubules called **spermatogonia**. Before puberty, these diploid cells undergo repeated rounds of mitosis to increase their numbers. As puberty begins, each spermatogonium divides into two different cells—one cell that stays a spermatogonium and another cell that becomes a **primary spermatocyte**.

The primary spermatocyte then undergoes meiosis I and divides into two **secondary spermatocytes** that migrate closer to the lumen of the tubule. The two secondary spermatocytes undergo meiosis II and give rise to four haploid **spermatids**. The spermatids move to the lumen of the seminiferous tubule, at which point they are called spermatozoa, or sperm cells. These cells then move to the epididymis to mature into functional gametes. Mature sperm contain three parts: the **head**, in which the DNA resides, the **midpiece**, which contains an axomere and mitochondria, and the **flagellum** (**Figure 27.11**).

Note in **Figure 27.12** that most of the stages of spermatogenesis are identifiable on a microscope slide of the seminiferous tubules. On the inner edge of the tubule near the lumen are small, round cells with little cytoplasm. These are the spermatozoa and spermatids. As we move to deeper layers of the tubule wall, we can see primary spermatocytes, and on the outer edge we can see the cuboidal spermatogonia. In between the tubules are small clusters of cells called **interstitial** (or **Leydig**) **cells**. These cells make testosterone, required for spermatogenesis to take place. Note that interstitial cells and secondary spermatocytes are not visible in **Figure 27.12**.

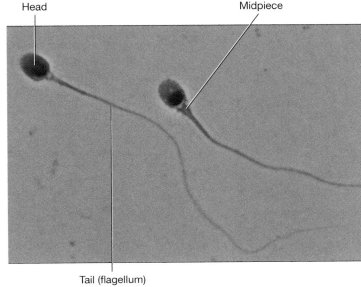

Head

Midpiece

Tail (flagellum)

FIGURE **27.10** Sperm cells

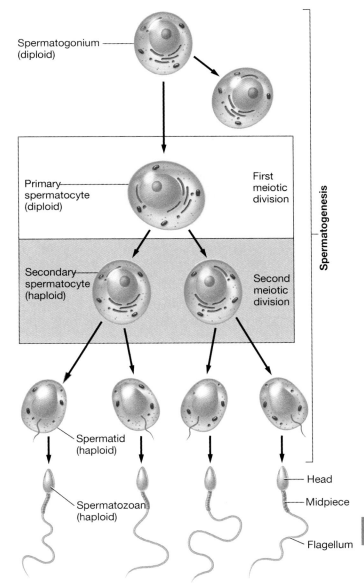

Spermatogonium (diploid)

Primary spermatocyte (diploid)

First meiotic division

Secondary spermatocyte (haploid)

Second meiotic division

Spermatid (haploid)

Spermatozoan (haploid)

Head

Midpiece

Flagellum

Spermatogenesis

FIGURE **27.11** Spermatogenesis

27

FIGURE **27.12** Seminiferous tubules, photomicrograph

Like spermatogenesis, oogenesis proceeds through meiosis to yield a haploid gamete, the **ovum** (**Figure 27.13**). But the two processes differ in some notable ways:

▌ *The number of oocytes is determined before birth.* During the fetal period, stem cells called **oogonia** undergo mitosis, increasing their numbers to about 500,000 to 700,000. This is the total number of oocytes a woman will ever produce. This is in sharp contrast to spermatogenesis, which begins at puberty and continues throughout a male's lifetime.

▌ *Meiosis I begins during the fetal period but is arrested.* Still during the fetal period, the oogonia become encased in a **primordial follicle**, enlarge, and become **primary oocytes**. The primary oocytes begin prophase I but are arrested at this stage. Meiosis I does not resume until puberty.

▌ *The first meiotic division results in one secondary oocyte and one polar body.* At puberty, one primary oocyte enlarges and becomes encased in a **primary follicle**. This primary oocyte then completes meiosis I to produce a **secondary oocyte** and a small bundle of nuclear material called a **polar body**. The formation of a polar body allows the oocyte to conserve cytoplasm, which will have to sustain the cell if fertilization occurs. The secondary oocyte, initially encased in a **secondary follicle**, enlarges to form a **vesicular** or **Graafian follicle**, which contains a large, fluid-filled space called the **antrum**.

▌ *Meiosis II completes only if fertilization takes place.* The secondary oocyte begins meiosis II

FIGURE **27.13** Oogenesis

27

Secondary oocyte

Antrum of vesicular follicle

Primordial follicles with primary oocytes

Secondary follicle with secondary oocytes

Vesicular (Graafian) follicle

Primary follicles with primary oocytes

FIGURE **27.14** Ovary, photomicrograph

and is released when the vesicular follicle ruptures during ovulation. Note that the ruptured follicle then becomes an endocrine organ called a **corpus luteum**. But the secondary oocyte only completes meiosis II to form an ovum and a second polar body if fertilization occurs. If fertilization does not occur, the secondary oocyte degenerates without ever completing meiosis II.

The ovary offers a nice opportunity to view oogenesis and meiosis in action (**Figure 27.14**). Though you typically cannot see the chromosomes of the oocytes, you can determine the stage of the oocyte by looking at the surrounding follicle: Primary oocytes are encased in primordial and primary follicles, and secondary oocytes in secondary and vesicular follicles. Note that we cannot see oogonia in the ovary because these cells begin meiosis I and become primary oocytes during the fetal period.

## Procedure Microscopy of Male Reproductive Structures

In this exercise you will examine prepared slides of the testes and a sperm smear. You will want to examine both slides on high power, and you may wish to examine the sperm smear with an oil-immersion lens. Because sperm cells are so small, the slide will likely have a thread or some other marker to help you find them. As you examine the slides of the seminiferous tubules and the sperm smear, use colored pencils to draw what you see, and label your drawings with the terms indicated.

### Testes

1. Seminiferous tubule
2. Spermatogonia
3. Primary spermatocytes
4. Spermatids
5. Interstitial cells

**Sperm**

1. Head
2. Midpiece
3. Flagellum

 **Procedure** Microscopy of the Ovary

Obtain a prepared slide of an ovary. Use your colored pencils to draw what you see, and label your drawing with the following terms. Please keep in mind that you may have to examine multiple slides to see all the follicular stages.

1. Primordial follicle
2. Primary follicle
   a. Primary oocyte
3. Secondary follicle
4. Graafian (vesicular) follicle
   a. Secondary oocyte
   b. Antrum

## Procedure Time to Trace

In this exercise, you will trace the male and female gametes from their production in the gonads to the point at which they meet in the uterine tube, where fertilization takes place.

### Step 1: Male Gamete

Trace the male gamete from its earliest stage—the spermatogonium in the seminiferous tubule—through the stages of spermatogenesis to a mature sperm, and then through the male reproductive tract until it exits the body and enters the uterine tube of the female reproductive tract. Trace the pathway using **Figure 27.15**, and also fill in the space below.

FIGURE **27.15** Male reproductive tract

**Start:** Spermatogonium

**End:** Uterine tube

## Step 2: Female Gamete

Trace the female gamete from its earliest stage, the oogonium in a primordial follicle, to the time at which it is ovulated and enters the uterine tube. Trace the pathway using **Figure 27.16**, and also fill in the space below.

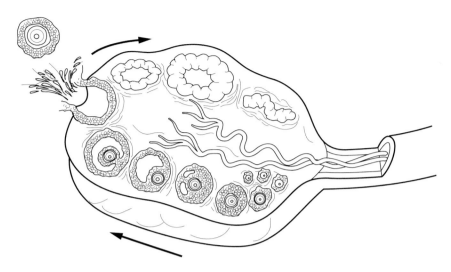

FIGURE **27.16** Ovary and follicles

**Start:** Oogonium

**End:** Uterine tube

## Check Your Recall

**1** Label the following structures on **Figure 27.17**.

☐ Epididymis (head)
☐ Epididymis (tail)
☐ Seminiferous tubules
☐ Spermatic cord
☐ Tunica albuginea
☐ Tunica vaginalis
☐ Vas (ductus) deferens

FIGURE **27.17** Midsagittal section through the testis

**2** Label the following structures on **Figure 27.18.**

- ☐ Bulbourethral gland
- ☐ Corpora cavernosa
- ☐ Corpus spongiosum
- ☐ Ejaculatory duct
- ☐ Epididymis
- ☐ Glans penis
- ☐ Prostate gland
- ☐ Scrotum
- ☐ Testis
- ☐ Vas (ductus) deferens

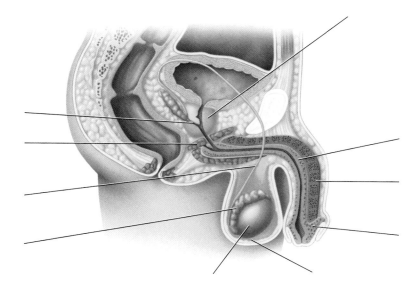

FIGURE **27.18** Midsagittal section through the male pelvis

**3** Label the following structures on **Figure 27.19.**

- ☐ Cervical os
- ☐ Labium major
- ☐ Labium minus
- ☐ Ovary
- ☐ Uterine tube
- ☐ Uterus
- ☐ Vaginal canal

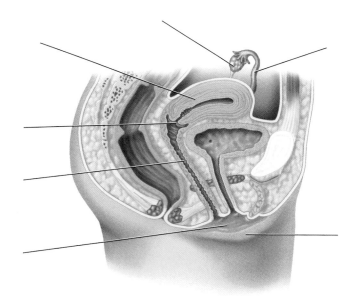

FIGURE **27.19** Midsagittal section of the female pelvis

**4** Label the following structures on **Figure 27.20.**

☐ Body of uterus      ☐ Fimbriae      ☐ Ovary
☐ Broad ligament      ☐ Fundus of uterus      ☐ Perimetrium
☐ Cervical canal      ☐ Myometrium      ☐ Uterine tube
☐ Endometrium      ☐ Ovarian follicle

FIGURE **27.20** Posterior view of the female reproductive organs

**5** Which of the following statements about meiosis is *false*?
a. Gametes proceed through two rounds of cell division.
b. Crossover occurs during meiosis, which increases genetic diversity.
c. The DNA replicates before both meiosis I and meiosis II.
d. Meiosis is carried out only in the gonads to produce gametes.

**6** *Fill in the blanks:* The daughter cells produced by mitosis are _____ with _____ set(s) of chromosomes, whereas the gametes produced by meiosis are _____ with _____ set(s) of chromosomes.

**27**

**7** Number the following events of spermatogenesis in the proper order, with number 1 next to the first event and number 4 next to the final event.

_____ Primary spermatocytes undergo meiosis I to produce two secondary spermatocytes.

_____ The two secondary spermatocytes undergo meiosis II to produce four haploid spermatids.

_____ Spermatogonia divide into more spermatogonia and primary spermatocytes.

_____ Spermatogonia undergo repeated rounds of mitosis.

**8** Which of the following statements about spermatogenesis and oogenesis is *false*?
a. Meiosis II does not complete in oogenesis unless fertilization takes place.
b. Meiosis I begins during the fetal period but is arrested in oogenesis.
c. Spermatogenesis begins at puberty and continues throughout the male's lifetime, whereas the total number of oocytes a woman will produce is determined before birth.
d. Spermatogenesis results in one spermatid and two polar bodies, whereas oogenesis results in four ova.

**9** Spermatids migrate to the _____ to mature.
a. vas (ductus) deferens
b. epididymis
c. seminal vesicle
d. prostate gland

**10** Label the following structures on **Figure 27.21**.

☐ Corpus luteum
☐ Primary follicle
☐ Primary oocyte

☐ Secondary follicle
☐ Secondary oocyte
☐ Vesicular follicle

Figure **27.21** Ovary

27

## Check Your Understanding

### Critical Thinking and Application Questions

**1** A condition called *testicular torsion* results when the spermatic cord becomes twisted. Why would this condition be a surgical emergency?

_____

_____

_____

_____

_____

_____

**2** What structure is sectioned in a vasectomy? What are the effects of this procedure? Will it affect the number of sperm or the volume of semen produced?

_____

_____

_____

_____

_____

_____

_____

**3** The condition *benign prostatic hypertrophy*, in which the prostate is enlarged, often results in urinary retention—the inability to completely empty the bladder. Considering the anatomy of the male genitourinary tract, explain this symptom.

_____

_____

_____

_____

_____

_____

27

**4** A tubal (or ectopic) pregnancy results from implantation of a fertilized ovum in the uterine tube instead of the uterus. Why is this dangerous?

_____

_____

_____

_____

_____

**5** One of the most common complaints of pregnant women is the need to urinate often. Explain this, considering the arrangement of the pelvic organs.

_____

_____

_____

_____

_____

**6** What happens if one of the chromosomes doesn't separate fully during meiosis II? Give an example of a common disorder caused by this phenomenon. What are the symptoms of this disorder?

_____

_____

_____

_____

_____

27

# Human Development and Heredity

# 28

## OBJECTIVES

*Once you have completed this unit, you should be able to:*

▌ Identify and describe structures and membranes associated with the fetus.

▌ Identify and describe the stages of development and structures of the fetal circulation.

▌ Trace a developing zygote (or conceptus) from fertilization to implantation.

▌ Define the terms phenotype and genotype, and perform analyses to determine a subject's phenotype and genotype.

▌ Describe dominant, recessive, and sex-linked traits.

# PRE-LAB EXERCISES

Complete the following exercises prior to coming to lab, using your textbook and lab manual for reference.

## Pre-Lab Exercise **28-1**

### ✎ Key Terms

You should be familiar with the following terms before coming to lab.

| Term | Definition |
|------|------------|

**Development**

Fertilization _____

_____

Zygote _____

_____

Blastocyst _____

_____

Embryo_____

_____

Fetus _____

_____

**Fetal Structures**

Chorion _____

_____

Amnion _____

_____

Placenta _____

_____

**Inheritance**

Allele _____

_____

Phenotype _____

_____

Genotype _____

_____

**28**

Autosomal dominant _____

_____

Autosomal recessive _____

_____

Sex-linked trait _____

_____

# Pre-Lab Exercise **28-2**

Embryonic Membranes

Label and color the embryonic membranes and structures depicted in **Figure 28.1** with the terms from Exercise 28-2 (p. 687). Use your text and Exercise 28-2 in this unit for reference.

FIGURE **28.1** Embryo and membranes (**A**) embryo in uterus; (**B**) chorionic villi

Label and color the structures of the fetal cardiovascular system depicted in **Figure 28.2** with the terms from Exercise 28-2 (p. 687). Use your text and Exercise 28-2 in this unit for reference.

FIGURE **28.2** Fetal cardiovascular anatomy

# EXERCISES

When two gametes unite in a process known as **fertilization**, a diploid **zygote** forms, and the incredible process of development begins. In this unit we examine human development and the unique anatomy and structures of the developing human. We conclude by exploring the basics of **heredity**—the passing of traits from parents to offspring.

## Exercise **28-1**

### Fertilization and Implantation

The first event in human development is **fertilization**, which takes place in the uterine tube when sperm cells encounter a secondary oocyte. Fertilization begins with a process called the **acrosomal reaction**, in which the acrosomes in the heads of the sperm release digestive enzymes by exocytosis. The acrosomal enzymes digest the corona radiata and the zona pellucida, the secondary oocyte's outer coverings.

When the zona pellucida has been cleared, the head and midpiece of a single sperm enter the oocyte. As the sperm enters, the secondary oocyte completes meiosis II, yielding an ovum and a second polar body. The sperm and ovum nuclei swell to become **pronuclei,** and their nuclear membranes rupture. The two sets of exposed chromosomes then fuse to form a single-celled structure called a **zygote** (**Figure 28.3**).

The zygote, also called a **conceptus,** undergoes multiple changes during the period of time known as **gestation.** Gestation is considered a 40-week period extending from the mother's last menstrual period to birth. The changes the conceptus undergoes can be divided into three processes:

1.  an increase in cell number,

2.  cellular differentiation, and

3.  the development of organ systems.

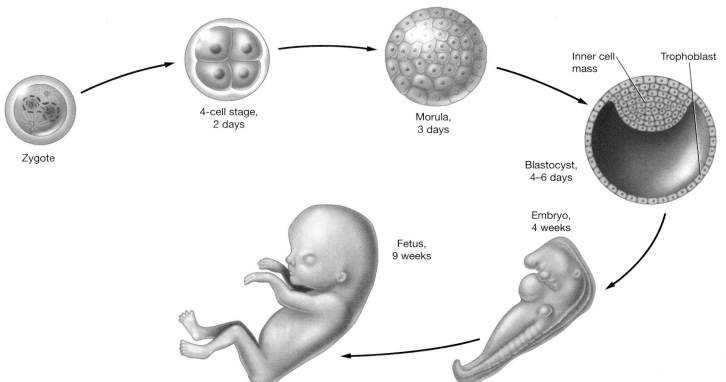

FIGURE **28.3** The stages of development

28

The increase in cell number begins about 30 hours after fertilization as the zygote travels down the uterine tube. At this time a process called **cleavage** occurs, in which the cell undergoes a series of mitotic divisions to produce two cells, then four, and so on. At about day 3, the conceptus reaches the uterus as a 16-cell ball called a **morula** (which means "little mulberry"). The morula floats around the uterus for another 2 to 3 days and continues to divide until it becomes a hollow sphere called a **blastocyst**. Note in **Figure 28.3** that the blastocyst has two populations of cells: the rounded **inner cell mass** and an outer layer of cells called the **trophoblast**.

At about day 6, the blastocyst adheres to the endometrium and begins the process of **implantation**. The implanting blastocyst secretes digestive enzymes that eat away the endometrial lining, and the endometrium reacts to the injury by growing over and enclosing the blastocyst. The process of implantation is generally complete by the second week, about the time a woman's menstrual period would begin.

## Procedure Time to Trace

FIGURE **28.4** Female reproductive tract

In this exercise you will be tracing the gametes from the point at which the male and female gametes meet to form a zygote to the point at which it implants into the endometrium. Trace the pathway using **Figure 28.4**, and also fill in the space below.

**Start:** Sperm and ovum meet in uterine tube _____

_____

_____

_____

_____

_____

_____

_____

_____

**End:** Implanted blastocyst

# Exercise 28-2

## Embryogenesis, Fetal Development, and Fetal Cardiovascular Anatomy

### MATERIALS
☐ Sequence of embryonic development model

☐ Female reproductive system with fetus model

☐ Fetal circulation model

As the blastocyst implants, it begins a process known as **embryogenesis** during which cellular differentiation takes place. During embryogenesis, the inner cell mass differentiates into three primary tissue layers known as **germ layers**, from which all other tissues arise. The outermost germ layer is called **ectoderm**, the middle germ layer is called **mesoderm**, and the innermost germ layer is called **endoderm**. The three primary germ layers are formed by the end of the second week, at which point the conceptus is considered an **embryo**.

The next six weeks of the embryonic period are marked by development of the **placenta**, formation of the extraembryonic membranes, and differentiation of the germ layers into rudimentary organ systems (**Figure 28.5**). The placenta begins to form at day 11, when the trophoblast forms the **chorion**, the fetal part of the placenta. Note in **Figure 28.5** that the chorion develops elaborate projections called the **chorionic villi** that eat into uterine blood vessels to create a space filled with maternal blood called the **placental sinus**. Nutrients and oxygen from maternal blood diffuse from the placental sinus to the chorionic villi and are delivered to the embryo by the large **umbilical vein**. Note that the umbilical vein is red, which reflects the fact that it carries oxygenated blood. Wastes are drained from the embryo via the paired **umbilical arteries**. All three vessels travel through the **umbilical cord**.

FIGURE **28.5** Embryo and membranes (**A**) embryo in uterus; (**B**) chorionic villi

During this period the extraembryonic membranes develop. The innermost membrane is called the **amnion**. The amnion completely surrounds the embryo (and later the fetus) and suspends it in **amniotic fluid** within the **amniotic cavity**. The amniotic fluid protects the embryo from trauma by allowing it to remain buoyant and also protects it from fluctuations in temperature. In addition, the cavity provides the embryo ample space in which to move, which is critical to early muscle development.

As the placenta and extraembryonic membranes form, the primitive organ systems develop. This process completes by the end of the eighth week, at which point the conceptus is considered a **fetus**. For the duration of gestation, the organ systems become progressively more specialized, and the fetus continues to grow and develop.

The fetus's cardiovascular system differs significantly from that of the neonate. Within the fetal cardiovascular system are three **shunts** that bypass the relatively inactive lungs and liver and reroute the blood to other, more metabolically active organs. The three shunts include the following (**Figure 28.6**):

1. **Ductus venosus.** The ductus venosus is a shunt that bypasses the liver. The umbilical vein delivers a small amount of blood to the liver but sends the majority of the blood through the ductus venosus to the fetal inferior vena cava. It closes around the time of birth and becomes the ligamentum venosum.

2. **Foramen ovale.** The foramen ovale is a hole in the interatrial septum that shunts blood from the right atrium to the left atrium. This allows the blood to bypass the collapsed fetal lungs. The foramen ovale closes about the time of birth, leaving a permanent indentation in the interatrial septum called the **fossa ovalis**.

3. **Ductus arteriosus.** The ductus arteriosus is a vascular bridge between the pulmonary artery and the aorta that bypasses the pulmonary circuit. It also closes around the time of birth and leaves behind a remnant called the **ligamentum arteriosum**.

- Arch of aorta
- Superior vena cava
- **Ductus arteriosus** (becomes ligamentum arteriosum)
- Pulmonary artery
- Pulmonary veins
- **Foramen ovale** (becomes fossa ovalis)
- Liver
- **Ductus venosus** (becomes ligamentum venosum)
- Hepatic portal vein
- **Umbilical vein** (becomes ligamentum teres)
- Umbilicus
- Inferior vena cava
- Abdominal aorta
- **Umbilical arteries** (become medial umbilical ligaments)
- Umbilical cord
- Placenta

■ Oxygenated blood
■ Mixed oxygenated and deoxygenated blood
■ Deoxygenated blood

FIGURE **28.6** Fetal circulatory system

## Procedure  Model Inventory of Fetal Development

Identify the following structures of development on models and diagrams, using your textbook and this unit for reference. As you examine the anatomical models and diagrams, record on the model inventory in **Table 28.1** the name of the model and the structures you were able to identify. When you have completed the activity, answer Check Your Understanding questions 1–4 (pp. 701–702).

1. Stages of development
   a. Zygote
   b. Morula
   c. Blastocyst
   d. Implanted blastocyst
   e. Embryo
   f. Fetus

2. Fetal membranes
   a. Amnion
      (1) Amniotic cavity
      (2) Amniotic fluid
   b. Chorion
      (1) Chorionic villi
   c. Yolk sac

3. Vascular structures
   a. Placenta
      (1) Placental sinus
      (2) Maternal arteries and veins
   b. Umbilical cord
      (1) Umbilical arteries
      (2) Umbilical vein
   c. Foramen ovale
   d. Ductus venosus
   e. Ductus arteriosus

TABLE **28.1** Model Inventory for the Stages of Development and Fetal Structures

| Model/Diagram | Structures Identified |
|---|---|
|  |  |
|  |  |
|  |  |
|  |  |
|  |  |

28

# Exercise 28-3

## Heredity

### MATERIALS

- ❏ Well plate
- ❏ Simulated blood of Ms. Rose, Ms. Magenta, Col. Lemon, Professor Purple, and Mr. Olive
- ❏ Anti-A, anti-B, and anti-Rh antisera
- ❏ PTC tasting papers
- ❏ Ishihara chart for color blindness

**Heredity** is defined as the passing of traits from parents to offspring. We refer to two properties of the cell when discussing traits:

1. **Phenotype** is the physical expression of a specific trait, such as hair color or blood type. Traits such as "blonde hair," "brown eyes," and "type A blood" refer to an individual's phenotype.

2. **Genotype** is the genetic makeup of an organism, or what a cell contains in its DNA. When someone performs DNA analysis, he or she is examining the genotype. We have two copies of each of our genes—one from the maternal side and one from the paternal side. The varying forms of a gene are known as **alleles**. The genotype is described using letters to stand for each of the alleles inherited. It is generally not possible to determine a person's precise genotype without doing DNA typing; however, if we know the phenotype, we can determine *possible* genotypes. Let's look at an example using blood groups:

a. From each parent, we can inherit one of the following ABO genes:
   (1) A allele: codes for A antigens
   (2) B allele: codes for B antigens
   (3) O allele: codes for neither A nor B antigen (please keep in mind that there is no "O" antigen)

b. From each parent, we can inherit one of the following Rh factor genes:
   (1) + allele: codes for the Rh antigen
   (2) − allele: does not code for the Rh antigen

We can put all of this information into a table to determine which phenotypes accompany which possible genotypes. Fill in the remainder of **Table 28.2**, following the pattern from the first two examples.

A person who inherits the same allele from both parents is said to be **homozygous** (e.g., the genotype AA is homozygous). A person who inherits different alleles from each parent is said to be **heterozygous** (e.g., the genotypes AB and BO are heterozygous).

TABLE **28.2** Phenotype and Genotype

| Genotype(s) | Phenotype |
|---|---|
| AA + +, AA + −, AO + +, AO + − | A + |
| AA − − , AO − − | A − |
| BB + +, BB + −, BO + +, BO + − | |
| | B − |
| OO + +, OO + − | |
| | O − |
| | AB + |
| AB − − | |

## **Procedure** Determining Paternity with Blood Typing

Blood typing is a simple and quick way to determine a child's potential heredity. Following is an example:

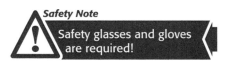

Child's phenotype:  B+

Mother's phenotype:  O−

Potential father 1's phenotype:  AB+

Potential father 2's phenotype:  O−

We start by first determining the possible genotypes of each person:

▌ Child's potential genotypes: BB+ +, BB+ −, BO+ +, BO+ −

▌ Mother's potential genotypes: OO− −

▌ Father 1's potential genotypes: AB+ +, AB+ −

▌ Father 2's potential genotypes: OO− −

Now what we have to do is attempt to determine which genes the child received from his mother. For the ABO gene, the mother can only give her child an O allele. For the Rh gene, the mother can only give her child a − allele. So, we can say that the child received an O allele and a − allele from mom.

The final step is to determine which gene(s) the child got from his father. The child has a B allele and a + allele that mom couldn't have given him, so these B and + alleles had to come from dad. Now let's look at both fathers' potential genotypes. We see that father 1 does have a B allele, but father 2 does not. We see also that father 1 has a + allele, but father 2 does not. Therefore:

We cannot exclude father 1 as a potential father of the child.

We *can* exclude father 2 as a potential father of the child.

Notice here we can't say for certain father 1 is indeed the child's father. All we can say is he cannot be excluded, and further testing is necessary. We can say for certain, however, that father 2 is completely excluded and does not have to be tested further.

For this procedure we return to our favorite cast of characters to sort out some sticky paternity issues using the genetics of blood typing. While investigating the murders from Unit 19, the investigators discovered that Ms. Magenta was hiding a secret: She had a 20-year-old daughter named Ms. Rose. The investigators are uncertain as to Ms. Rose's paternity, but they are certain that there are three possible fathers—Col. Lemon, Mr. Olive, and Professor Purple. You must determine if any of these three men may be ruled out as Ms. Rose's father on the basis of their blood types.

Following is the procedure for testing the blood samples. When you have completed the activity, answer Check Your Understanding question 5 (p. 702).

**1** Obtain a well plate, simulated anti-A, anti-B, and anti-Rh antisera and samples of simulated blood from Ms. Magenta, Ms. Rose, Col. Lemon, Mr. Olive, and Professor Purple.

**2** Place two drops of Ms. Magenta's blood into three separate wells.

**3** Place one drop of anti-A antiserum into the first well, and watch for a reaction. Record your results in **Table 28.3**.

**4** Place one drop of anti-B antiserum into the second well, and watch for a reaction. Record your results in **Table 28.3**.

**5** Place one drop of anti-Rh antiserum into the third well, and watch for a reaction. Record your results in **Table 28.3**.

**6** Repeat this procedure with the blood of Ms. Rose, Col. Lemon, Mr. Olive, and Professor Purple. Record your results in **Table 28.3**.

TABLE **28.3** Blood Type Results

| Blood Sample | Reaction with Anti-A? | Reaction with Anti-B? | Reaction with Anti-Rh? |
|---|---|---|---|
| Ms. Magenta | | | |
| Ms. Rose | | | |
| Col. Lemon | | | |
| Mr. Olive | | | |
| Prof. Purple | | | |

**7** Use your data from Table 28.3 to complete Table 28.4.

TABLE **28.4** Phenotypes and Potential Genotypes

| Blood Sample | Blood Type (Phenotype) | Potential Genotypes |
|---|---|---|
| Ms. Magenta | | |
| Ms. Rose | | |
| Col. Lemon | | |
| Mr. Olive | | |
| Prof. Purple | | |

**8** Interpret your results:

**a** Can any of the men be excluded as Ms. Rose's potential father? Explain.

**b** Can any of the men *not* be excluded as Ms. Rose's potential father? Explain.

## Procedure Testing the Ability to Taste PTC

The genetic effect known as **dominance** refers to the effect different alleles have on an individual's phenotype. An allele expressed as the phenotype is known as the **dominant allele**, and an allele not expressed that has little to no measurable effect on phenotype is known as the **recessive allele**. Dominant alleles are nearly always represented with capital letters and recessive alleles with lowercase letters.

The example of free and attached earlobes can be used to show the effect of dominant and recessive genes on phenotype. The gene that codes for free earlobes is dominant and written as *E*, whereas the gene that codes for attached earlobes is recessive and written as *e*. Three different combinations of genes and two phenotypes are possible:

*EE*, or homozygous dominant individuals, will have free earlobes;

*Ee*, or heterozygous individuals, will have free earlobes;

*ee*, or homozygous recessive individuals, will have attached earlobes.

Another dominant allele is one that determines the ability to taste **PTC** (phenylthiocarbamide), a bitter chemical found in foods such as brussels sprouts and coffee. About 75% of the population has the dominant allele, *T*, and can therefore taste PTC. The remaining 25% of the population is homozygous for the recessive allele, *t*, and cannot taste PTC. In the following procedure, you will test your ability and that of your classmates to taste PTC. You will then pool your data to determine the percentage of tasters and nontasters in your class.

**1** Obtain a piece of PTC paper, and place it on your tongue.

**2** If you can taste the PTC right away, you may remove the paper, and dispose of it. If you do not taste it right away, leave the paper on your tongue for up to 1 minute. If you cannot taste the PTC after 1 minute, remove the paper, and dispose of it.

Were you able to taste PTC? _____

**3** Pool the data from your class, and record it below:

_____ Number of tasters                    _____ Number of nontasters

_____ Percentage of class who are tasters     _____ Percentage of class who are nontasters

**4** Interpret your results:

**a** Which genotype(s) do the tasters have?

_____

**b** Which genotype(s) do the nontasters have?

_____

**c** How do the results from your class compare with the general population?

_____

_____

_____

## Procedure Testing Color Vision

Recall that humans have 23 pairs of chromosomes. Chromosomes 1–22 are called **autosomes**, and the last pair are the **sex chromosomes**. The two sex chromosomes are known as **X** and **Y**. Females have the genotype XX, and males have the genotype XY. Until now we have been discussing **autosomal traits**, or those passed down on autosomes. Certain traits, however, are passed down on the sex chromosomes, and these are known as **sex-linked traits**.

The inheritance of autosomal traits is not generally influenced by gender; however, this is not the case with the inheritance of sex-linked traits. We can use the example of the blood disease hemophilia to see how this works. Hemophilia is the result of a recessive allele ($h$) found on the X chromosome. The normal allele is the dominant allele $H$.

- A female with the genotype $X^H X^H$ will *not* have the disease.
- A female with the genotype $X^H X^h$ will *not* have the disease; however, she will be a "carrier" of the disease, capable of passing the allele to her offspring.
- A female with the genotype $X^h X^h$ will have the disease.
- A male with the genotype $X^H Y$ will *not* have the disease.
- A male with the genotype $X^h Y$ will have the disease.

Notice here that because males have only one X chromosome, they require only one copy of the recessive allele to express the trait and have the disease. Conversely, females still require two copies of the recessive allele to express the trait. For this reason, hemophilia is much more common in males than in females.

Another example of a sex-linked trait is color blindness, or the inability to distinguish between certain colors. The allele for the disease is a recessive allele $c$, and the allele for normal color vision is the dominant allele $C$. The following procedure will allow you to test your vision for abnormalities in color vision, which may result from the presence of this allele. The test uses a chart called an *Ishihara chart* that has a number embedded among a group of colored disks (**Figure 28.7**). A person with normal color vision will be able to read the number, whereas a person with color blindness will only be able to see randomly arranged colored dots. When you have completed the activity, answer Check Your Understanding question 6 (p. 702).

**1** Obtain an Ishihara chart. Hold the chart at a comfortable reading distance. Record what you see below:

_____

Was the test negative or positive for color blindness? (Note that "negative" means that your color vision was normal.)

_____

**2** Pool the results from your class, and record the data below:

_____ Number of females with normal color vision

_____ Number of females with impaired color vision

_____ Number of males with normal color vision

_____ Number of males with impaired color vision

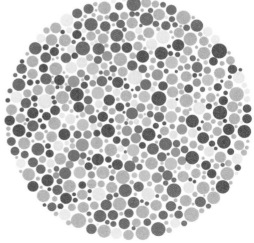

FIGURE **28.7** Ishihara chart for color blindness

28

**3** Interpret your results:

**a** What is/are the genotype(s) for females with normal color vision?

_____

_____

**b** What is the genotype for females with impaired color vision?

_____

_____

**c** What is the genotype for males with normal color vision?

_____

_____

**d** What is the genotype for males with impaired color vision?

_____

_____

**e** How did the number of males with impaired color vision compare with the number of females with impaired color vision? Explain this finding.

_____

_____

_____

_____

_____

_____

_____

_____

**UNIT 28 REVIEW**

## Check Your Recall

**1** *Matching:* Match the following terms with the correct definition.

_____ Chorion

_____ Blastocyst

_____ Fetus

_____ Gestation

_____ Zygote

_____ Morula

_____ Amnion

_____ Embryo

_____ Acrosome

_____ Germ layer

A. A single-celled structure formed after fertilization

B. The end of a sperm that contains digestive enzymes

C. A hollow sphere of cells that implants into the endometrium

D. The conceptus from two weeks to eight weeks of development

E. A 16-cell ball that reaches the uterus

F. The fetal membrane that contributes to the placenta

G. The early tissues that will go on to form all tissues in the body

H. The period that extends from the mother's last menstrual period to birth

I. The innermost fetal membrane

J. The conceptus from nine weeks to birth

**2** Which of the following is *not* one of the three basic changes a conceptus undergoes during development?

a. An increase in cell number

b. Cellular differentiation

c. Development of organ systems

d. Neoplasia

**3** The placenta is formed from

a. the chorion.

b. maternal tissues.

c. the amnion.

d. Both a and b are correct.

e. Both b and c are correct.

**28**

**4** Label **Figure 28.8** with the terms below.

☐ Ductus arteriosus
☐ Ductus venosus
☐ Foramen ovale
☐ Placenta
☐ Umbilical arteries
☐ Umbilical vein

FIGURE **28.8** Fetal circulatory system

**5** Label **Figure 28.9** with the terms below.

- ☐ Amnion
- ☐ Amniotic cavity
- ☐ Chorion
- ☐ Chorionic villi
- ☐ Maternal arteries and veins
- ☐ Placenta
- ☐ Placental sinus
- ☐ Umbilical arteries
- ☐ Umbilical vein
- ☐ Yolk sac

FIGURE **28.9** Embryo in uterus

**6** *Fill in the blanks:* The umbilical _____ carries oxygenated blood to the fetus, and the

_____ carries deoxygenated blood away from the fetus. The _____

bypasses the fetal liver, and the _____ and _____ bypass the fetal lungs.

**7** Variant forms of a gene are known as
   a. genotypes.
   b. alleles.
   c. phenotypes.
   d. antigens.

**8** *Fill in the blanks:* The _____ is the physical manifestation of a trait, and the _____
   is the genetic makeup of an organism.

**9** How do dominant and recessive alleles differ?

_____

_____

_____

_____

_____

_____

**10** Which of the following statements is *false*?
   a. Sex-linked disorders tend to affect females more than males.
   b. Chromosomes 1–22 are called autosomes.
   c. A person with two identical genes is homozygous.
   d. Blood typing may be used to conclusively exclude paternity, but it cannot conclusively prove paternity.

Name _____

Section _____ Date _____

# Check Your Understanding

**Critical Thinking and Application Questions**

**1** Predict the effects of failure of the foramen ovale and/or the ductus arteriosus to close shortly after birth.

_____

_____

_____

_____

_____

_____

_____

_____

**2** The use of certain medications, such as nonsteroidal anti-inflammatories (e.g., ibuprofen), in the seventh to ninth months of pregnancy increases the risk for premature closure of the ductus arteriosus. What problems might this potentially cause for the fetus and the neonate?

_____

_____

_____

_____

_____

_____

**3** A condition called *placenta previa* occurs when the placenta is too low in the uterus. Why do you think this is a cause for concern?

_____

_____

_____

_____

_____

_____

**4** Active labor generally is considered to begin about the time the "water breaks." What is the "water," and which membranes rupture to release it? What could be the potential consequences if the membranes were to rupture too early?

_____

_____

_____

_____

_____

**5** Further investigation has revealed that Ms. Magenta is in fact *not* Ms. Rose's mother. Because of a freak baby-switching accident, Mrs. Feather is really Ms. Rose's biological mother. Mrs. Feather's blood type is determined to be AB + . Col. Lemon, Professor Purple, and Mr. Olive remain Ms. Rose's potential fathers.

   **a** What is Mrs. Feather's phenotype? What are her possible genotypes?

_____

_____

_____

   **b** Can any of the potential fathers be excluded? Explain.

_____

_____

_____

   **c** What if Ms. Rose's blood type were O + ? How would this change the scenario, both *maternally* and *paternally*?

_____

_____

_____

**6** You have just tested two patients' color vision, one male and one female, and found that both have color blindness. Can you conclusively determine the male's genotype? Can you conclusively determine the female's genotype? Explain why or why not for both patients. What are each patient's potential genotypes?

_____

_____

_____

_____

_____

_____

# The Big Picture of Human Anatomy and Physiology

## 29

## OBJECTIVES

*Once you have completed this unit, you should be able to:*

■ Find physiological connections between different organs and systems in the body.

■ Review the physiology of the organ systems we have covered.

■ Trace the pathway of the physiological response to a stimulus through various body systems.

At this point in our study of anatomy and physiology, we have examined each organ system and its individual organs and have studied how the parts fit together to make the whole. Along the way, we also have developed an appreciation for how the form of the system, organ, tissue, and cell all follow the primary functions of each part. Now, in this last unit, we will develop an appreciation for the big picture of how the organs work together to maintain homeostasis.

# EXERCISES

## Exercise 29-1

### Physiological Connections

**MATERIALS**
- ❑ Anatomical models: human torso
- ❑ Laminated outline of human body
- ❑ Water-soluble marking pens

In anatomy and physiology the anatomical connections between organs and system are usually easy to see. All organs are connected by the cardiovascular system, bones are united by ligaments and cartilage, and accessory organs of the digestive system are joined to the alimentary canal by common ducts. What is sometimes difficult to see is the *physiological* connections that exist between all organs and organ systems.

Your final task in this lab manual is to find physiological connections between seemingly unrelated groups of organs. The physiological connection between the salivary glands, stomach, and duodenum is easy to find, but the connection between the heart, bone, and thyroid may be a bit less obvious. Following are some hints, and an example, to make this task easier.

▌ Begin each exercise by listing the functions of each organ listed. Some functions are obvious, and others may require a bit of digging on your part. Take, for example, the skin. The skin has the obvious functions of protection and thermoregulation. But don't forget about its other tasks: The skin also synthesizes vitamin D when it is exposed to certain wavelengths of ultraviolet radiation, it excretes small amounts of metabolic wastes, and it serves immunological functions. These less-discussed functions are easy to overlook but are vitally important.

▌ Beware of the tendency to fall back to anatomical connections. When given the combination "lungs, kidneys, and blood vessels," you may be tempted to say that the lungs breathe in oxygen, delivered by the blood vessels to the kidneys. That's all true, but this is actually an anatomical connection. With some creative thinking on your part, you could come up with something much better. Remember that lung tissue makes angiotensin-converting enzymes, and, having said that, the connection between the blood vessels and the kidneys becomes more obvious (at least I hope it does).

▌ When in doubt, look to the endocrine system. Remember that many different organs, not just the organs discussed in the endocrine unit, make hormones. The heart, kidneys, and even adipose tissue all function in some respects as endocrine organs. For the organs that have a primary endocrine function, don't overlook some of the hormones they make. For example, the thyroid gland makes T3 and thyroxine but also calcitonin, which students tend to forget.

*Here's an example of physiological connection:*

## Lungs, Kidney, and Bone

**Connection:** The lungs have the function of *gas exchange*. When this exchange is inadequate, and oxygen intake is insufficient, the kidneys detect the decreased partial pressure of oxygen in the blood. The kidneys, in turn, make the hormone *erythropoietin*, which acts on the bone marrow to stimulate increased *hematopoiesis* and an increased rate of red blood cell maturation.

Wasn't that easy?

**29**

Now it's your turn.

## Part 1: Heart, Parathyroid Glands, Bone

*Hint:* Think electrophysiology of the heart.

**Connection:** _____

_____

_____

_____

_____

_____

_____

_____

## Part 2: Small Intestine, Lymphatic Vessels, Liver

*Hint:* How does the body process dietary fats?

**Connection:** _____

_____

_____

_____

_____

_____

_____

_____

## Part 3: Spleen, Liver, Large Intestine

*Hint:* Think about the functions of the spleen with respect to erythrocytes.

**Connection:** _____

_____

_____

_____

_____

_____

_____

29

**Part 4:** Adrenal Glands, Blood Vessels, Kidney

*Hint:* What do the different regions of the adrenal gland make?

**Connection:** _____

_____

_____

_____

_____

_____

_____

**Part 5:** Pancreas, Sympathetic Nervous System, Liver

*You're on your own for this one!!*

**Connection:** _____

_____

_____

_____

_____

_____

_____

29

## Exercise 29-2

### Tracing—The Big Picture

Throughout this manual, you have done many different tracings of both anatomical and physiological processes. Let's now put many of those tracings together to get a bigger picture of how the body functions as a whole. We are going to trace the response to a stimulus you have likely experienced: an angry wasp coming toward you. The tracing has been broken up into parts so that it is easier to manage. As with other tracings, hints have been placed throughout to help you.

### Part 1: Sight and Sound of Wasp to Perception and Interpretation in the Brain

In this first part, Professor Purple has just seen a wasp. You will trace the pathway light and sound takes from when he first hears and sees the wasp to the perception of these stimuli in the brain. You will trace the anatomical path the light and sound takes from the point where it enters his eye and ear to its interpretation in his brain. For the first pathway, don't forget about the liquids in the eyeball through which light passes. You can use Unit 15 for reference.

**Start:** See the wasp → light enters the _____ → _____ humor → _____ →

_____ → _____ → depolarization of _____ →

propagation of _____ along the _____ → _____ →

_____ → _____ lobe → _____ →

**End**

**Start:** Hear the wasp → sound enters the _____ → _____ →

_____ → malleus → _____ → _____ →

_____ window → _____ → _____ depolarize →

_____ propagated along the _____ → _____ lobe →

perception and interpretation → **End**

### Part 2: Initiation of the Sympathetic Nervous System Response

Now let's trace Professor Purple's natural response to an angry wasp: fear. As you've learned, fear triggers activation of the sympathetic nervous system, which prepares the body for a fight or flight response. For help, see Unit 14 on the PNS and ANS.

**Start:** Hypothalamus and cerebral cortex trigger the SNS response → sympathetic postganglionic neurons release

_____ onto target organs → heart rate and force of contraction _____ →

smooth muscle in bronchioles _____ → digestive and urinary functions _____ →

blood vessels to skeletal muscles _____ → blood glucose _____ →

body primed for fight and flight → **End**

## Part 3: Sting and Inflammatory Response

In this next part of the exercise, poor Professor Purple has gotten stung by the wasp on his right hand. As with any tissue injury, a wasp sting leads to an inflammatory response, so you will be tracing the steps of this response. You may wish to consult Unit 20 and your textbook for reference.

**Start:** Wasp stinger damages tissue → damaged cells release _____ → local blood vessels

_____ → capillary permeability _____ → _____

(*hint*: a cell type) are attracted and activated → _____ attract additional _____

(*hint*: another cell type) → _____ ingest _____ → warmth, redness, swelling,

and pain of the inflammatory response → **End**


## Part 4: Ingestion and Delivery of Medication

Now Professor Purple has taken an acetaminophen tablet for the pain of his aching right hand. You will first trace the path the tablet takes through the organs of the alimentary canal to its absorption into the bloodstream and then to its delivery to the lateral hand by arterioles and capillary beds via the radial artery. Remember that you will need to first go through a capillary bed, then through the veins (don't forget the hepatic portal system) and the heart, and finally through arteries. See Units 17 and 24 for reference.

**Start:** Ingested at the mouth → _____ → _____ → _____ →

_____ → enter the enterocyte → diffuse into _____ →

_____ vein → _____ vein → _____ system →

_____ veins → inferior vena cava → _____ → _____ →

_____ → _____ → _____ capillaries →

_____ → _____ → _____ → ascending aorta →

_____ arch → _____ → _____ →

_____ → _____ → radial artery → **End**

## Part 5: Excretion of the Medication

In this final part of the tracing, you are following the path Professor Purple's metabolized acetaminophen molecule takes as it leaves the body via the urinary system. You will start from the capillary beds of the liver, where the acetaminophen is metabolized, and trace it through the blood vessels to the kidneys. Once it reaches the kidneys, you will trace its pathway through the nephron and out of the body. You can use Units 17 and 25 for reference.

**Start:** Capillary beds of the hepatic portal system → _____ veins → _____ →

right _____ → _____ → _____ →

_____ → _____ capillaries →

_____ → _____ → _____ →

ascending aorta → _____ arch → _____ →

_____ → _____ artery → segmental artery →

_____ → _____ →

_____ → _____ → filtration at the

_____ → _____ → _____ →

_____ → _____ →

_____ → _____ →

minor calyx → _____ → _____ →

_____ → _____ → urethra → **End**

---

# Congratulations!

*You've done what you probably thought was impossible—
completed this manual! I truly hope it has been fun
(at least a little!) and that you have enjoyed it as much
as I have. Best wishes with your future plans,
down whichever path they may take you.*

# Photo Credits

Morton Publishing expresses thanks to the following sources for allowing us to use their photos.

**Unit 1**
Opener: SPL/Photo Researchers
Fig. 1.7, Fig. 1.8: Kent M. Van De Graaff and John Crawley, *A Photographic Atlas for the Anatomy and Physiology Laboratory*, 6E © Morton Publishing
Fig. 1.11A–B: Neil Borden/Photo Researchers
Fig. 1.11C: SPL/Photo Researchers

**Unit 2**
Opener: Ken Eward/Photo Researchers
Fig. 2.5: SPL/Photo Researchers
Fig. 2.6: Justin Moore
Fig. 2.7: SPL/Photo Researchers

**Unit 3**
Opener: SPL/Photo Researchers
Fig. 3.1: Courtesy of Olympus America

**Unit 4**
Opener: SPL/Photo Researchers
Fig. 4.5: SPL/Photo Researchers
Fig. 4.7: GIPhotoStock/Photo Researchers
Fig. 4.9: Michael Abbey/Photo Researchers

**Unit 5**
Opener: SPL/Photo Researchers
Fig. 5.2A: Biophoto Assoc./Photo Researchers
Fig. 5.2B: Biology Pics/Photo Researchers
Fig. 5.2C: Spike Walker/Photo Researchers
Fig. 5.2D: David Phillips/Photo Researchers
Fig. 5.2E: Biophoto Assoc./Photo Researchers
Fig. 5.2F–G: SPL/Photo Researchers
Fig. 5.2H: Biophoto Assoc./Photo Researchers
Fig. 5.3A: Biophoto Assoc./Photo Researchers
Fig. 5.3B: Visuals Unlimited
Fig. 5.3C: Erin Amerman
Fig. 5.3D–F: Biophoto Assoc./Photo Researchers
Fig. 5.3G: Erin Amerman
Fig. 5.3H: Spike Walker/Photo Researchers
Fig. 5.3I: Chuck Brown/Photo Researchers
Fig 5.3J: SPL/Photo Researchers
Fig. 5.3K: Biophoto Assoc./Photo Researchers
Fig. 5.4A: Eric Grave/Photo Researchers
Fig. 5.4B: SPL/Photo Researchers
Fig. 5.4C: Getty Images
Fig. 5.4D: Biophoto Assoc./Photo Researchers
Fig. 5.5: Spike Walker/Photo Researchers

**Unit 6**
Opener: SPL/Photo Researchers
Fig. 6.5A: SPL/Photo Researchers
Fig. 6.5B: Garry Delong/Photo Researchers
Fig. 6.6: Fabrication Enterprises

**Unit 7**
Opener: SPL/Photo Researchers
Fig. 7.3: SPL/Photo Researchers
Fig. 7.6: Justin Moore
Fig. 7.8: A. Travelogue/Photo Researchers

**Unit 8**
Opener: SPL/Photo Researchers
Fig. 8.3: VideoSurgery/Photo Researchers
Fig. 8.4: David Bassett/Photo Researchers
Fig. 8.5: Getty Images
Fig. 8.6: SPL/Photo Researchers

Fig. 8.7: VideoSurgery/Photo Researchers
Fig. 8.8A–B: Martin Shields/Photo Researchers
Fig. 8.9: SPL/Photo Researchers
Fig. 8.11: SPL/Photo Researchers
Fig. 8.13A–B: Kent M. Van De Graaff and John Crawley, *A Photographic Atlas for the Anatomy and Physiology Laboratory*, 6E © Morton Publishing
Fig. 8.16A–B, Fig. 8.17, Fig. 8.20A–B: Video Surgery/Photo Researchers
Fig. 8.23A: Kent M. Van De Graaff and John Crawley, *A Photographic Atlas for the Zoology Laboratory*, 6E © Morton Publishing
Fig. 8.23B: VideoSurgery/Photo Researchers
Fig. 8.24: VideoSurgery/Photo Researchers
Fig. 8.36: SPL/Photo Researchers

**Unit 9**
Opener: SPL/Photo Researchers
Fig. 9.5A: VideoSurgery/Photo Researchers

**Unit 10**
Opener: SPL/Photo Researchers
Fig. 10.4B: VideoSurgery/Photo Researchers
Fig. 10.6B: Mike Leboffe
Fig. 10.7B: Biophoto Assoc./Photo Researchers
Fig. 10.12A: SPL/Photo Researchers
Fig. 10.12B: Biophoto Assoc./Photo Researchers

**Unit 11**
Opener: SPL/Photo Researchers

**Unit 12**
Opener: Hybrid/Photo Researchers
Fig. 12.2: Spike Walker/Photo Researchers
Fig. 12.7: Biophoto Assoc./Photo Researchers

**Unit 13**
Opener, Fig. 13.9: SPL/Photo Researchers
Fig. 13.12C: Martin Rotker/Photo Researchers
Fig. 13.14, Fig. 13.15, Fig. 13.16: Kent M. Van De Graaff and John Crawley, *A Photographic Atlas for the Anatomy and Physiology Laboratory*, 6E © Morton Publishing

**Unit 14**
Opener: 3D4Medical/Photo Researchers
Fig. 14.11: SPL/Photo Researchers

**Unit 15**
Opener: VideoSurgery/Photo Researchers
Fig. 15.8, Fig. 15.9, Fig. 15.10: Kent M. Van De Graaff and John Crawley, *A Photographic Atlas for the Anatomy and Physiology Laboratory*, 6E © Morton Publishing

**Unit 16**
Opener: SPL/Photo Researchers
Fig. 16.3: VideoSurgery/Photo Researchers
Fig. 16.5, Fig. 16.6, Fig. 16.7: Justin Moore
Fig. 16.9: SPL/Photo Researchers

**Unit 17**
Opener, Fig. 17.17: SPL/Photo Researchers

**Unit 18**
Opener: BSIP/Photo Researchers

Fig. 18.2: Bruce Levy/*Journal of Bone and Joint Surgery* 92(12), September 15, 2010, Fig. 2, copyright © *JBJS*

**Unit 19**
Opener: SPL/Photo Researchers
Fig. 19.1: Biophoto Assoc./Photo Researchers and Michael Ross/Photo Researchers
Fig. 19.2: Justin Moore
Fig. 19.4: Biophoto Assoc./Photo Researchers

**Unit 20**
Opener: SPL/Photo Researchers
Fig. 20.4A: Michael Abbey/Photo Researchers
Fig. 20.4B–C: Biophoto Assoc./Photo Researchers

**Unit 21**
Opener: SPL/Photo Researchers
Fig. 21.3, Fig. 21.4B: VideoSurgery/Photo Researchers
Fig. 21.6A: Erin Amerman
Fig. 21.6B–D: Biophoto Assoc./Photo Researchers
Fig. 21.7A–B: Justin Moore

**Unit 22**
Opener: Carolyn McKeone/Photo Researchers
Fig. 22.3: SPL/Photo Researchers
Fig. 22.5: Justin Moore

**Unit 23**
Opener: A. Travelogue/Photo Researchers
Fig. 23.5B: Visuals Unlimited
Fig. 23.7: VideoSurgery/Photo Researchers
Fig. 23.10B: Visuals Unlimited
Fig. 23.12, Fig. 23.13, Fig. 23.14, Fig. 23.15: Biophoto Assoc./Photo Researchers
Fig. 23.16B: Spike Walker/Photo Researchers

**Unit 24**
Opener: SPL/Photo Researchers
Fig. 24.10: VideoSurgery/Photo Researchers
Fig. 24.11: Garry Delong/Photo Researchers
Fig. 24.12: SPL/Photo Researchers
Fig. 24.13A: Biophoto Assoc./Photo Researchers
Fig. 24.13A: Erin Amerman
Fig. 24.13B: SPL/Photo Researchers
Fig. 24.13C: Biophoto Assoc./Photo Researchers

**Unit 25**
Opener, Fig. 25.3: Scott Camazine/Photo Researchers

**Unit 26**
Opener: A. Travelogue/Photo Researchers
Fig. 26.3: Michael Ross/Photo Researchers
Fig. 26.4, Fig. 26.5: Biophoto Assoc./Photo Researchers

**Unit 27**
Opener: SPL/Photo Researchers
Fig. 27.10: Michael Abbey/Photo Researchers
Fig. 27.12: Biophoto Assoc./Photo Researchers
Fig. 27.14: Erin Amerman

**Unit 28**
Opener: Petit Format/Photo Researchers

**Unit 29**
Opener: SPL/Photo Researchers

# Index

iron-deficiency anemia, 480
ischial tuberosities, 178
ischium, 176, 178
Ishihara chart, 695
islets of Langerhans, 567, 637
isopropyl alcohol, 45
isovolumetric contraction, 447

jaw bones, 165
jejunum, 559
joint(s), 159
  classification, 207
  distal radioulnar and proximal radioulnar, 175
  proximal and distal tibiofibular, 178
  synovial, 207, 209–210, 214
joint capsule, 209
jugular veins, 424, 491, 572
juxtaglomerular (JG) apparatus/cells, 596

keratin/keratinocytes, 89, 123, 124
keratinized papillae, 560
kidneys, 420, 424, 452, 593–595, 600, 601, 602, 621, 635, 636, 704
knee cap/joint/muscles, 146, 178, 212–213, 260, 262

labia majora/minora, 662
lacrimal bones, 165
lacrimal apparatus/gland, 165, 363–364
lacteal, 491, 572
lactiferous sinuses, 663
lacunae, 97, 142
lambdoid suture (cranium), 159
lamellae, 98, 141, 143
lamellated (Pacinian) corpuscles, 124
*lamina reticularis*, 89
laminae, 168
laryngopharynx, 517, 522
larynx, 517, 522
lateral collateral ligaments (LCL), 212
lateral rectus muscle, 338
lateral term and view of body cavity, 6, 9
latissimus dorsi, 257, 259
leg
  bone, 178
  muscles, 262, 263
  nerves to, 345
length-tension relationship, 238
leptin, 635
leukocytes, 98, 467, 470
Leydig cells, 637, 667
ligaments, 209, 212, 314, 364, 662, 704
ligamentum arteriosum/venosum, 688
light microscopes, 52–53, 54
lingual tonsil, 491
lipase, 41
lipid absorption/digestion, 41, 571
lipid synthesis, 66
lips, muscles, 257
liver/lobules, 424, 560, 567, 568, 571, 593, 637
local potentials, 291
long bones, anatomy of, 148
longissimus, 259
loop of Henle, 596
lower limb bones, 178–179

Lugol's iodine, 573
lumbar nerves/lumbar plexus, 343, 345
lumbar vertebrae, 168, 170
lungs, 515, 517, 537, 539, 704
luteinizing hormone, 635
lymphatic system/vessels, 335, 491, 495, 559, 572
lymphocytes (B and T), 467, 491, 495, 498, 508
lymphoid organs, 492
lysosomes, 66
lysozyme, 560

macrophages, 495, 498
macula densa (kidney), 596
macula lutea (eye), 365
magnification, 52
male reproductive system, 635, 637, 659–660, 667, 685, 695
malleolus, lateral, 178
malleus (hammer), 370
mammary glands, 635, 662–663
mandibular angle/condyle/rami, 165
manubrium, upper, 171
mastication, muscles of, 257, 337
mastoid region and process, 159
maxillae/maxillary sinuses, 165
mechanical digestion, 557, 560
medial collateral ligaments (MCL), 212
median nerve, 345
mediastinum, 10, 393, 515, 517, 637
medulla (lymph node), 495
medulla oblongata, 308, 313
medullary cavity/pyramids, 148, 593
meibomian glands, 363
meiosis, 665, 666, 667, 668, 685
melatonin, 308, 635
membranous labyrinth, 370
memory cells, 498
meningeal branch, 343
meninges/meninx, 309, 310, 313
menisci, 209, 212
Merkel disks, 128
mesenchyme, 97
mesenteric arteries/veins, 420, 424
mesentery, 561
mesoderm, 687
mesovarium, 662
metabolic rate, 636
metacarpals, 175
metaphase, 78
metatarsals, 178
methylene blue, 68
metopic (frontal) suture, 165
micelles, 571
microglial cells, 288
micrograph diagrams, drawing, 69
microscopy, hints and tips, 56
microtubules, 65, 66, 78, 285
microvilli, 65, 89, 571, 601
micturition, 596
midbrain, 308
midsagittal sections, 16, 17
mineralocorticoids, 637
mitochondria, 66, 667
mitochondrial cytopathies, 83
mitosis, 77–78, 665, 668, 686
mitotic spindle, 78
mitral valve, 394, 447
monocytes, 467
monofilament, 128
monoglycerides and monosaccharides, 571

morula, 686
motor end plate, 233, 235
motor nerves, 337–338
motor neuron, 233, 235, 314
mouth, 557, 571
mouth muscles, 257
mucosal-associated lymphoid tissue (MALT), 491, 495, 564
mucous membranes/mucosae, 498, 522
mucus/mucosa
  alimentary canal/digestive, 564, 565, 566, 567
  female reproductive, 662
  salivary gland, 560
  urinary bladder, 601
multipolar neuron, 286
murder mysteries (blood typing), 473
muscle action potential, 235
muscle fibers, 229, 557
muscles/muscle tissue, 89, 90, 106, 107, 123, 229, 257, 264–265, 291
  abdominal, 259
  actions (origin and insertion), 268, 270
  adductor group, 260
  back, 257, 259
  cardiac, 393
  extraocular, 364
  eye/eyeball, 337
  forearm, 259–260, 261
  head, neck, and face, 257
  hip, 260
  intercostal, 258, 259, 343, 537, 538
  knee, 260
  leg, 262, 263
  lower limb, 262
  *Myasthenia gravis*, 247
  pulmonary, 537–538
  shoulder, 259
  stomach, 557
  trunk, 257–259
  upper limb, 257
  wrist, 259–260. *See also* cardiac muscle tissue; skeletal muscles; smooth muscle
muscle spindles, 347
muscularis externa/mucosa, 563, 564
musculocutaneous nerve, 344
myelin sheath, 287, 288
myocardial infarction, 396
myocardium, 394, 396
myocytes, 106, 403
myofibrils, 229, 230
myofilaments, 106, 230
myometrium, 662
myosin, 230, 235

nail anatomy, 123, 124, 125
nasal bones/cavity/conchae/septum, 165, 375, 515
nasopharynx, 491, 515, 522
neck muscles, 257
negative feedback loop, 642
nephrons/loops, 593, 594, 601, 621
nerve(s), 335, 561
  cranial, 335, 337–338, 375
  mixed, 343
  spinal, 313, 314, 335, 343–344
nerve impulse. *See* action potentials
nerve plexus, 333
nervous system/tissue, 89, 90, 110, 123, 285, 287, 291, 635

neurilemma, 287
neurofilaments, 285
neuroglial cells, 110, 285, 287, 307
neurohypophysis, 635
neuromuscular junction (NMJ), 233, 287
neurons, 110, 285, 286, 287, 291
neurotransmitters, 285, 287, 291, 348, 635
neutrophils, 467, 498, 499
nipples, 663
Nissl bodies, 110, 285
nodes of Ranvier, 287
nonbuffered solutions, 40
nonpolar covalent bonds, 36, 43
norepinephrine, 348, 638
normal sinus rhythm, 457
nose, 165
nuclear envelope/pores, 66
nuclear membrane, 78, 685
nucleic acids, 46, 66
nucleus/nuclei, 65, 66, 110, 285, 308
nutrients, 557, 571

objective lens (microscope), 52, 53
obliques, internal and external abdominal, 259
oblique section, 16
obstructive diseases (respiratory), 540
obturator foramen, 176
occipital bone, 159
ocular lens (microscope), 52, 53
oculomotor nerve, 337
odontoid process, 169
oil-immersion lens (microscope), 52
olecranon process and fossa, 174, 175, 268
olfactory epithelium/foramina/nerves/ receptor cells, 165, 286, 337, 375
oligodendrocytes, 288
oligosaccharides, 571
omentum, greater and lesser, 561
oocytes, 637, 662, 668, 669, 685
oogenesis, 667, 668
oogonia, 668, 669
optic chiasma/disc/nerve/tracts, 337, 365
orbicularis oculi/oris, 257
orbit (facial bone), 165
organ(s), 19, 20, 141, 685
  embryonic, 687, 688
organelles, 65, 66, 68, 110
organology, 111
origin, muscle, 257, 268
oropharynx, 491, 517, 522
osmosis, 65, 74
osseous tissue, 98, 141, 144
osteoblasts/osteoclasts, 141, 142, 636
osteocytes, 98, 142
*Osteogenesis imperfecta*, 152
osteogenic cells, 141
osteomalacia, 152
osteons, 141–142
osteoporosis, 652
ovary/ovaries/ovulation/ovum, 635, 637, 659, 662, 663, 668, 669, 672, 685
oxygen, 518, 543. *See also* gas exchange
oxytocin, 635

pacemaker/nonpacemaker cells, 403, 404

pain
  neurons transmitting, 286
  sensing, 363, 377, 498
palatine bones/processes, 165
palatine tonsils, 491
palpebrae, 363
pancreatic duct/enzymes/islets/juice, 561, 567, 571, 635, 637, 639, 640
papillae (tongue), 375, 560
papillary ducts (kidney), 596
papillary layer (dermis), 124
papillary muscles, 394
parafollicular cells, 636
paranasal sinuses, 159, 165, 515
parasagittal sections, 16
parasympathetic nervous system (PSNS), 334, 335, 338, 348, 449
parathyroid glands/hormone (PTH), 635, 636, 652
parietal bones, 159
parietal layer (serous membranes), 9
parietal pericardium, 393
parietal pleura, 10, 515
parotid glands, 560
patella, 146, 178
patellar reflex, 347
pectineus, 260
pectoral girdle, 159, 171, 173
pectoralis minor/major, 257, 259
pedicles, 168
pedis artery, 420
pelvic bones/cavity, 10, 159, 173, 176–178, 659, 660, 662, 663, 674
penis, 660
pepsin, 571, 575
pepsinogen, 571
perception, 363
perforating canals/fibers, 141, 142
pericardial cavity, membranes, and pericardium, 10, 393
perilymph, 370
perimetrium, 662
perimysium, 229
perineurium, 335
periosteum, 141
peripheral blood smear, 468
peripheral nervous system, 285, 314, 335, 348
peripheral process, 286
peripheral resistance, 452, 453
peripheral vascular disease, 455
peripheral vision, 365
peristalsis, 557, 564
peritoneal membranes and cavity, 10, 561
peritubular capillaries, 594
perivascular feet, 287
peroneal nerve, 345
peroxisomes, 66
perpendicular plate, 165
petrous region of temporal bone, 159
Peyer's patches, 495
pH
  acidic and alkaline, 37
  blood, 543
  buffers and, 40
  definition and measurement, 35, 37, 38, 39
  stomach, 39
phagocytes, 66, 288, 491, 495, 498
phalanges, 176, 178
pharyngeal tonsils, 491

pharyngitis, 508
pharyngotympanic tube, 370
pharynx, 338, 370, 515, 517, 557
phenol group, 45
phenol red indicator, 41, 577
phenotype, 690, 694
phenylthiocarbamide (PTC), 694
phospholipids, 43
photoreceptors, 365
phrenic nerve, 343
physiological connections between organs, 704
pia mater, 311, 314
pineal gland, 308, 635
pinna, 370
pisiform bone, 176
pituitary gland, 159, 308, 635, 636
pivot joints, 210
placenta/sinus, 687, 688
planes of motion, synovial joints, 209, 210
planes of section, anatomical, 16, 17
plantarflexion, 262
plasma, 98, 467, 498
plasma membrane, 62, 65, 66, 71
  in digestive system, 571
  in endocrine system, 635
  in nervous system, 285, 287, 291
  See also axolemma
platelets, 98, 467
pleural cavities and membranes (thoracic), 10, 515
pleural effusion, 531
pleural membranes (lung), 515
plicae circulares, 566
podocytes, 594, 596
polar body, 668, 685
polar covalent bonds/compounds/solutes, 36, 43
polymorphonucleocytes, 467
polypeptides, 571
polysaccharides, 571
pons, 308
popliteal artery, 420
popliteal vein, 426
portal triads, 567
portal vein, 567
posterior cruciate ligament (PCL), 212
posterior view of body/posterior body cavity, 7, 8, 9
postsynaptic neuron, 287, 291
potassium ions in neuronal action potential, 291
potassium permanganate dye, 72
potassium-sparing diuretics, 652
power stroke (muscle), 235
pregnancy, 637
presynaptic neuron, 287, 291
progesterone, 637
prolactin, 635
prolapse, 394
pronuclei, 685
prophase, 78
prostate gland, 659
proteases, 50
protein(s)
  antimicrobial, 498
  complement system, 498
  digestion, 571
  enzymes and, 41, 46
  filaments, 65
  muscle, 229, 235
  See also collagen
protein fibers, 89, 97, 98, 144

protein synthesis, 65, 66
pseudostratified ciliated columnar epithelium, 91, 515, 522
pseudounipolar neurons, 286
pterygoid processes, 159
puberty, 667, 668
pubic arch/bone/symphysis, 176, 178
pudendal nerve, 345
pulmonary system arteries (trunks)/capillaries/ gas exchange/veins/ventilation (inspiration and expiration)/volume, 394, 419, 447, 518, 522, 537, 538, 539, 540, 543
pulp, red and white, 495
pulse and pulse palpation, 449
Purkinje system, 403, 404
pus, 498
P wave, 457
pyloric antrum, 447
pylorus, 557
pyrogens, 498

QRS complex, 457
quadriceps femoris muscle group, 260

radial arteries/pulse/veins, 174, 419, 425
radial nerve, 344
radius (bone), 159, 174, 175, 268
rami, superior and inferior, 178
range of motion, 209
recruitment, 410
rectum, 559, 662
rectus abdominis muscle, 259
red blood cells, 98, 467
reflex/arc, 346, 347
regional terms, 7
regurgitation, 447
renal arteries/capsule/columns/corpuscle/cortex/medulla/tubule/veins, 420, 424, 593–594, 596, 601
repolarization. See depolarization
reproductive system. See female reproductive system; male reproductive system
residual volume (RV), 539, 540
respiratory system capacities/disorders/epithelium/membrane/rate/tract/volumes, 515, 522–523, 539, 540, 543
restrictive diseases (pulmonary), 540
rete testis, 659
reticular cells, 97
reticular layer (dermis), 124
retroperitoneal organs, 10
Rh factor, 476, 690
ribosomes/ribosomal RNA, 65, 66
ribs, 159, 171, 259
rickets, 152
rigor mortis, 247
Rinne test, 372
Romberg test, 374
rotator cuff muscles, 259
rough endoplasmic reticulum (RER), 66
rugae, 557, 596

saccule, 370
sacral foramina/sacral crest, median, 171
sacral nerves and plexus, 343, 345
sacroiliac joints, 171

sacrum, 168, 171
saddle joints, 210
sagittal plane, 16
sagittal suture (cranium), 159
salicylic acid, 43, 45
saliva, 560
salivary amylase, 560, 571
salivary glands, 338, 557, 560
saphenous vein, 426
sarcolemma, 229, 233, 235
sarcomere, 231, 238
sarcoplasm, 229, 235
sarcoplasmic reticulum (SR), 229
sartorius muscle, 260
satellite cells, 287
scapula, 173, 259, 268
Schwann cells, 287, 288
sciatic notch and nerve, 178, 345
sclera, 364
scrotum, 659
sebaceous glands, 124, 125
sebum, 125
secondary active transport mechanisms, 571
selectively permeable membrane, 74
sella turcica, 159
semicircular canals and ducts, 370
semilunar valves, 394
semimembranosus muscles, 260
seminal vesicles, 659
seminiferous tubules, 659, 667, 668
semitendinosus muscles, 260
sensation, 363
senses
  brain origins of, 308
  neurons associated with, 286
sensorineural hearing loss, 372
sensory axons, 314
sensory nerves, 337–338
sensory receptors, 128, 377
sensory tunic, 365
septum, interatrial and interventricular, 394
seromucous glands, 522
serosa, 564
serous fluid, 9, 515, 561
serous membranes, 9, 393, 515
serous pericardium, 393
serratus anterior muscle, 259
sesamoid bones, 146
sex chromosomes and sex-linked traits, 695
Sharpey's fibers, 141
Sheep
  brain, 316, 318
  heart, 398–400
shoulder joint, 173
shoulder muscles, 259
simple columnar epithelium, 89, 564
simple cuboidal epithelium, 89, 630
simple squamous epithelium, 89, 522, 594, 601
sinoatrial (SA) node, 403–404
sister chromatids, 665
skeletal muscle(s)/fiber, 106, 229, 233, 235, 241, 257, 564
skeletal system, 141, 159
skeleton, 157–158
  appendicular, 159, 173
  axial, 159, 168–171
  and gender determination, 178
skin, 123, 498, 704
  accessory structures, 123
  nerves, 343
  neurons, 286